AF850727

WITHDRAWN

063320

COASTAL GEOMORPHOLOGY

HOMERTON
COLLEGE
LIBRARY

First published in 1973
First published by Allen & Unwin in 1980

This book is copyright under the Berne Convention. All rights are reserved. Apart from any fair dealing for the purpose of private study, research, criticism or review, as permitted under the Copyright Act, 1956, no part of this publication may be reproduced, stored in a retrieval system, or transmitted, in any form or by any means, electronic, electrical, chemical, mechanical, optical, photocopying, recording or otherwise, without the prior permission of the copyright owner. Enquiries should be sent to the publishers at the undermentioned address:

GEORGE ALLEN & UNWIN LTD
40 Museum Street, London WC1A 1LU

© Donald R. Coates 1973, 1980

British Library Cataloguing in Publication Data

Coastal geomorphology.
1. Coasts - Congresses
I. Coates, Donald Robert
551.4'5 GB450

ISBN 0-04-551038-5

Dedicated to Cuchlaine King — esteemed friend and leader in modern geomorphology

Cover photographs of Alaska and Massachusetts from "Suzanne's Lament." Courtesy of Miles Hayes.

COASTAL GEOMORPHOLOGY

Donald R. Coates, Editor

A proceedings volume of the Third Annual Geomorphology Symposia Series, held at Binghamton, New York
September 28–30, 1972

London
GEORGE ALLEN & UNWIN
Boston Sydney

HOMERTON
COLLEGE
LIBRARY

HOMERTON COLLEGE LIBRARY
16/2/09 ✓
ML
551.457 COA
693320 NL
ACC 063320
~~48420~~

CONTENTS

HOMERTON
COLLEGE
LIBRARY

PREFACE

This proceedings volume presents the collection of papers that were given at the Third Annual Geomorphology Symposium conducted in the Department of Geology, State University of New York at Binghamton, September 28-30, 1972. All elements that contributed to a successful conference cannot be bound in writing, and one of these were the performances of "Suzanne's Lament". This beautifully created audio-visual experience of music and slides, developed by the University of Massachusetts Coastal Research Center, set the proper mood for speakers and participants. In this symposium on "Coastal Geomorphology" a particular type of environment forms the basis of discussion. This environment is such that scientists who classify themselves as geomorphologists, ecologists, engineers, geographers, sedimentologists, and stratigraphers can all make contributions of great value. The necessity for interdisciplinary cooperation in research in order to understand the processes and the terrain of the littoral zone is clearly reflected in the contributions to the symposium. The central theme that emerges from all papers is the dynamic aspect of the coastal environment, and the way geomorphic principles can be used to solve problems (in a problem-solving context). Thus all papers examine some aspect of the geomorphic situation, such as erosion, transportation, deposition, and/or morphology of coastal terrain.

For a period of almost two decades little additional work was done on the geomorphology of coasts after D.W. Johnson (1919) published his monumental work. During and since World War II, the study of beaches and coasts has grown rapidly. Their investigation has been both for pure scholarly research, such as elucidating the origin of certain features, and for study of practical problems that directly concern man. The increased insight gained from both types is necessary in today's world.

Coastlines comprise a major element of the earth's landscape, and processes that shape this terrain are exceptionally complex. To have attempted to cover all of the myriad items about coastlines with their great variability would have been impossible and would have produced a smorgasbord of disjointed papers. Instead the volume largely concentrates on beaches, with particular influence to processes operating on barrier islands, and it provides insight into a variety of different locales.

The selection of authors was made to obtain as wide a geographic and topical forum as possible. They represent many academic institutions, from the United States, Canada and England, and also include workers from governments and industry. Coastal areas in the United States that are examined include Alaska, Delaware, Florida, Maryland, Massachusetts, New York, North Carolina, Oregon, Virginia and Washington. Also coasts of Bermuda, Canada, England, and Zululand are analyzed. The authors are specialists in several different fields of science and engineering, and many are the outstanding leaders of their discipline.

They indicate the current status of research and provide some examples of new methods that are being used to obtain greater precision in studying this fragile and changeable environment.

The volume begins with the introductory statement that officially opened the symposium. The papers that were presented are called chapters and are divided into three main topical areas, or Parts. Each of the Parts contains a general statement by the editor which links the chapters and hopefully gives some perspective on the general theme. Some chapters could have been included in more than one section. They have been arranged to provide a balance between the themes of the three sections. For example, nearly all chapters contain ideas related to coastal processes, and geomorphic applications, in addition to those comprising the first and the final parts of the book.

Part 1 Coastal Processes contains 7 chapters that specifically deal with the dynamics of the littoral zone and the resulting landforms. Methods of investigations are reviewed and emphasis is placed on erosional and accretional phases of shoreline development.

Part 2 Barrier Islands has 4 chapters devoted to this important landform that is so common on many of today's coastlines. The chapters range from historical accounts and re-evaluation of their origin, to investigation of areas previously unexamined. Man's alteration of the natural system provides valuable data for increasing geomorphic knowledge and using it to remedy a particular situation.

Part 3 Applications of Geomorphology includes 5 chapters that illustrate how geomorphic principles may be used. Two of the chapters show the relation of geomorphology to man and engineering, and three chapters stress the practical aspect of geomorphic input towards providing greater understanding of other academic disciplines, including archaeology and sedimentary geology.

The intent of this volume is to provide an up-to-date synthesis of research areas concerning the geomorphic aspect of the coastal environment. As such, it should be suited for special readings in a variety of courses. Many of the applications are relevant to professionals working on shorelines and for managers who must design utilization of the coastal corridor.

Appreciation is expressed to the Geomorphology Group at Binghamton, the students, Marie Morisawa and Gary Carver, for their logistical support during the symposium. Special thanks are given to my colleague, Cuchlaine A.M. King for her invaluable reading and correction of manuscripts so that this volume could be increased in accuracy and polish.

Donald R. Coates
Editor

INTRODUCTION TO COASTAL GEOMORPHOLOGY

Donald R. Coates

Welcome to the Third Annual Geomorphology Symposium. We are slightly overwhelmed, yet greatly encouraged, by the turnout of more than 300 registrants who have taken time off from busy schedules to participate in this year's conference. *Environmental Geomorphology* (Coates, 1971) started the symposia series because of its timeliness and cross-cutting nature of many sub-disciplines. *Quantitative Geomorphology,* (Morisawa, 1972) was the second year theme and stress was placed on methods that are now used in all aspects of geomorphology. The topic this year *Coastal Geomorphology* represents the first in a continuing set of meetings that will analyze the different geomorphic processes that shape the landscape. Perhaps you are now wondering – Why should we start with coastal processes? Allow me to mention at least three reasons:

1. We are extremely fortunate to have as Visiting Professor on our Binghamton campus for the 1972-73 academic year, Professor Cuchlaine A. M. King, Department of Geography, University of Nottingham, England. Her leadership and expertise in this marine-land interface is well established, and has again been documented by publication of the second edition (1972) of her classic *Beaches and Coasts.* We are grateful to the National Science Foundation for her sponsorship, so we wanted to maximize her talents at this meeting.

2. Dr. Marie Morisawa and I are joint recipients of a grant from the U.S. Department of Commerce (National Oceanic and Atmospheric Administration, NOAA) under the Sea Grant Program. The purpose of this award is to study, up to a 5-yr period, the environmental geomorphology of the south shore of Long Island. Although I operated on South Pacific shores for 2 yr during World War II, I readily acknowledge my noviceship of coastal knowhow. Thus, we thought that all of you coastal experts could provide us with a quick cram course, that will help orient our fluvial background into littoral knowledge.

3. For many years Douglas W. Johnson's publication on coasts was considered the definitive work, and until fairly recently study of this environment was neglected by geomorphologists. In the past few years, however, there has been much renewed interest by geomorphologists in coastal processes and landforms. Their efforts are greatly augmented by workers in allied disciplines. This introduces what is probably the central purpose of this symposium, namely, to bring together specialists in a variety of fields, including ecologists, engineers, geographers, geomorphologists, and sedimentary geologists toward a common effort to learn more about this very fragile, dynamic and beautiful environment . . . the coastal terrain.

A large array of different topics will be presented but the common thread that unites all papers is the dynamics of the coastal zone and documentation of the interdisciplinary character of research. For convenience and organizational structure the symposium has been divided into several serially-conducted sessions. I wish to take this advance opportunity to thank the chairmen of the sessions for their leadership which will help make the conference a success .... Dr. Jerry Galvin of CERC, Dr. Les Sirkin of Adelphi University, and Dr. Warren Yasso of Columbia University. Thanks are also due to the speakers, many of whom have come great distances to share their work with you. They represent many different colleges and universities, as well as government agencies and industry. It is recognition of their ability which has attracted such a large gathering. Now sit back and enjoy learning of regions from the arctic to tropics, in the United States from the Atlantic to the Pacific, and other areas from England to Africa.

Our entire staff joins me in your welcome to Binghamton. We hope to see you all next year when the theme will be "fluvial geomorphology."

REFERENCES

Coates, D.R. 1971. *Environmental Geomorphology:* Publications in Geomorphology, State Univ. of N.Y., Binghamton, 262 p.

Morisawa, M. 1972. *Quantitative Geomorphology: Some Aspects and Applications:* Publications in Geomorphology, State Univ. of N.Y., Binghamton, 315 p.

PART 1

COASTAL PROCESSES

Shorelines are one of the most rapidly changing landforms on the earth. Throughout history man has held in awe the power of the oceans and wondered about the dynamics of this fragile environment. Some understanding of coastal processes has been necessary ever since man first ventured onto the seas because safe harbors were a necessity in all sea-going ventures. Although some general facts about tidal changes and of wind-driven waves have been known for 1,000's of years, owing to the complexity of the total marine system many of the processes involved are still not completely understood and constitute an area of continuing basic research. The purpose of Part 1 is to emphasize the geomorphic components that constitute the coastal system developed by the dynamic processes of this land-water interface. Basic guidelines for study at different scales are provided in Chapter 1, and the following chapters exemplify the type of current research that is necessary to increase knowledge of the coastal environment. Many of the processes discussed in these chapters also occur in lakes, but this volume is devoted to the land adjacent to marine waters.

Coastal processes constitute one of the major sets of geomorphic forces that sculpture the landscape. They have in common with other processes, such as running water, glaciers, groundwater, gravity movement and wind, the ability to erode, transport, and deposit materials, and in this chain of events landforms are created. A variety of these coastal landforms are considered in the several chapters of the book.

Seven of the 16 chapters for this volume are included in Part 1, since they emphasize processes, however, all chapters contain some data that are also relevant to this section. The following listing provides such an overview:

A Erosional processes—chapters 3, 7, 10, 12.

B Depositional processes—chapters 4, 6, 15, 16.

C Morphological processes—chapters 5, 8, 9, 13, 14

D Field techniques, process theory, botanical processes—chapters 1, 2, 11.

Coastal processes differ from subaerial processes in that water movements in the nearshore environment are more complex. They are oscillatory, not unidirectional as in fluvial and gravity processes, and act at variable levels owing to tidal and other changes of sea level. Feedback operates in all geomorphological systems, but the relationship between the processes and responses in the coastal zone are particularly complex, and include both negative, or self-regulating, and positive, or self-generating loops. As a specific process creates a particular form, that form in turn modifies the process. Chapters 4 and 5 provide examples of this type of interaction.

Since all authors were asked to assume a general knowledge by the reader of typical littoral forces a summary statement of such processes is now appropriate. Movement of ocean waters constitutes a complex series of interactions powered by energy derived from the sun and moon. These bodies provide gravitational attraction that causes the tides. The different type of tidal regime is related to the tide producing forces set up by the motions of the sun-moon-earth system, and the response of the oceans to these forces. Rising and falling tides permit coastal processes to affect a wider zone of the shoreline. In areas with embayed coastlines of suitable dimensions tidal forces can produce ranges exceeding 50 ft as a result of resonance. The Bay of Fundy illustrates the situation where the natural period of oscillation of the water body is equal to that of the semidiurnal lunar tide producing force, and hence a very large tidal range occurs.

Wind waves are oscillatory water movements that are initiated by the wind-generated frictional drag on the upper water surface. Wave height, length, and period are thus a function of the wind duration, velocity, and the fetch. Wind can also pile up water and create currents, and during storm periods surges can be initiated in some areas. When combined with high tides, sea level along coasts can be raised 20 ft or more, for example during hurricanes, causing great damage to coastal structures and changes in beach characteristics. When waves strike the land obliquely or where beaches have complex configurations, the waves will have a vector of motion parallel to the beach and thus initiate a longshore current. The littoral drift refers to the sediment moving alongshore under the influence of this longshore current. The combination of tides, waves, and currents provides a continuing dynamo of energy that is constantly working to alter coasts.

Eroded materials, removed by the hydraulic force of water or by solution processes, is transported for ultimate deposition in a new locale. Beach sediments may be moved back and forth countless times until attrition of particles has ground them sufficiently small to be permanently lost far offshore. Sand grains may be moved laterally along the shore in the littoral drift, which transports more than 100,000 yd^3 per yr along many United States beaches, such as in New York and New Jersey. Other changes in beaches can occur on a seasonal base, since summer and winter regimes often create different beach profiles. Ocean swells are generally considered to build beaches, while storm waves are usually destructive, moving sand to deeper water. Thus many beaches build up during calmer summer weather and are lowered by the destructive action of steep storm waves which are often made even more destructive by onshore winds. The mobile beach is continuously adjusting its profile to the variations in wave size and character, the coarser the beach material the more rapid and the greater the changes are on the foreshore. Indeed coarse beach material, in the form of shingle, reacts differently to storm waves than sand, in that storm waves can construct the longest lasting features of a shingle shore. These are the beach ridges that build up 10's of ft above the high water level. The pattern of such ridges reveals the longer term developments of the coastline. The beach can reach a short term dynamic equilibrium under reasonably constant wave action, but the coast, which is on a larger scale and which develops at a slower tempo, reveals its youthful nature. All coasts are immature because sea level, following the rapid Flandrian transgression, only reached approximately its present level about 4000 yr ago.

The youthfulness of present coastlines accounts for the wide variety of coastal forms. As the rising sea has flooded a diverse landscape and the waves and tides have shaped the abundance of sediment available to them, a great range of coastal forms has developed. The forms can be related to the variations in energy level, the structural stability, wave climate, and many other influences. The youthfulness of many coastlines means they are not yet adjusted to the waves and tides, thus longshore drift is active and zones of accretion and erosion related to the net drift often occur. Within the nearshore and foreshore zones a wide variety of forms develop including berms, scarps, terraces, ridges and runnels, spits, barriers, and a broad spectrum of different forms of bars. Where the berms are well adjusted to the waves, they illustrate clearly in their alignment the relationship between the offshore relief and the wave character. The beaches become adjusted to the direction of approach of the long refracted swell waves. Thus the offshore relief, which affects the wave refraction pattern, plays its part in accounting for the outline of the beach in plan. This adjustment can only take place fully where rocky headlands do not restrict the movement of the material alongshore. Adjustments between rocky headlands do occur, however, and take the form of log spiral beaches, where wave conditions are sufficiently stable. The emphasis in this book is on the mobile beaches rather than the rocky shores.

Chapters included in Part 1 illustrate many different aspects of the processes operating in the coastal environment. Chapter 1 provides introductory material, general guidelines for study and evaluation of shorelines, and an overview of basic principles. Emphasis is placed on methodologies for coastal interpretation using several different scales of investigation, ranging from reconnaissance mapping to the complex interrelationship of elements at a single part of the beach over time. Alaska and Massachusetts are used as field sites to document these approaches. Chapter 2 introduces the idea of using mathematical equations to develop and test a coastal model. In the model of a concave beach from headland to bay, five significant energy components exist that determine beach configuration and its rate of change. Florida beaches are used as field examples. Chapter 3 discusses the coastal processes that are responsible for changes in beach position of the eastern Long Island coastline. Here increase in sea level is not considered to be a major factor affecting present shore configuration. Instead the characters of inlets through barrier beaches are producing rapid changes in the region and determine the advance or retreat of the shore position.

Chapters 4 and 5 deal with sedimentation in the coastal zone and some of the moderate size landforms that are developed. Large size landforms, barrier islands, are discussed in Part 2. King relates dynamics of littoral transport and depositional processes to the resulting coastal landforms. Trend analysis reveals the marked regularity and magnitude of accretion in terms of beach accretion and ridge size, and the movement and growth sequences of a small spit on the east coast of England. Ridge development occurs where there is a surfeit of sand on the foreshore, producing a gradient less than the equilibrium swash slope. The development of special types of bars that previously have received little attention constitutes the theme for Chapter 5. The cases sited, especially the transverse bars, provide excellent examples of the feedback mechanism. The wave forms that are instrumental in bar development in turn become altered by the shape of the sandy protuberance. The parallel and transverse bars in this preliminary report seem more likely to develop along low energy beaches.

Chapters 6 and 7 provide somewhat special cases for consideration in beach processes. Chapter 6 shows complexities that can occur in the coastal zone where small rivers empty onto the Oregon shoreline. This investigation of the interaction of fluvial and marine processes indicates the importance of seasonal variations on small-scale depositional forms. An orderly and systematic pattern of evolutionary changes occurs in such an environment and this unique assemblage of structures can be used to interpret similar stream mouth-littoral interface deposits in the geologic record. Chapter 7 calls attention to the special characteristics of polar shorelines. In such regions the process regimes differ in magnitude and style from those of lower latitude beaches. Freeze-up conditions cause wave dampening, but in turn ice shove may be important in providing littoral movements with unusual beach features resulting.

Donald R. Coates

Chapter 1

THE INVESTIGATION OF FORM AND PROCESSES IN THE COASTAL ZONE

Miles O. Hayes, Edward H. Owens, Dennis K. Hubbard, and Ralph W. Abele

ABSTRACT

Studies undertaken in the coastal zone (by the Coastal Research Group) can be classified into three categories, dependent upon the scale and detail of the study: (1) reconnaissance of a large section of coast (2) studies at the intermediate level which involve some systematic process measurements, and (3) detailed time-series studies of a small area.

At the reconnaissance scale, a method of study (*zonal method*) has been developed which is based upon a systematic sampling program and upon the quantitative investigation of representative geomorphic units or zones. This approach has provided a rapid method for synthesizing critical environmental properties of large coastal areas, such as the glacial outwash plain shoreline of southeastern Alaska and the southern half of the Gulf of St. Lawrence. The intermediate scale of investigation is by far the most common and usually provides the explanation of coastal phenomena at the local level. A study of the sedimentation and hydrography of the entrance to the Merrimack River, Mass., a tidal inlet that persistently resists engineering efforts to stabilize it, is an example of this level of study.

Continuous detailed studies provide the most effective approach for gaining an understanding of the complex and interrelated problems of process and response within the littoral zone. Without this kind of data and explanation of specific processes, generalizations at the higher level are weakened considerably. A 660 m section of the barrier island at Plum Island, Mass. was studied in detail for 43 days during the summer of 1971 and 27 days during the winter of 1972 to provide a continuous three-dimensional picture of the processes and response for a variety of wave and tidal conditions.

Coastal Research Division, Department of Geology, University of South Carolina, Columbia, South Carolina 29208.

As an example of the type of results that may be obtained by integrating the three levels of study, five generalizations pertaining to sedimentation and morphology of mesotidal barrier islands are presented:

1. The cycle of beach response to the passage of coastal storms in mesotidal areas consists of: (a) eroded, flat post-storm profile (b) early-recovery profile of landward-migrating ridge-and-runnel systems, and (c) late-recovery profile of wide, welded berm.
2. Nearshore bar morphology is exceedingly complex and much more work is needed to determine their variety and distribution before meaningful classification and interpretation of nearshore bars can be accomplished.
3. The migration of tidal inlets is an important process in the formation of barrier island deposits. Historical surveys of Nauset Inlet, Mass. and a geological study of Sandy Neck, Mass. (Redfield, 1965) provide ample evidence of this process.
4. Ebb-and flood-tidal deltas in mesotidal areas are remarkably similar in morphological development. Important in shaping their morphology is the time-velocity asymmetry of tidal currents.
5. Mesotidal barrier islands also have a characteristic shape and morphology, which is dominated by a large bulbous form at the updrift end of the island that results, in part, from sediment accretion caused by wave refraction around the ebb-tidal delta.

INTRODUCTION

The study of form and processes in the coastal zone is undertaken at a variety of different scales. Over the past 7 years the Coastal Research Group[2] has carried out a large number of individual studies. On the basis of the approaches and methods used in these investigations, it is possible to group them into three classes: (1) reconnaissance studies of the morphology and sediment dispersal patterns of large areas (average length of shoreline approximately 500 km) (2) more localized, problem-oriented studies (e.g., study of the circulation pattern of an estuary; study of the sand bodies at a single tidal inlet), and (3) detailed time-series studies of small segments of the shoreline ($<$1 km in length). This paper describes field methods and techniques that have been used by our group at these different scales and presents some generalizations derived from the integration of results obtained by the three types of study.

RECONNAISSANCE STUDIES

Introduction

The purpose of this level of study is to determine regional patterns of coastal morphology and sedimentation. To accomplish such studies quickly, a simple method of rapid reconnaissance was developed (the *zonal method*). The first area studies by this method was the glacial outwash plain shoreline of

---

[2] Formerly at the University of Massachusetts and now at the University of South Carolina.

southeastern Alaska. The bulk of the field work on the project was completed in the summer of 1970. Since then other studies of this type have been conducted on the shoreline of Cape Cod and on the southern shoreline of the Gulf of St. Lawrence. Presently, a study of the northern half of the west coast of Baja California is in progress.

It is important to point out that to date this method has been applied only to relatively straight shorelines between major rivers. These studies have been primarily of depositional coasts with an emphasis on beach morphology and processes. Some coastal regions, such as large deltaic systems (e.g., Mississippi, Niger), shorelines highly indented by estuaries (e.g., Chesapeake Bay area), or large tide-dominated embayments (e.g., upper Gulf of California, Bay of Fundy), are not amenable to study by this method in its present form. However, we feel the concept of the zonal method can be applied in these areas by modification of the data-collection techniques.

Zonal Method

The *zonal method* consists of the following steps:

1. A single large physiographic unit is chosen as the region of study. For example, it may be a connecting series of glacial outwash plains, as was the area studied in Alaska, or some other relatively homogeneous physiographic unit, such as the southern coast of the Gulf of St. Lawrence.
2. Study of aerial photographs, maps and charts of the area chosen, as well as of the available literature, precedes the field work.
3. Field work begins by aerial reconnaissance of the entire area. This initial flight should be flown at low tide to get maximum exposure of the intertidal zone.
4. After the first flight and some preliminary ground inspection, a sediment-sampling interval is chosen. This interval is dependent upon the detail desired, the variation of sediment type, practical considerations such as amount of time and manpower available for field and laboratory work, or certain statistical considerations that the investigator may wish to apply. Our sampling intervals have ranged 4-10 km.

   At each sampling station, the intertidal zone is sampled by a simple ordered-random technique. Samples are collected at three positions, one in the middle of the intertidal zone (sample B), one midway between B and the last high-tide swash line (sample A), and one midway between B and the low-tide swash (sample C). At each station, a sample of approximately 200 grams is scraped from the surface (attempting to collect only the last sedimentation unit). A core sample 15-20 cm in length is also collected at station B. Landward of the last high-tide swash, surface and core samples are collected in each of the major Holocene environments, such as dunes and washover fans, according to a sampling plan designed to fit each area studied. Adjacent pre-Holocene materials, such as bedrock or till outcrops may also be sampled.
5. Next, the study area is divided into physiographic subdivisions. These

subdivisions are defined on the basis of a change in morphology or in one of the major process parameters (e.g., deltas, barrier-island systems, or cliffed shorelines). Figure 1 shows the location of the 10 subdivisions of the southern Gulf of St. Lawrence area. The location of sampling stations within a part of that area is shown in Figure 2. The sampling interval is 10 km.

6. Within each of the subdivisions, specific sites thought to be representative of that particular type of coastal morphology are chosen for detailed study. There may be several of these sites within each subdivision, depending upon the complexity of the unit. On the Alaskan coast, 18 of these detailed studies, called *zonal studies*, were carried out by a 5-man crew.

   Each *zonal study* includes the following:

   (a) Construction of a three-dimensional block diagram of a zone that extends from the low-tide line to the landward extent of active Holocene sedimentation.[3] In Alaska, these zones were usually 200 m wide. The block diagram is constructed by measuring topographic profiles perpendicular to the shore along the two margins of the zone, using standard beach-profiling methods (Figs. 3A & 3B). These two profile lines are connected by surveying critical topographic units between them.

   (b) Grain size estimates are made at 3 to 12 m intervals along the two lateral profile lines, the interval chosen being dependent upon the complexity of the area. On an average, between 30 and 50 grain size estimates are made along each profile line. The grain size estimates are made by comparing the sediment with sand fractions sieved at ¼ phi intervals (Fig. 3C).

   (c) Detailed topographic surveys of features within the zone, such as spits, ridge-and-runnel systems, and dune ridges.

   (d) Statistical sampling and measurements of features such as bedforms, pebble distribution and imbrication, internal structures of bedforms, and other features that may occur in the zone.

   (e) Detailed photographs to illustrate sediment characteristics and topographic features within the zone, both from the ground and from the air.

   (f) Sediment samples are collected along one of the profile lines, following the method described above (section 4).

   (g) Detailed sketches are made of the zone. These are important because they force the observer to carefully inspect all aspects of the morphology and sediments within the zone. Also, they can generally be more inclusive (of the entire zone) than single photographs, and they permit correlation of sampling sites with local morphology (on the sketches, see Fig. 4B).

---

[3] There are some instances where it is not possible to include all of the Holocene sediments (e.g., wide dune plain). In those cases, a few hundred meters of the area landward of the beach zone is profiled and the rest is mapped from aerial photographs.

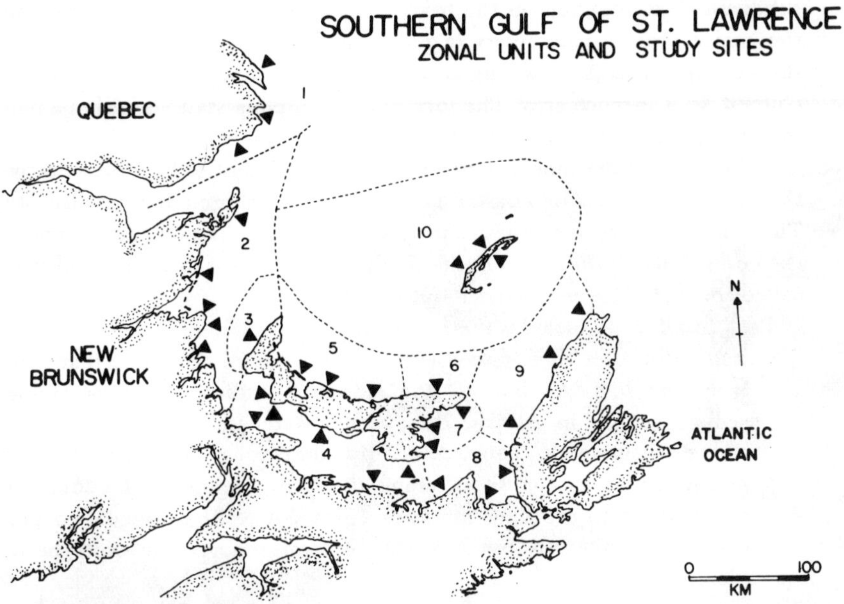

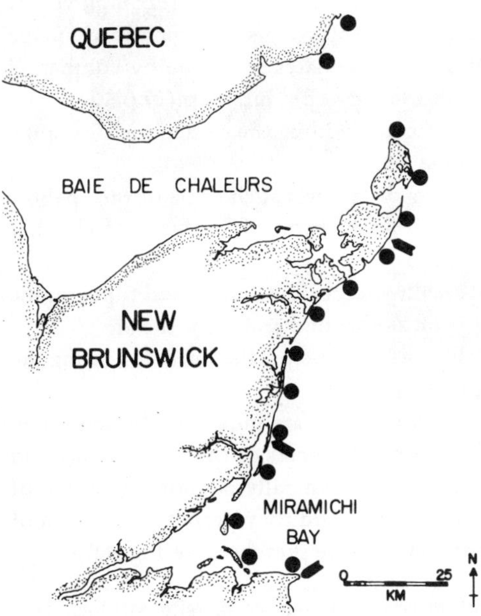

Figure 1. Location of 10 physiographic divisions within the southern Gulf of St. Lawrence study area. Triangles indicate locations of detailed study sites.

Figure 2. Sample stations (circles) and zonal study sites (arrows) within part of units 1 and 2 of the southern Gulf of St. Lawrence study area.

Figure 3. Some field techniques used during detailed zonal site studies. (A) Demonstration of beach-profiling technique, a modification of a method described by Emery (1961). (B) The change in elevation between the two vertical staffs is measured by leveling on the horizon, as is demonstrated in this photograph. (C) Estimation of mean grain size by comparison of grains with sieve fractions sealed in plastic (phi-size estimator).

7. In addition to the *zonal studies*, other short-term projects can be undertaken outside the zonal sites, such as mapping of major features and sediment sampling at a more detailed level.
8. While the sampling and zonal studies are under way, the area is flown again and is described carefully from the air, the observations being recorded on tape. From the information gathered during this flight and the ground studies at the sampling and zonal sites, it is possible to construct a detailed map of the coastal morphology of the entire area.
9. In the laboratory sediment samples are analyzed for grain size and mineralogical variations.
10. Finally, in summarizing the characteristics of the major study area, other sources of information such as data on tides, waves and regional tectonics, and historical surveys are used to supplement the field data.

The end product of the zonal study is a general summary and map of the morphology of a large coastal region, with some details on the local morphology and sediment dispersal trends. Little of the world's coastline has been investigated at even the reconnaissance level; therefore, this method represents a valuable tool for the collection and interpretation of information about regional coastal processes and geomorphology. This basic type of information should be an essential part of meaningful environmental planning and coastal zone management, yet for the most part, it is not available even for the developed eastern seaboard of the United States, let alone other less populated areas.

Alaskan Study

Some preliminary results of the Alaskan study illustrate the types of information that can be obtained by the zonal method. The 600 km of glacial outwash plain was divided into 9 major morphological types (or zones), in which 18 site investigations (*zonal studies*) were carried out. Samples were collected at 90 stations, using a 4 km spacing. From analysis of these samples, it was possible to determine regional variations in textural parameters of the sediments. For example, the average percent gravel for the 3 samples at each station is summarized by the histogram in Figure 5. This plot shows that the samples were composed largely of gravel or coarser-sized material where glaciers or rocky cliffs were adjacent to the coast.

The information from the individual *zonal studies* provides data on some of the general characteristics of the beach morphology. Beach profiles from six of the zonal sites are presented in Figure 6, along with a ratio scale representation of mean grain size. These data indicate the variety of beach profiles in the region and the close correlation between beach gradient and mean grain size. The beach profile at station MAL-5, located at the terminus of the Malaspina Glacier (Fig. 5), has a short, steep profile and is composed of extremely coarse material ($\bar{X}$= 9.0 ϕ; Figs. 4A and 4B). By contrast, the beach at station EG-8 is characterized by a wide, flat profile with a mean grain size of 2.07 ϕ. At this profile, it is common to find ridge-and-runnel systems migrating across a wide, flat low-tide

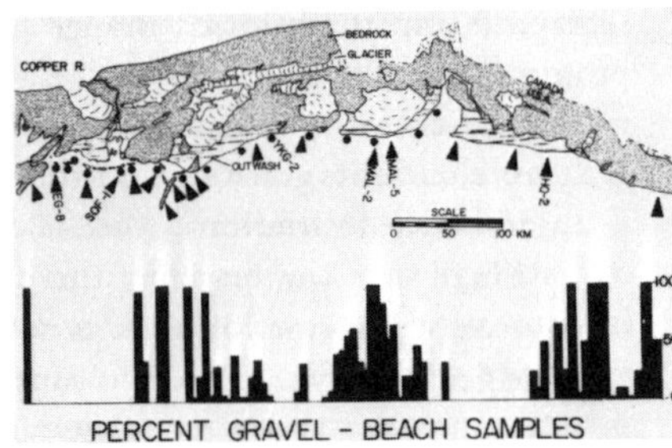

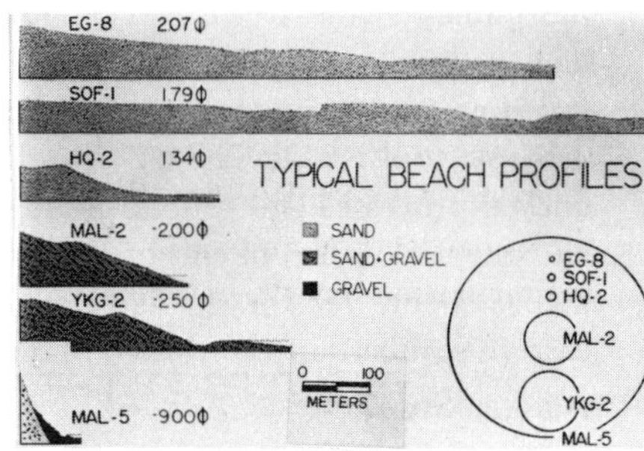

Figure 4. Zonal study site MAL-5 in southeastern Alaska. This site is located at the terminus of the Malaspina Glacier (Fig. 5). (A) Ground view looking east. (B) Field sketch. Arrows indicate location of same boulder in photograph and sketch.

Figure 5. Glacial outwash plain shoreline of southeastern Alaska. Triangles indicate location of detailed zonal study sites; the circles show the location of permanent beach profiles. The histogram is a plot of the average percent gravel occurring in the 3 samples at each of the 90 sediment-sampling localities. The sampling stations were plotted on the east-west abscissa of the histogram by dropping perpendicular lines from the arc of the main shoreline trend. The high gravel percentages occur in areas where glaciers or rocky cliffs border on the shore.

Figure 6. Typical beach profiles from the southeastern Alaska zonal study area (profiles located on Fig. 5). Note the correlation between beach gradient and mean grain size of the beach sediment. Profiles are plotted at a 5:1 vertical exaggeration. The grain sizes indicated are average values for the means of the three samples collected at each locality. The circles in the lower right are ratio comparisons of the sizes of the sediments at each profile (using phi scale).

terrace (Figs. 7A and 7B).

In addition to the zonal investigations and the sediment sampling program, several auxiliary one-day projects were undertaken to investigate specific geomorphological features. A set of beach samples, spaced at 250 m intervals, were collected along the 8 km length of Duktoth Spit. Analysis of these sediments shows a marked change in maximum grain size from $>$ 64 mm at the proximal end to $<$ 8 mm at the distal end (Figs. 8A and 8B). The sediment transport in this area is from east to west, which illustrates the importance of longshore currents generated by the dominant southeasterly storm winds.

In addition to studies in the field, it is valuable to investigate historical data and surveys that can be related to coastal changes. An example of this approach is a study of the Icy Bay area, where the glacial ice has retreated more than 40 km since 1900 (Fig. 9). At the turn of the century, sediment was fed into the littoral zone at the glacier terminus and transported to the west by longshore drift to build up a series of prograding beach ridges. Since the retreat of the glacier, this sediment supply has been eliminated and the beach ridges have been eroded about 1.5 km.

These few examples of the kinds of data collected during the Alaskan project indicate some of the characteristics of the coastal geomorphology in this region. A great deal of information has been obtained from this and other areas using the zonal method and from these data basic patterns of coastal morphology and sedimentation have been delineated within each region.

STUDIES AT THE INTERMEDIATE LEVEL

Introduction

A field investigation of the Merrimack River Inlet, Massachusetts, by Hubbard, provides an example of a study at the intermediate level. This type of study yields further insights into local problems that, in some cases, were originally defined by the zonal method. The "problem" at the Merrimack Inlet is continuing erosion in the vicinity of the south jetty.

The Merrimack River Estuary (Figs. 10A and 10B) is a type B estuary, using the criteria of Pritchard (1955). Usually, a well-developed slightly tilted contact exists between the high density saline water on the botton and the fresh river water on the top. The normal discharge of the Merrimack River is an excess of 2000 c.m.s. Mean tidal range for the Merrimack River entrance is 2.53 meters.

Between 1966 and 1971, the Coastal Research Group studied 8 inlets (and estuaries) in the New England area. Many of the methods and techniques developed for those studies were applied to the Merrimack.

Methods of Study

Twenty-one beach profiles (Fig. 11) were measured at regular intervals throughout the summer (approximately once a week) in order to determine rates of change along the shorelines adjacent to the inlet. Twenty-nine hydrographic

Figure 8. Duktoth spit, southeastern Alaska. (A) Aerial view (looking east) of the prograding beach-ridge plain near Yakataga. The maximum grain size of the beach sediments were measured at 250 m intervals along the outer spit (Duktoth), which trails west in response to strong westerly longshore currents. (B) Plot of maximum grain size vs distance along the 8 km length of the spit.

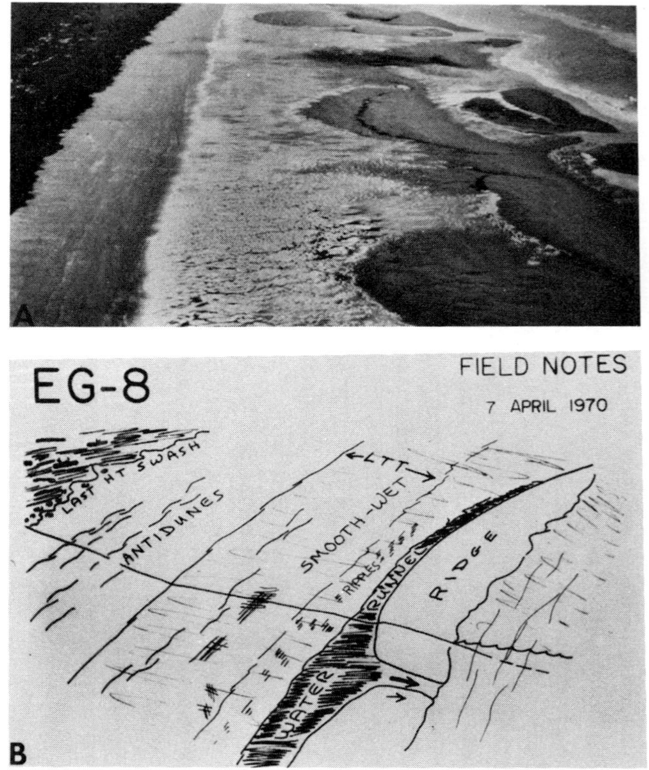

Figure 7. Aerial photograph (A) and field sketch (B) of detailed zonal study locality EG-8 (located in Fig. 5). This is an exceptionally flat profile (Fig. 6). The beach zone consists of a moderately steep high-tide beach face, which is covered with anitdunes, and a wide low-tide terrace, which has numerous ridge and runnel systems on its seaward margin.

stations (Fig. 12) were monitored for tidal current velocity and direction, salinity and temperature. Measurements were taken hourly at 1 m intervals through the water column for 13 hours. This procedure was repeated during spring, mean and neap tide periods for all stations except those associated with the system of offshore bars. At these locations tidal currents were negligible during mean and neap tide periods.

Bathymetric maps were constructed from over 100 bottom profiles. The exact location of any feature is fixed from high altitude (2500 m) aerial photographs and the morphologic character of each feature is determined from the carefully controlled fathograms. At a smaller scale, maps were constructed of minor topographic features in the area (e.g., bars, spits, intertidal bedforms) using a Wild T-16 theodolite. The accuracy of this instrument permits contouring at a 10 cm interval with little difficulty.

Wave Refraction

Preliminary evidence indicates that a complex relationship between the morphology of the ebb-tidal delta, wave refraction and the intervention of man (in the form of jetties and groins) is responsible for the erosion problem along the beach front at Plum Island. This relationship was considered by Hayes, Goldsmith and Hobbs (1970), who concluded that refraction of northeast waves

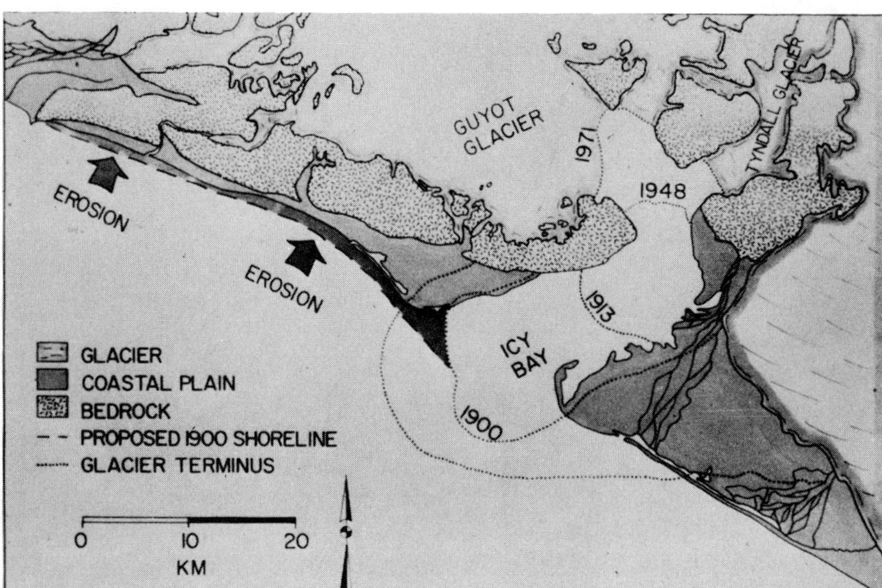

Figure 9. Icy Bay, Alaska. Dotted lines indicate the locations of the end of the glacial ice in Icy Bay at the years indicated. Since retreat of the ice, the shoreline west of the bay has undergone severe erosion because of the depletion of the sediment supply formerly provided by the ice. Dominant longshore drift is from east to west.

systems placed on intertidal sand bodies, it was possible to construct maps showing bedform scale, type and orientation. These maps show the dominant local processes and help explain the relationship between tidal currents and wave parameters in the area. A generalized distribution of bedform type and orientation is given in Figure 14.

The above data were combined to determine the pattern of sediment dispersal at the Merrimack River Inlet (Fig. 15). Waves refracting around the ebb-tidal delta transport sand southward along the offshore bar until a point is reached where a part of the sand is transported toward the shore in the form of swash bars and ridges (King, 1972). Four sequential maps of an offshore bar were constructed (Fig. 16) to illustrate this process. The bar was mapped at the following stages of development:

(1) 12 March 1972 (Fig. 16A): At this stage, two slipface orientations indicate a complex interrelationship between waves and longshore currents.

(2) 15 April 1972 (Fig. 16B): The bar started to develop a marked landward directed slipface, although it was still a subtidal feature.

(3) 21 June 1972 (Fig. 16C): The bar had become a transverse bar, but had not yet welded to the beach.

(4) 4 July 1972 (Fig. 16D): Welding was complete. The bar was oriented obliquely to the beach and was starting to migrate northward. This mechanism

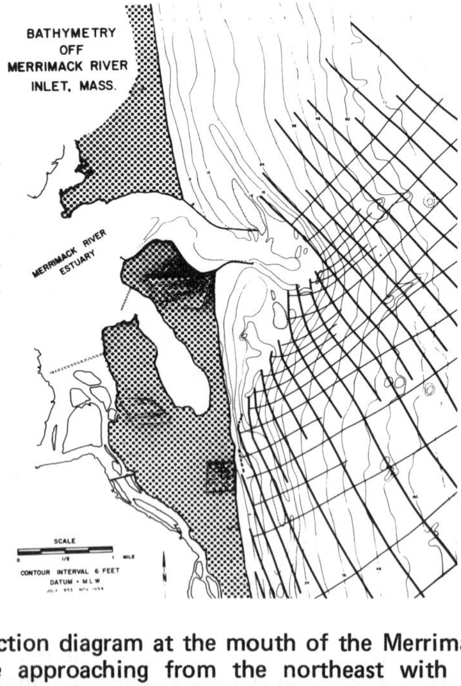

Figure 13. Wave-refraction diagram at the mouth of the Merrimack Inlet. Wave fronts are approaching from the northeast with a period of 9 seconds. Note that refraction of the dominant northeast waves around the ebb-tidal delta has resulted in waves approaching from the southeast at the south (i.e., downdrift) side of the inlet. Diagram constructed by Jon C. Boothroyd.

of bar migration and the concommitant transport of material to the beach is therefore an important factor in the overall sand budget of this inlet area. Thus, the understanding of the processes and morphology within the estuary and around the inlet leads to an explanation for the distribution of sediments and the associated patterns of erosion and deposition.

DETAILED STUDIES

Introduction

The data on the Merrimack Inlet were presented to illustrate the intermediate level of investigation. To explain specific processes and morphological features in more detail, it is necessary to monitor smaller sections of the littoral zone. The nature of process and response within the littoral zone can only be studied successfully on a continuous basis. Weekly or monthly observations provide a glimpse of some of the processes and their effects, but are not sufficient to enable a clear understanding of these complex interrelationships.

Plum Island Study

In order to study one 660 m section of a barrier beach at Plum Island, Massachusetts, two detailed investigations were completed by Abele in July and

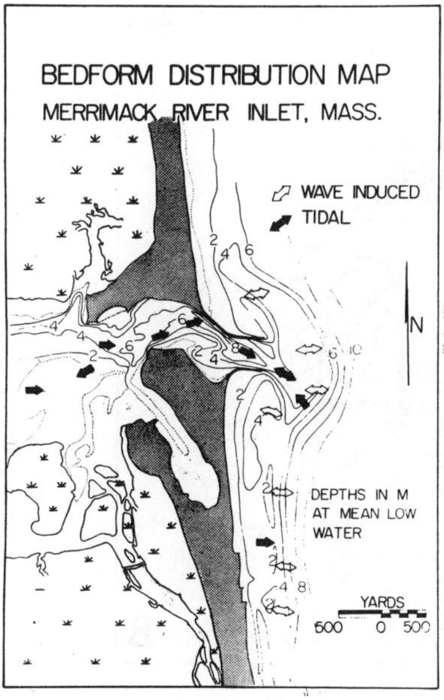

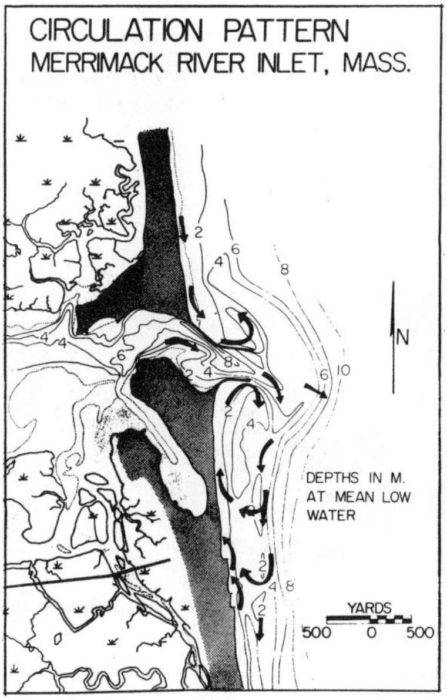

Figure 14. Bedform distribution and orientation in the Merrimack Inlet area. Arrows indicate dominate orientations of bedform slipfaces.

Figure 15. Pattern of sediment circulation in the Merrimack Inlet area deduced from hydrographic, morphologic and bedform orientation data.

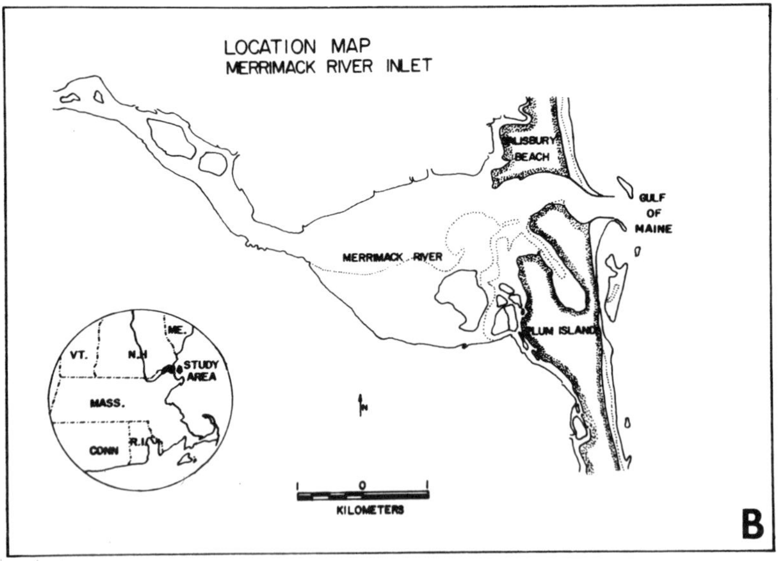

Figure 10. Merrimack Inlet, Massachusetts. (A) Aerial view looking north on 5 July 1972. (B) Location map.

around the ebb-tidal delta (Fig. 13) was responsible for a reversal of sediment transport south of the inlet. This northerly transport of sediments created a shoaling problem in the harbor before rehabilitation of the south jetty in 1967.

Sediment Distribution

Bottom samples were collected by SCUBA divers at over 200 locations. Sample locations were chosen on the basis of significant topographic changes, in order to include all morphologic features. To assure uniformity of conditions, sampling was confined to the period between 2 hours before and 2 hours after high water.

This method is thought to be superior to grab sampling as it eliminates loss of a portion of the sample in transport to the surface as well as blind sampling without regard to bottom morphology. This approach affords the investigator a much clearer view of the morphologic and hydrologic parameters responsible for a given grain-size distribution. For example, it can now be shown that the coarse sand at the 18 m contour, thought by Anan (1971) to be relict Pleistocene deltaic sediments, occurs in bedforms generated by storm waves. During periods of relative calm between storms, these bedforms are partially obscured by fine sediments that settle from suspension.

From these underwater observations and from observations along 30 m grid

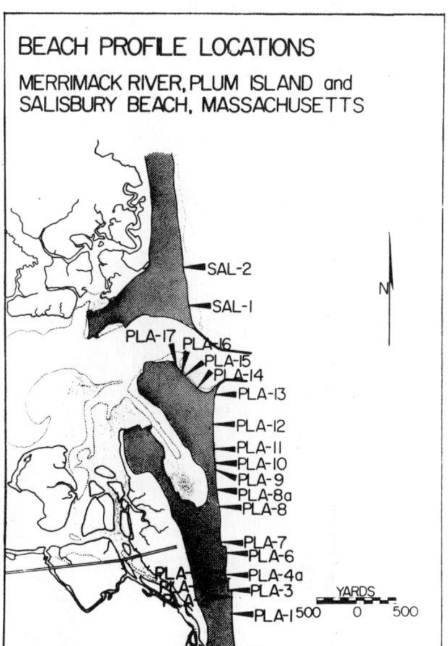

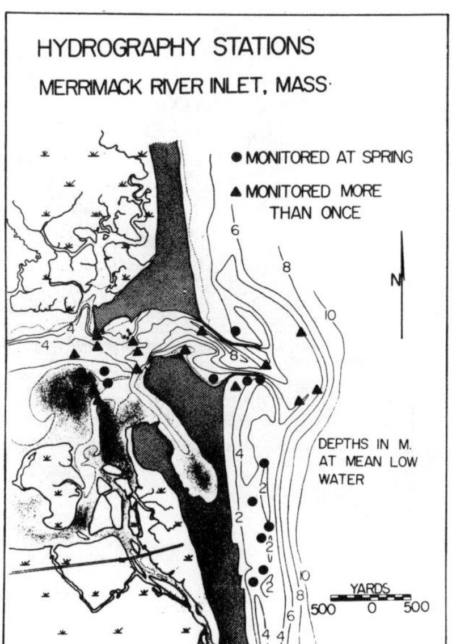

Figure 11. Location of beach profile stations for the Merrimack Inlet study.
Figure 12. Location of hydrography stations for the Merrimack Inlet study.

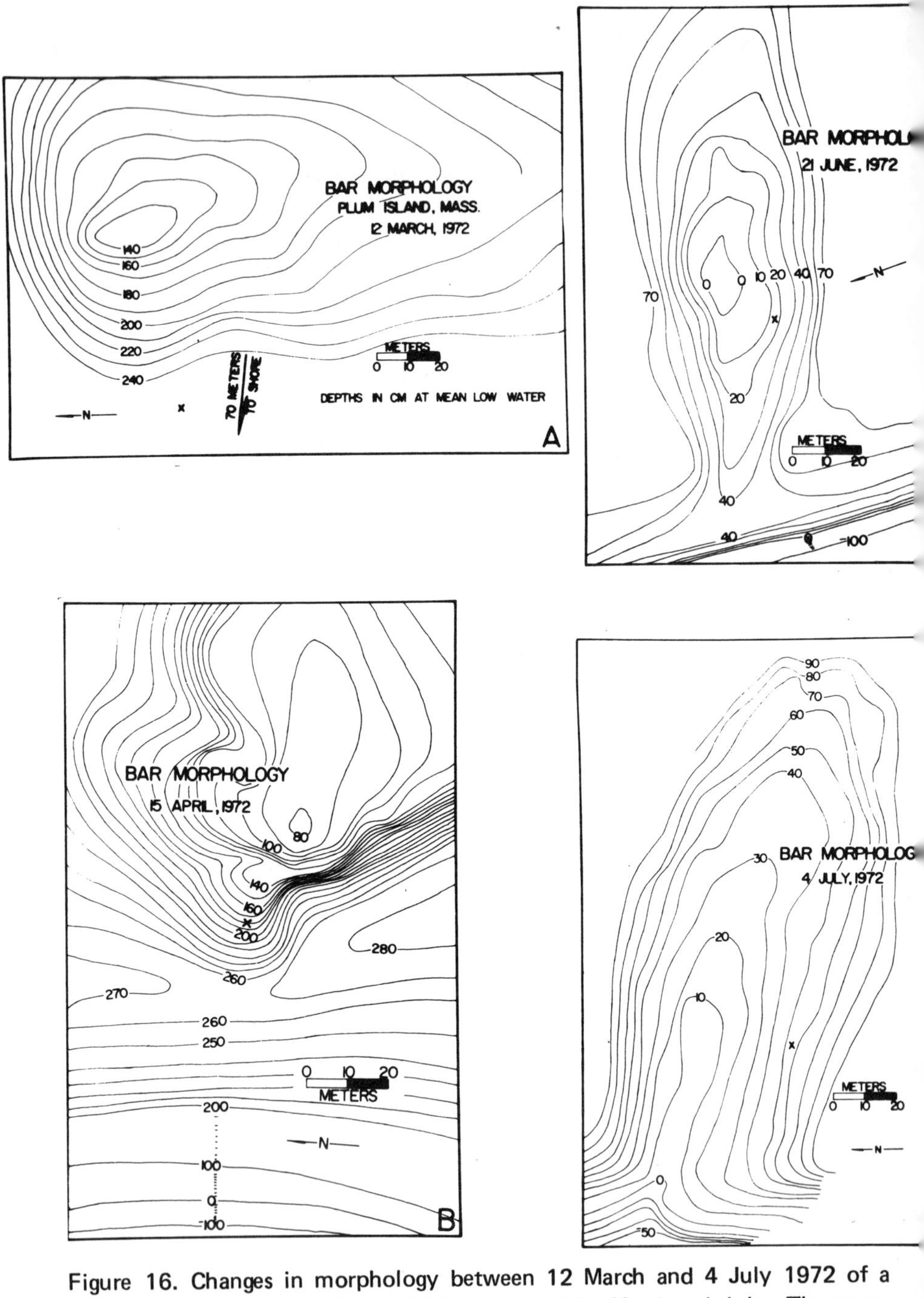

Figure 16. Changes in morphology between 12 March and 4 July 1972 of a nearshore bar located 2 km south of the Merrimack Inlet. The arrow in the photograph in Figure 10A indicates the location of the bar. Water depths are expressed in cm at mean low water. Contour interval is 10 cm. The X is in the same position on all maps.

August 1971 and in January 1972 (Figs. 17A and 17B). Following procedures developed by Fox and Davis (1970), bi-hourly measurements were recorded. These included monitoring of meteorological variables (barometric pressure, wind speed and direction), and wave parameters (breaker height, breaker depth, breaker type, wave period, breaker angle and longshore current velocity). In the summer, groundwater elevation was measured. Wave steepness and breaker power were calculated for each set of wave measurements. A series of twelve equally spaced profile lines extending from base stakes on the foredune ridge to a point seaward of spring low water were run on a daily basis for 42 d in July and August 1971 and for 24 d during January 1972. In addition to profiles run during the summer and winter studies, twelve daily profiles were run for an 8-day period following a severe northeaster of February 19, 1972.

These closely spaced observations of changes in beach morphology in response to dynamic processes during summer and winter periods reveal that changes in beach morphology are directly related to the passage of high-and low-pressure systems over Plum Island. The path of the pressure system relative to the study area is critical in terms of changes in beach process variables. Figure 18 illustrates barometric pressure, breaker height, and breaker power measurements for the summer period. Breaker power was calculated by the method suggested by Harrison and Fausak (1970). Low-pressure systems that

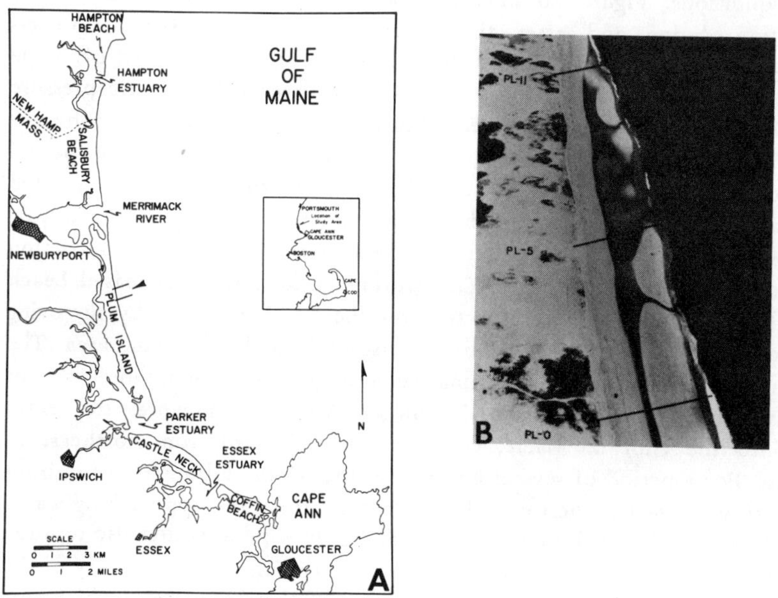

Figure 17. Detailed study area, Plum Island, Mass. (A) Location map. Area of study is a 660 m section of the central portion of Plum Island (arrow). (B) Aerial view of study area on 28 June 1971; profile PL-O marks the southern limit of the study area and PL-11 the northern limit. Photograph taken at low tide.

pass over Plum Island (e.g., July 7 and July 15) result in an increase in breaker height and an increase in breaker power per meter of wave crest. The increase in breaker height is related to the generation of long-period swells as the low-pressure system moves offshore.

As an example of one of the relationships derived from this type of study, it can be demonstrated that high-tide beach-face gradient measurements reflect changes in beach morphology due to differing levels of breaker power. Figure 19 illustrates changes in beach-face gradient along two profile lines for the summer study period. Profile PL-O was characterized by a wide welded berm formed by the landward migration of a large ($>$90 cm high) ridge during the early portion of the study. Between July 8 and 10, the ridge gradient at PL-O increased markedly due to a period of greater breaker power on July 7 and July 8 (compare Figures 18 and 19). The fact that the high-tide beach-face gradient did not show as much change as the ridge gradient is due to the greater period of time that waves may act on the ridge beach face. During the final stages of ridge welding, waves rarely act on the high-tide beach face for more than one hour per tidal cycle, whereas the ridge beach face may be exposed to wave action for as much as 7 h per tidal cycle.

During a severe northeasterly storm (extratropical cyclone), which occurred on February 19, 1972, beach process variables were measured under extreme energy conditions. Figure 20 illustrates some of the results of this 6-day investigation during and after the storm. At the beginning of process measurements on February 18, the low-pressure center was located off Cape Hatteras, North Carolina. As the low moved northward, barometric pressure steadily decreased over the study area. With the decrease in pressure, as the low neared Long Island, northeasterly wind velocities increased, wave period decreased and breaker height increased. The wind shift to the northeast resulted in a change in the direction of longshore drift from northward to southward. As the pressure continued to decrease, wind velocities increased to a maximum average of approximately 66 kph. This stormy onshore wind increased beach erosion by steepening the waves acting on the beach face and by removing considerable sediment through aeolian transport into the foredune area. The waves created by the increased wind velocity generated longshore current velocities measured in excess of 115 cm/sec. When the low-pressure center moved into the Gulf of Maine, wind direction changed from northeast to northwest. For a period of several hours after the winds shifted from onshore to offshore wave height increased; however, with time, the approaching waves were flattened by the offshore winds. The change in wind direction also resulted in a noticeable decrease in longshore current velocity. Post-storm recovery of the beach proceeded quickly, except for recovery of the dune scarp, which was eroded back 3 to 5 m; the first ridge welded onto the back beach 4 d after the storm.

These are examples of results obtained at the detailed level. The conclusions derived from investigations at this level can then be used in the interpretation

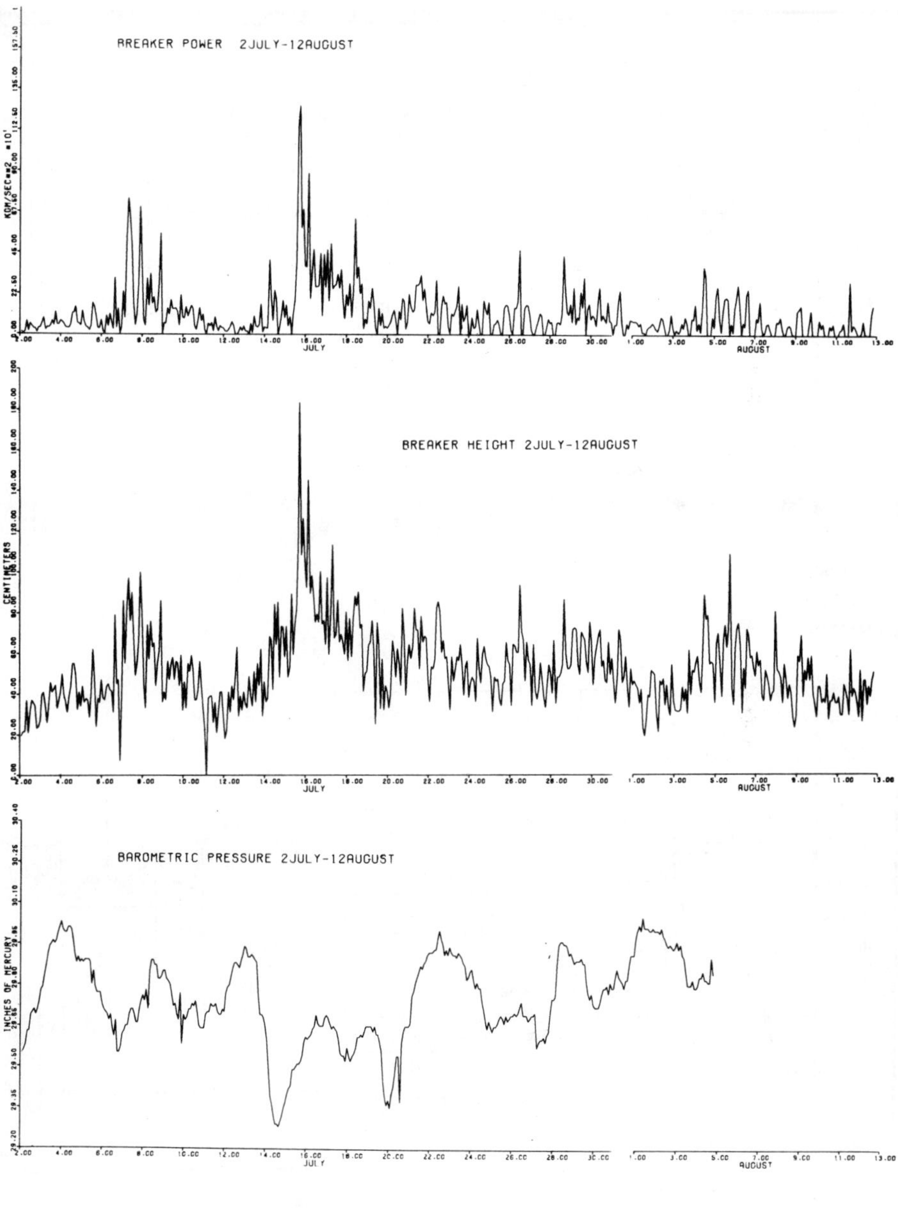

Figure 18. Relationship of barometric pressure to breaker height and breaker power for the summer period of the detailed Plum Island study.

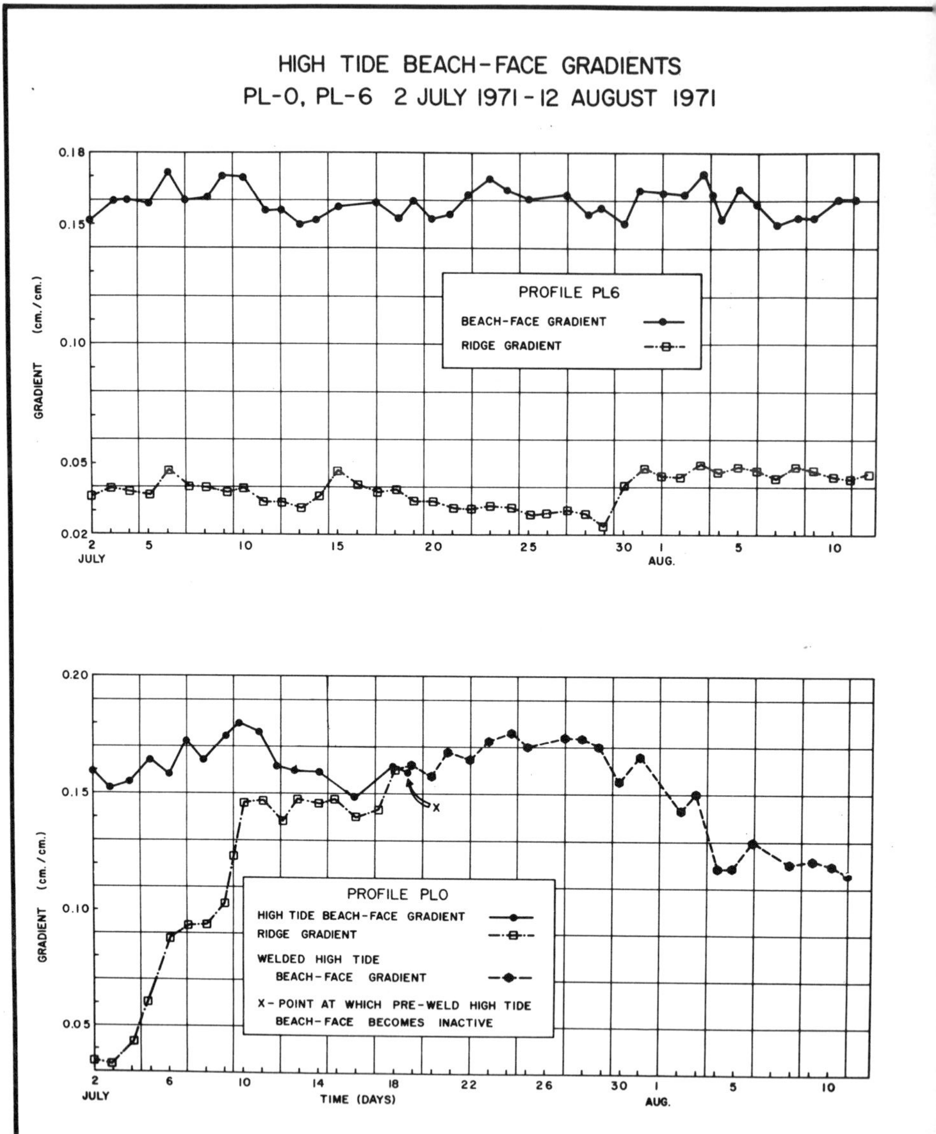

Figure 19. High-tide beach-face gradient for profiles PLO and PL6 for the period between 2 July and 12 August 1971.

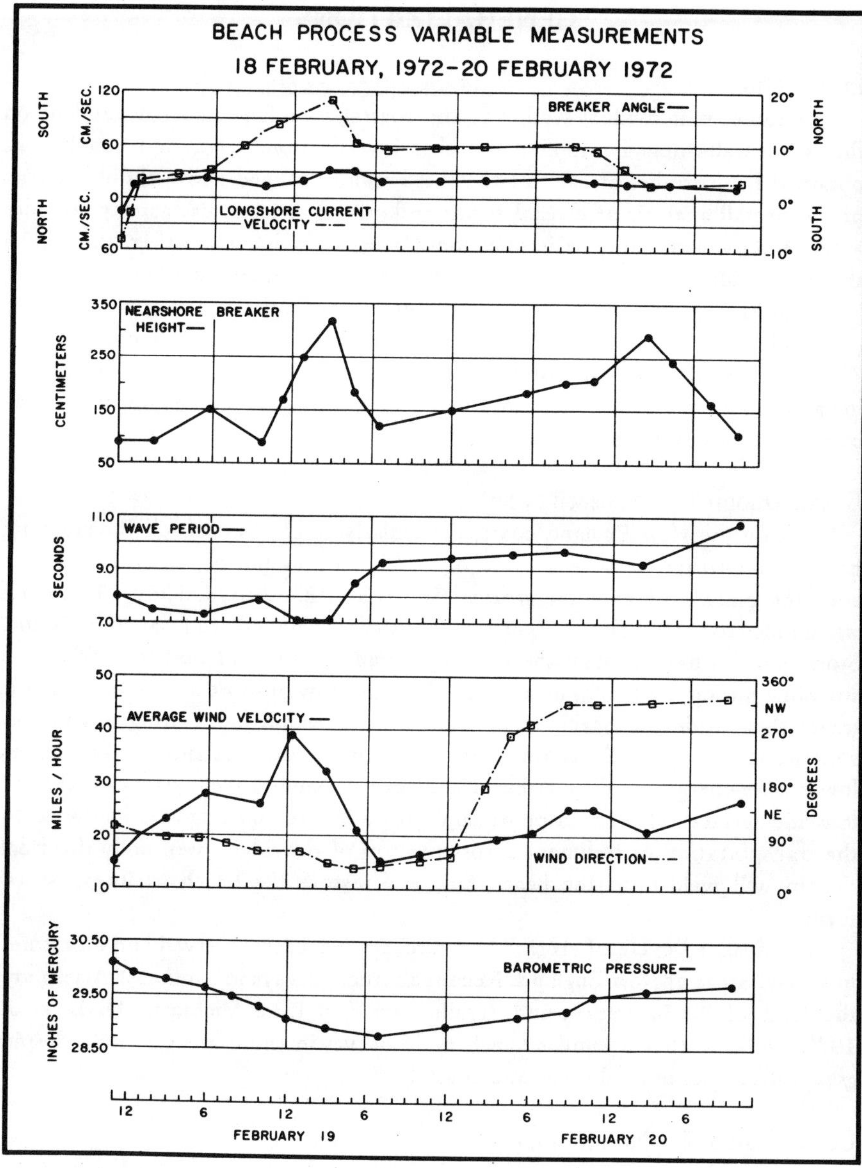

Figure 20. Beach process variable measurements (18 February–20 February 1972) during the passage of a major storm.

and explanation of problems defined at the intermediate and reconnaissance levels.

GENERALIZATIONS

Introduction

The value of individual studies in the coastal zone is partially dependent on the way results may be applied to other investigations. Data and conclusions obtained from studies at each of the three scales outlined above should be part of the overall analysis of coastal forms and littoral process. Each study provides a partial answer to some basic questions. When considered together, the results should provide a more realistic analysis of the coastal environment.

An an example of an effort to carry out such a synthesis, five generalizations (or models) of coastal morphology and sedimentation have been developed. These represent the integration of data collected by our group at the three levels in a variety of areas; they are based primarily on studies of mesotidal, barrier-island shorelines.

Generalization 1 – the beach cycle

Work on the New England coast, particularly in the Merrimack Embayment, has indicated that there is a cycle of beach erosion and deposition associated with the passage of cyclonic storms (Hayes and Boothroyd, 1969). This cycle was alluded to in the above discussion of Abele's detailed study at Plum Island. Storm centers pass through the region generally from southwest to northeast at any time during the year. The associated winds blow onshore and generate storm waves that erode the beach to produce a low, flat profile. During this period, sediment is moved from the intertidal zone offshore. Subsequent lower-wave-energy conditions build a small ridge-and-runnel system on the flat low-tide terrace. This envelope of sand grows in size and migrates landward by the transportation of sediment across the top of the ridge. Eventually this ridge system will weld onto the beach face and restore the beach to its pre-storm form.

This idealized cycle of erosion and recovery has been observed and monitored in several areas of New England. Reconnaissance surveys in southeast Alaska and the Gulf of St. Lawrence, and detailed work in Lake Michigan (Davis, et al, 1972) indicate that a similar beach cycle occurs in those areas as well, despite great differences in tidal range and coastal orientation.

Generalization 2 – bar morphology

Nearshore bar types are very complex even within relatively small geographic areas, and no simple classification of nearshore bars is yet possible. The ridge-and-runnel system described about (Fig. 17B) is only one type of bar formation found in New England. A zonal reconnaissance of Cape Cod in the spring of 1972 indicated that at least four major bar types occur in that area.

Apart from the typical ridge-and-runnel system, a series of swash bars and rip channels was observed on the outer Cape (Fig. 21A). In the more sheltered area of Cape Cod Bay, a series of up to 30 parallel, linear bars was present on the shallow nearshore terrace (Fig. 21B) studied by Nilsson (1972). By contrast, on the south shore of the Cape the bars were less numerous and more crescentic in form (Fig.21C).

Generalization 3 – inlet migration

The migration of tidal inlets is an important process in the formation and modification of barrier island deposits. One of the more dramatic examples is the migration of Nauset Inlet, on Cape Cod, which has moved over 6 km to the south in the last 100 yr (Fig. 22). This particular location was studied in detail by Goldsmith (1972) and Hine (1972). The migration of Nauset Inlet has been accompanied by the southerly growth of recurved beach ridges at the end of the spit updrift of the inlet. A similar situation is reported by Redfield (1965) at Sandy Neck in Cape Cod Bay, where the updrift spit has grown eastwards over the past 3300 yr (Fig. 23). At present, the deposition of sediments at the distal end of the spit is causing burial of the channel in the lee of the spit.

Generalization 4 – tidal delta models

Particularly important in the study of inlets and adjacent barrier beaches is the description and explanation of sand deposits in and adjacent to the inlet itself. Models have been developed from studies in New England to describe the ebb– and flood-tidal delta systems.* Although these sand bodies appear complex in form and origin, it is possible to describe the morphology and processes in relatively simple terms. In particular, each of the 8 inlets studies in New England have tidal-delta deposits that exhibit the same characteristic features. Reconnaissance work in Alaska and Canada has shown that these models derived to illustrate New England tidal deltas also in those areas.

Ebb-tidal delta model. The ebb-tidal delta usually consists of 5 major components (Fig. 24): (1) major ebb channel (2) channel-margin linear bars (3) terminal lobe (4) marginal flood channels and, (5) swash bars. These features are maintained by a balance of three hydrodynamic forces – wave action, flood currents, and ebb currents, the ebb currents being the dominant factor. The channel-margin linear bars are apparently built by the interaction of ebb currents and waves; swash bars are the result of wave action; the major ebb channel and terminal lobe are maintained by ebb currents; and marginal flood channels are a

---

**ebb-tidal delta* – lobate sediment accumulation deposited outside (seaward of) a tidal inlet by ebb-tidal currents. It is usually highly modified by wave action (see Fig. 24).

flood-tidal delta – wedge-shaped sediment accumulation deposited inside (landward of) a tidal inlet by flood-tidal currents (see Fig. 26).

Figure 21. Examples of different types of nearshore bars occurring in the Cape Cod, Mass. area. (A) Nearshore bar and rip-current system on the Atlantic Ocean shoreline of the outer Cape (near Truro, Mass.). (B) Multiple linear bars on Cape Cod Bay shoreline of the outer Cape (near Wellfleet, Mass.). A maximum of 30 of these parallel bars occur in this area. (See also Nilsson). (C) Irregular nearshore bars on the south shore of Cape Cod (near Hyannisport, Mass.).

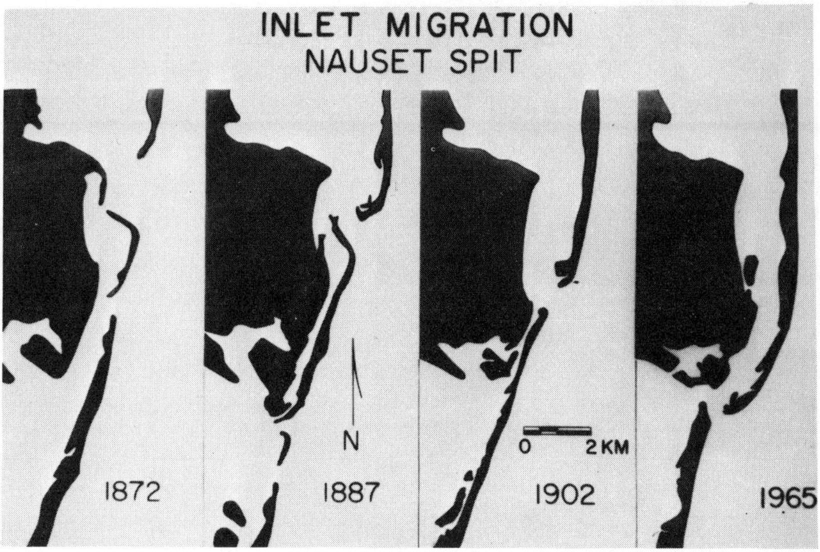

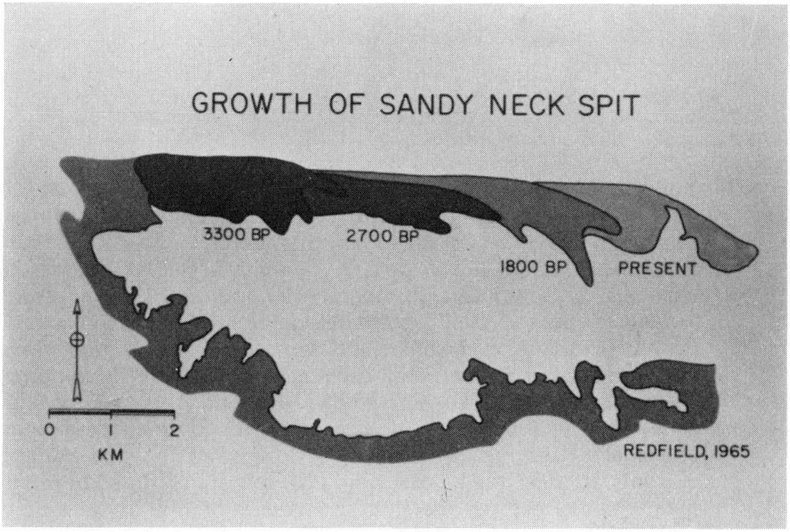

Figure 22. Migration of Nauset Inlet (near Chatham, Mass.) between 1872 and 1965 (from Goldsmith, 1972).

Figure 23. Growth of Sandy Neck Spit, Mass. between 3300 BP and the present (after Redfield, 1965).

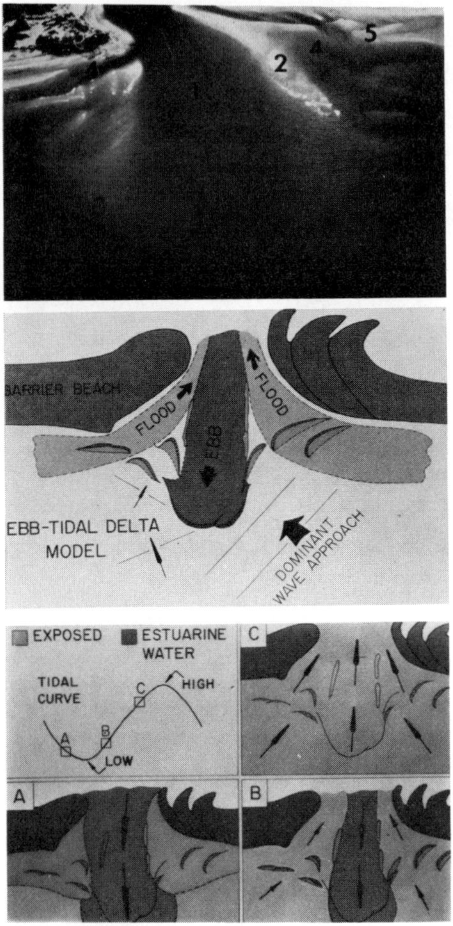

Figure 24. Ebb-tidal delta model. (upper) Ebb-tidal delta of Essex estuary, Mass. The numbers on the photograph indicate the following morphological subdivisions of the tidal delta: 1. Major ebb channel. 2. Channel-margin linear bars. 3. Terminal lobe. 4. Marginal flood channels. 5. Swash bars. Photograph by Jon C. Boothroyd. (middle) Model of the "typical" ebb-tidal delta in mesotidal areas. Note downdrift-offset of the inlet and the segregation of tidal channels into an ebb-dominant main channel and flanking flood-dominant marginal channels. (lower) Illustration of the relationship of time-velocity asymmetry of tidal currents to the morphology of ebb-tidal deltas: (condition A) Near low water (on tidal curve) ebb currents are near their peak (Fig. 25), but are confined primarily to main channel because marginal channels are either dry or extremely shallow. (condition B) Even though water level is rising, ebb currents are still flowing out the main channel. As the water level continues to rise, flood waters choose the paths of least resistance and flow into the flanking channels. (condition C) At high water the flood waters flow in as a sheet over the entire area.

response to flood currents. An important factor in the origin and the maintenance of this morphology is the *time-velocity asymmetry* of tidal currents that commonly occurs in the region of the ebb-tidal delta. Detailed hydrographic measurements in several areas show that maximum velocity of the tidal currents does not occur at midtide (Fig. 25). In fact, in many areas studies, the maximum ebb velocity occurs near low water and the maximum flood velocity usually occurs between midtide and high water. The relationship of the time-velocity asymmetry of tidal currents to the morphology of the typical ebb-tidal delta is illustrated in Figure 24.

Flood-tidal delta model. The flood-tidal delta is usually composed of 6 major components: (1) flood ramp (2) flood channels (3) high area (4) ebb shields (5) ebb spits and, (6) spillover lobes. The features are labelled on the aerial photograph of the flood-tidal delta in Essex estuary, Mass. (Fig. 26). The flood-tidal delta is formed essentially by flood currents. Note the abundance of flood-oriented bedforms on the Essex tidal delta. Ebb shields and ebb spits protect the flood ramp and flood channels from modification by ebb currents. Where the ebb currents break through these barriers, spillover lobes are formed. Furthermore, as a result of time-velocity asymmetry, maximum ebb currents occur after flow is confined to the major ebb channels that border the flood-tidal delta. The high area is usually salt marsh or clam flat areas that have been built above the limit of active tidal currents. The flood-tidal deltas of the Merrimack Embayment were described in detail in an earlier publication by the Coastal Research Group (1969). There is a remarkable similarity between the flood-tidal deltas of the Merrimack Embayment and those observed in other areas, such as on the coast of New Jersey, southeast Alaska, Baja California, and the Gulf of St. Lawrence.

Generalization 5 – mesotidal barrier islands

Much of our work has been carried out on mesotidal shorelines (tidal range = 6-12 ft (2-4 m); Davies, 1964) that have abundant barrier islands. From these studies it is apparent that mesotidal barriers frequently have a common shape, which consists of: (a) an accreting bulbous updrift end (b) a thin mid-section and (c) a series of recurves at the migrating downdrift end of the barrier. An example of a barrier island that exhibits this "typical" shape, from the Copper River Delta area of Alaska, is shown in Figure 27. Accretion at the updrift end of the island is thought to be aided by wave refraction around the ebb-tidal delta, which produces a net reversal and retardation of sediment transport downdrift of the inlet. The middle of the island is essentially a zone of sediment bypassing, whereas the downdrift end is also an accreting zone, which grows by the welding of a series of curving ridges to the end of the island as it migrates.

CONCLUSION

The compilation of results from investigations in different areas at a variety of

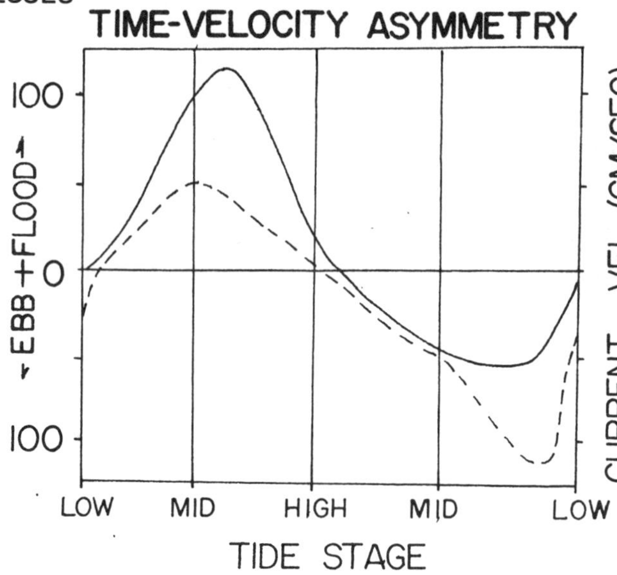

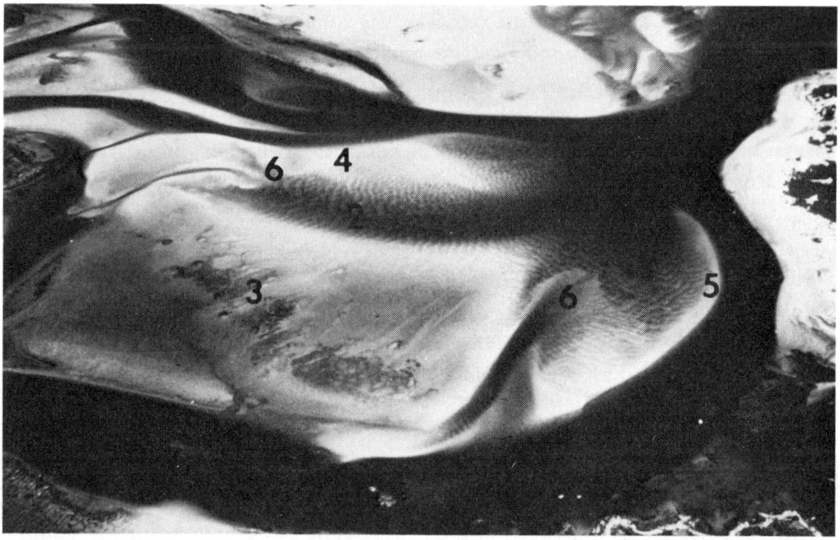

Figure 25. Illustration of the principle of time-velocity asymmetry of tidal currents. The solid line represents a velocity curve from a typical flood channel. Note that maximum flood and ebb velocities do not occur at mid tide. The dashed line represents a typical velocity curve from an ebb channel. Note that maximum ebb velocity occurs late in the tidal cycle, near low tide.

Figure 26. Flood-tidal delta of Essex estuary, Mass. The Gulf of Maine is at upper right. The numbers on the photograph indicate the following morphological subdivisions of the tidal delta: 1. flood ramp, 2. flood channels, 3. high area, 4. ebb shields, 5. ebb spits, 6. spillover lobes. From Boothroyd (1969).

scales enables comparisons of processes and forms in regions characterized by different environmental parameters (e.g., Davis, et al, 1972). The explanation of coastal phenomena at large and small scales requires an understanding of both regional and local processes and forms. In this respect, an initial reconnaissance provides an outline of particular problems for further analysis, while the more detailed studies enable the explanation of the features and problems recognized in the general survey. Although it is not yet possible to standardize methods and techniques for the investigation of the coastal zone, an attempt should be made to design projects so results may be available for comparative studies with work at different scales in different areas. The value of individual projects lies partly in the provision of data which can be used in the more general explanation of form and processes as well as providing models or anwers related to the particular problem under investigation. The study of any section of coast involves some consideration of all the three levels outlined in this paper. Each element of the littoral zone forms a part of the wider system, yet is itself dependent on smaller scale effects.

ACKNOWLEDGMENTS

Financial support for the work discussed in this paper was provided by a

Figure 27. Example of a mesotidal barrier island from the Copper River Delta area of Alaska (tidal range = 3 to 4 m). Dominant longshore drift from right to left. Note bulbous updrift end of the island. Photograph by Frank J. Raffaldi, Jr.

number of different sources. The Alaskan project was supported by the Geography Programs of the Office of Naval Research (Contract No. N00014-67-A-0230-0001). Support for the Gulf of St. Lawrence project was provided by the Geological Survey of Canada through the Bedford Institute of Oceanography (under direction of B.R. Pelletier). The Merrimack Inlet and Plum Island studies were supported by the Coastal Engineering Research Center of the U.S. Army Corps of Engineers [Contract DACW 72-72-C-0032 (Merrimack project); Contract DACW 72-71-C-0023 (Plum Island project)]. The Geology Departments at both the University of Massachusetts and the University of South Carolina provided additional financial support. Dennis Gimmel is acknowledged for his help in processing the photographs.

REFERENCES

Anan, Fayez S. 1971. Provenance and statistical parameters of sediments of the Merrimack Embayment, Gulf of Maine: Ph. D. dissertation, Univ. of Mass., 377 p.

Boothroyd, Jon C. 1969. Essex Bay sand bodies—a preliminary report: *in Coastal Environments: NE Massachusetts and New Hampshire*: S.E.P.M. Field Trip Guidebook, Cont. No. 1—CRG, Univ. of Mass. Department of Geology Publication Series, p. 128-146.

Coastal Research Group, 1969. *Coastal environments: NE Massachusetts and New Hampshire*: S.E.P.M. Field Trip Guidebook, Cont. No. 1-CRG, Univ. of Mass. Department of Geology Publication Series, 462 p.

Davies, J.L. 1964. A morphogenetic approach to world shorelines: Zeits. *fur* Geomorph., 8 (Sp. No.), p. 127-142.

Davis, R.A., Jr., W.T. Fox, M.O. Hayes, and J.C. Boothroyd. 1972. Comparison of ridge-and-runnel systems in tidal and non-tidal environments: J. of Sed. Pet., v. 2, p. 413-421.

Emery, K.O. 1961. A simple method of measuring beach profiles: Limnology and Oceanography, v. 6, p. 90-93.

Fox, W.T. and R.A. Davis, Jr. 1970. Fourier analysis of weather and wave data from Lake Michigan: Tech. Rept. No. 1, ONR Contract 388-092, 47 p.

Goldsmith, V. 1972. Coastal processes of a barrier island, Monomoy Island, Massachusetts: Ph. D. dissertation, Univ. of Mass., 467 p.

Harrison, W., and L.E. Fausak. 1970. A time series from the beach environment—II: Data Report No. 7, Virginia Institute of Marine Science, 96 p.

Hayes, M.O. and J.C. Boothroyd, 1969. Storms as modifying agents in the coastal environment: *in Coastal Environments: NE Massachusetts and New Hampshire*: S.E.P.M. Field Trip Guidebook, Cont. No. 1-CRG, Univ. of Mass. Department of Geology Publication Series, p. 245-265.

Hayes, M.O. and J.C. Boothroyd. 1971. Reconnaissance study of coastal processes on the glacial outwash plain shoreline of southeastern Alaska: (abs.) in Proceedings of Second Coastal and Shallow Water Conference, Univ. of Delaware, Oct. 1971.

Hayes, M.O., V. Goldsmith, and C.H. Hobbs, III. 1970. Offset coastal inlets: in Proceedings of 12th Coastal Engineering Conference, Washington, D.C., p. 1187-1200.

Hine, A.C., III. 1972. Sand deposition in the Chatham Harbor Estuary and neighboring beaches. Cape Cod, Massachusetts: M.S. thesis, Univ. of Mass.

King, C.A.M. 1972. *Beaches and Coasts*: Edward Arnold, Ltd., London, 570 p.

Nilsson, H.D. 1971. Coastal and subtidal morphology of eastern Cape Cod Bay, Massachusetts: M.S. thesis, Univ. of Mass., 174 p.

Pritchard, D.W. 1955. Estuarine circulation patterns: Amer. Soc. Civil Engineers, Proc. 81, 717/1-717/11.

Redfield, A.C. 1965. Ontogeny of a salt marsh estuary: Science , v. 147, p. 50-55.

Chapter 2

THE LITTORAL POWER GRADIENT AND SHORELINE CHANGES

James P. May and William F. Tanner

ABSTRACT

The beach from a headland to the bay head can be marked by the following five significant points:

a. Headland. Maximum total wave energy density (E). Minimum unidirectional littoral component of wave power (P_L). $dq/dx \rightarrow 0$ (where q is quantity of sand transported in the drift system, and x is a measure of distance along the beach).

b. dq/dx = maximum.

c. Maximum P_L. $dq/dx = 0$.

d. dq/dx = minimum (i.e., negative).

e. Bay head. Minimum E. Minimum P_L (probably zero). $dq/dx \rightarrow 0$.

The littoral drift along this beach is from (a) toward (e). E decreases systematically from a maximum at (a) (the headland) to (e), whereas P_L increases from a minimum at (a) to a maximum at (c) and then decreases to a minimum at (e). Furthermore, dq/dx is or approaches zero at three points: (a), (c) and (e), and is maximum at (b) and minimum at (d). Where dq/dx is maximum, the greatest amount of material is being put into transit, hence erosion is greatest in the vicinity of (b). Where dq/dx is minimum (and negative), the greatest amount of material is being taken out of transit, and deposition is greatest in the vicinity of (d). Where dq/dx is near zero, there is no change in the amount of material in transit, hence there is either no transportation (as at (a) and (e)) or there is plenty of transportation but it is neither augmented nor diminished (as at (c)).

A computer program has been written so that deep-water wave parameters (i.e., height, length, approach direction) can be modified across a shoaling bottom having known bathymetry, to produce an expression, at selected points along the beach, for the longshore component of wave power per unit length of shoreline:

Florida Bureau of Geology, Department of Natural Resources, Tallahassee, Florida 32302
Department of Geology, Florida State University Tallahassee, Florida 32306

$$P_L = 0.5\ E\ c\ \sin 2\beta$$

where E is wave energy density, c is wave phase velocity, and β is the angle formed by the intersection of the wave crest and the isobath, all measured at the breaker line. The local sand drift rate (immersed weight/time) can be related to the longshore component of wave power by the expression:

$$I_L = KP_L$$

where I_L is the immersed weight delivered per unit time. For the moderate energy, medium-to-fine sand beaches of the Florida panhandle, the value of K may be about 10^{-2} to 10^{-3}.

Determination of P_L at a series of points along a shoreline permits the computation of a longshore power gradient. A positive power gradient (increasing power) in the direction of longshore drift indicates shoreline erosion (i.e., in the sector from (a) to (c)); a negative gradient indicates deposition (in the sector from (c) to (e)). Calculation of P_L and I_L provides, for the input wave characteristics, volume rates of transport for various parts of the beach, and tends to confirm the theoretical analysis given above.

Studies such as this permit useful statements about where erosion, transportation, and deposition are taking place now (under specified wave conditions). They can be coupled with studies of past transport to show whether or not important time-dependent changes are taking place in the littoral drift system. With both kinds of data, more reliable predictions can be made about probable future trends.

The computer program necessarily involves a numerical approximation to the prototype bathymetry, and produces very small errors. A single small error may lead to cumulated effects at the surf zone: thus not all data points along the beach are equally reliable. Comparison with the theoretical model, however, coupled with more detailed analysis in questionable areas, can lead to the elimination of trouble spots. Within these limits, the theoretical model is confirmed.

INTRODUCTION

This chapter consists of three main parts: (1) a theoretical model developed from well-known facts or from principles, (2) a computer program designed to exploit the model,, and (3) a set of applications to idealized and to real coastal configurations and bathymetries. The respective parts of the research were carried out in that order, and the results are presented here in that order.

ENERGY

It is sometimes convenient to speak of the "energy" of a sediment transport system, such as the littoral drift mechanism, when only a rather loose concept is being treated. In this chapter, energy has the precise definition of physics: it is the capability, within a given system, of doing work, and has the dimensions ML^2T^{-2}.

The energy in a single wave length and a unit width, divided by the wavelength, is the energy content of a wave per unit area; this is the *energy density, E,* with the dimensions MT^{-2} (i.e., energy/area, or ML^2T^{-2}/L^2):

$$E = \rho\gamma H^2/8 \tag{1}$$

where ρ is mass density of water, γ is acceleration of gravity, and H is wave height (for full derivation, see Eagleson and Dean, 1966, p. 51 ff.). (See Table 1 for complete list of symbols and definitions.)

The energy density varies along the beach of half a reentrant (or bay), from a maximum at the headland, to a minimum at the bay head (Fig. 1). This is due to the well-known effect of wave refraction, which concentrates wave orthogonals, and hence wave energy, on those features (such as headlands and capes) having isobaths which are convex seaward, and which spreads wave energy more thinly on those features where the isobaths are concave seaward. In an ideal situation, the decrease in energy density from headlands to bay head is smooth and regular; along many real bays, it may be complicated by the presence of inlets, various bottom features outside the breaker zone, and variations in the materials exposed to wave attack.

At most places, the refracted waves do not strike the beach head on.There is generally s small angle between the wave crest and the beach (the angle β). The angle β can be measured, also, between the wave orthogonal and a line taken normal to the beach. Because β is rarely precisely zero, there is a small littoral component of motion in the breaker zone and in the swash. That is, the uprush of the waves is not precicely up the slope of the beach, but varies to one side or to the other by the angle β. The backwash, which is controlled by gravity, moves directly down the beach slope, and hence does not offset the effect of the angle β. As a result, a small net lateral transport of both water and sediment takes place: this is littoral transport, or drift.

If the wave refraction were such that $\beta = 0$, any sand grains in motion would oscillate back-and-forth along a line normal to the beach, and there would be no littoral transport. A finite non-zero value for β, however, provides for littoral transfer. The larger the angle β, up to roughly 45°, the greater the littoral motion of sand, with a constant energy density E. Hence the actual transport of sand along the beach cannot be a function of E alone.

The littoral transport of sand along the beach is best expressed in terms of energy (not energy density), per unit of distance along the beach, per unit time. Inasmuch as energy per unit time is power, the new expression will be power per unit distance along the beach. This term, if properly defined from the angle β, will be the littoral component of power, hence P_L, having the dimensions

MLT^{-3}.

The total wave power per unit width of crest is given by (Rayleigh, 1876):

$$P = nEc \qquad (2)$$

where $n = (0.5 + kh/\sinh 2kh)$, and expresses the ratio of group (energy) velocity to phase velocity. The total wave power per unit distance along the shoreline delivered at the breaker line is:

$$P_b = Ec \qquad (3)$$

(at the breaker line $n = 1$). The longshore component of wave power that is effective in transporting sediment in the longshore direction is (Watts, 1953; Komar, 1971):

$$P_L = 0.5\, P_b \sin 2\beta \qquad (4)$$

On the cape, or headland, there is a locality where the wave approach direction is 90° from the local coast line. At that point, $\beta = 0$, 2β is still zero, and $\sin 2\beta$ is also zero; hence $P_L = 0$. Likewise, at the bay head there must be another point where $\beta = 0$ and hence $P_L = 0$.

Somewhere between these two "zero power" points, along the side of the bay, the angle β has a series of finite non-zero values, and hence P_L must be greater than zero. The parameter E also changes along the side of the bay, but the changes in E do not negate the changes in β. Therefore, P_L increases from the zero-point on the headland, to a maximum along the side of the bay, and then decreases to the other zero point at the bay head. It is convenient to provide formal designations for these three points: the zero P_L on the headland is at point (1), the maximum P_L locality along the side of the bay is (c), and the zero P_L at the bay head occurs at point (c) (Fig. 1).

LITTORAL TRANSPORT

The quantity of sand moved, per unit time, in the littoral drift system can be designated as q; it has the dimensions L^3T^{-1}. If a mass-rate of transfer is required, the expression $\rho_s q$ can be used (where ρ_s is the mass density of the loose sediment, immersed in sea water). q is a removal rate (where erosion is taking place), a transfer rate, a delivery rate, or a deposition rate; it is the quantity of sand which would cross, per unit time, a line painted at right angles to the beach.

q is also a direct function of P_L, increasing from zero at (a) to a maximum value at (c), and then decreasing to zero at (e). Under high energy conditions, such that E is large, P_L is nevertheless zero, or very close to zero, at (a) and (e), and q is likewise zero or nearly zero at these two points. Obviously, q cannot be constant along the stretch of beach between (a) and (e).

Consider two points, x_0 and x_1, the former located precisely at (a) and the latter located a small distance toward (c). The value of q at x_1 must be finite and non-zero: something larger than at x_0. Hence erosion, however small, must take place at x_1. Designating the quantity of sand in littoral transport by the same subscripts, $q_0 = 0$, but q_1 is greater than zero, and hence at x_1 a supply of sand must be picked up and put in transit. This is, specifically, erosion. Likewise q_2 (observed at x_2) must be greater than q_1, and hence additional sand is picked up from the beach and added to the littoral transport system. The quantity of sand in motion along the beach (q) thus increases, away from (a), for some significant distance. At some point the rate of change of q, along the beach, must begin to decrease; this point can be designated as (b). It is located between (a) and (c).

Changes in q along the beach are easier to understand if examined in terms of a change per distance, x, along the beach: dq/dx. dq/dx is identically the slope of the curve $q = f(x)$, and is zero precisely at (a) = x_a, and likewise must be zero wherever q again becomes zero (at (e) = x_e). dq/dx must be zero where P_L and q are at a maximum: at (c) = x_c.

From x_a to x_c, q is increasing. From x_c to x_e, q is decreasing. Therefore, precisely at x_c, dq/dx must be zero. At this point, dq/dx changes from positive (q is increasing) to negative (q is decreasing), and hence is zero. It is clear that dq/dx must have a maximum value (this location is at point (b) and a minimum value (at point (d)); (see Fig. 1).

The concept of dq/dx is extremely important to our understanding of the littoral drift system. Suppose, for example, that q is larger than zero along a given stretch of beach, and that dq/dx is zero (q is constant). *No* erosion takes place along this stretch of beach, and *no* deposition. For whatever reasons involved in combining E and β, P_L is constant throughout the stretch. The amount of sand moving at the up-drift end is the amount transported past any one point within the reach, and is likewise the amount finally delivered at the terminus. Sand is passing through the system, but the beach is unaffected by that fact, except for the possible exchange of individual grains between the beach and the drift.

The linear rate of change of the delivery rate, dq/dx, must be zero at three points: (a), (c), and (e). It must be maximum (positive) at point (b) where erosion is great, and it must be minimum (negative) at point (d) where deposition is great.

If there are no gains or losses by the complete system from headland to bay head, the volume of sand picked up between (a) and (c) must equal the volume deposited between (c) and (e). Obviously, net gains or losses, such as provided by leakage from or to an adjacent system, or through an inlet, or from a river, will complicate the pattern somewhat.

There is no *a priori* reason for knowing the locations of points (b), (c), and (d), or even for knowing their spacing. Although points (a) and (e) can be located with fair assurance, from maps or air photos, the other three points can be found only as a result of detailed investigation. Hence the distances between points – whether expressed in kilometers or in percentages – cannot be stated in advance. In fact, as wave parameters (including especially approach direction) change from time to time, the locations of points (b), (c), and (d) may be subject to large changes.

COASTAL INDEX

The sand volume delivery rate, q can be rewritten as I_L (the "dynamic transport rate" of Inman and Bagnold (1963), the immersed weight of the sand mass being transported along the beach past a given observation point:

$$I_L = q\rho_s\gamma \quad (5)$$

where ρ_s is the mass density of the moving sand, submerged in sea water as individual grains, and γ is the acceleration of gravity (and hence the product of the two is the weight of the submerged sand in motion). I_L has the dimensions MLT^{-3} (the same as P_L).

The dynamic transport rate can be related to the littoral component of power as follows:

$$I_L = KP_L \quad (6)$$

where K is a dimensionless numerical coefficient which can serve as an index to the maturity of the littoral drift system.

The coefficient K can be obtained by careful measurement, under various conditions, along a given beach (Komar and Inman, 1970); or it can be computed from the estimated mean value of P_L and from longterm measures of q obtained by map differencing techniques (Stapor, 1971). The two concepts of K are slightly different. In the former, K describes the instantaneous condition, which may or may not be significant over long time spans. The latter concept of K describes a long term condition that is important over a specific time span. This latter concept is used here. P_L, in this case, is obtained by wave-hindcasting for the area under consideration. This is accomplished by a computer program which approximates shoreline wave conditions (given bathymetry and deep water conditions) based on the shoaling wave model described below.

Within the study area (Fig. 2), we have determined that the value of K (based on the time interval from 1874-1942) is approximately 10^{-2} to 10^{-3}. This cannot be compared directly with the study of Komar and Inman (1970), who determined a value of K (in the instantaneous sense) to be 0.77.

SHOALING WAVES

As a deep water wave enters shoal depths, a point is reached at which the interaction between the wave-caused fluid motion and the bottom results in a significant drain of energy from the wave. Shoalward from this point (defined as the nearshore zone), the energy drain continues until either the breaker line is reached or the total energy possessed by the wave has been expended. In the former instance, the remaining energy is dissipated through turbulence due to breaking and through bottom friction. In the latter case, the wave form disappears before reaching a breaking condition.

The changes in wave characteristics due to bottom friction were first considered by Keulegan (1948) and Putnam and Johnson (1949). Savage (1953) performed laboratory experiments to evaluate the Putnam and Johnson theory. Additional treatment was given the subject by Bretschneider and Reid (1954), who applied the theory, slightly modified, to field examples on the Texas coast. None of these studies had the advantage of the use of a high-speed, digital computer, necessary for the detailed evaluation of the shoaling wave model.

A completely separate line of work has evolved regarding the longshore drift of sediment due to waves. This type of study has attempted to relate the amount of longshore sediment drift to the longshore component of wave power. Early work included studies by Watts (1953), Caldwell (1956), and Savage (1959), who derived empirical relations. This relationship was placed on theoretical grounds by Inman and Bagnold (1963), Komar and Inman (1970), and Komar (1971).

A logical next step was to combine these two types of studies into a single model so that, given deep water wave characteristics and nearshore bathymetry of an area, the wave power and potential for longshore drift can be determined. We developed a model for the determination of wave energy dissipation due to bottom friction independent of the work of Putnam and Johnson (1949) and Bretschneider and Reid (1954). Although this model turned out to be nearly the same as that of Bretschneider and Reid, it was designed specifically for computer application whereas theirs was presented in graphic form that required previous determination of refracted ray paths and spacing. Our model, furthermore, incorporates the longshore wave power component concept of Watts and later workers.

The model presented below assumes there is no significant energy loss due to internal friction, friction at the free surface, adverse currents, or percolation into the bottom. Lamb (1945) has shown that internal friction as a cause of wave energy loss is negligible for low viscosity fluids such as water. The effect of free surface friction is eliminated by the assumption of the absence of wind. The nearshore sediments along the northwest Florida coast study area (Fig. 2) are composed of fine sand, with low permeability, and hence offer negligible percolation effects. Finally, it is assumed that bottom friction is negligible in depths greater than one-half wavelength. This has been found valid by empirical methods in a series of experiments which will be described elsewhere.

Linear theory (Airy, 1842) is used to compute mean values of wave orbital velocity near the bottom. Linear theory does not accurately describe the *distribution* of water particle orbital velocities through either time or space at shoal depths. Yet, fortuitously, it has been found experimentally that linear

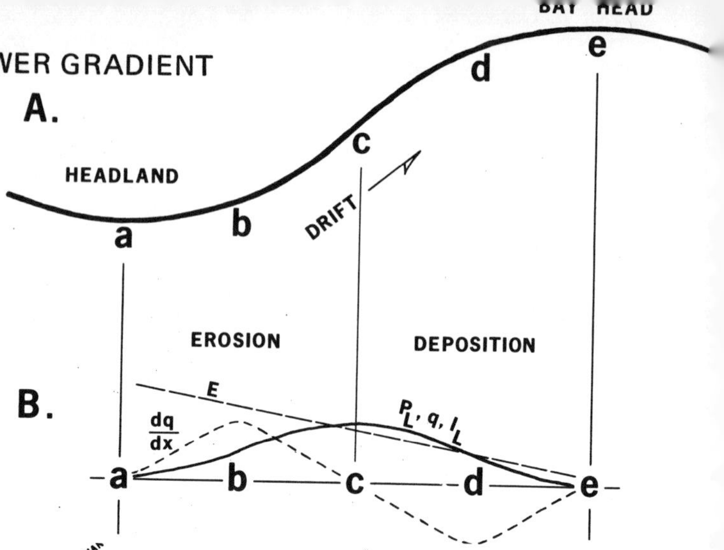

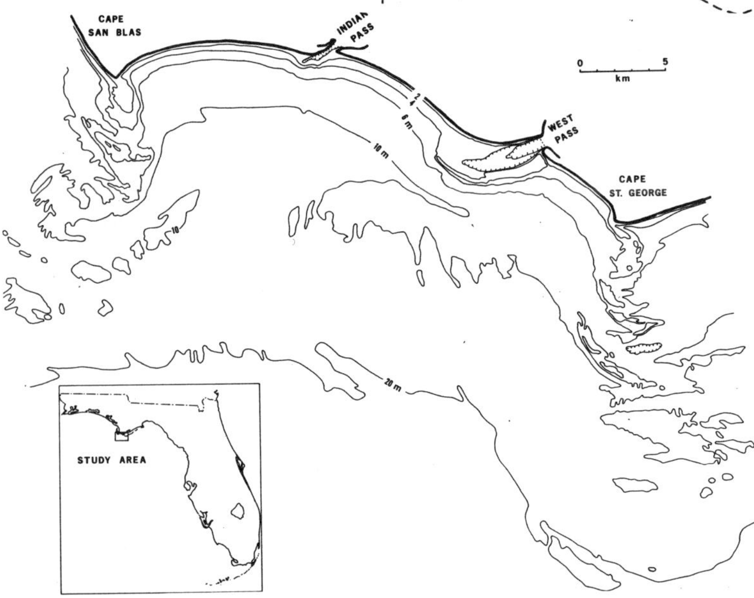

Figure 1. (A) Map of half of a re-entrant, showing headland and bay head, with littoral drift direction indicated. The other half of the bay would be roughly a mirror image of this half. Five important points, discussed in the text, are labelled: (a) through (e). Their positions are relative, rather than absolute, and there is no scale. These points are used to identify parts of the lower diagram (B), which likewise has no scale. Three important curves are shown in (B): the decreasing energy density (E) in the down-drift direction; the unimodal littoral power (P_L) curve which does not differ in shape from a q curve or an I_L curve; and the doubly-reversing dq/dx curve. Where dq/dx is positive (above the baseline), erosion takes place, and where negative (below the baseline), deposition takes place. For fuller discussion, see text.

Figure 2. Index map of Cape San Blas-Cape St. George study area of northwest Florida. Bathymetry is given in meters.

theory provides a reasonably accurate estimation of mean (and maximum) horizontal orbital velocity at the bottom (Le Mehauté, et al, 1968). Furthermore, if the mean orbital velocity at the bottom is calculated according to (1) third order Stoke's theory (Wiegel and Johnson, 1951), (2) cnoidal theory (Wiegel, 1960), and (3) linear theory (Airy, 1842), the value obtained by (3) falls between those obtained by (1) and (2), both of which supposedly better describe the shoal depth situation.

The mechanism of energy dissipation under waves is presumed to be similar to that occurring in rivers, tidal flow, and wind (Taylor, 1919), yet there is a basic difference. Wave induced flow is periodic and oscillatory over a short time base. As a consequence, the flow near the bottom tends to be largely laminar, except for long period waves. In unidirectional fluid flow, a general turbulence obtains after a short time interval. This difference between the two types of flow is significant in the determination of a coefficient of friction (i.e., the coefficient of friction is a function of the Reynold's number). This is developed further in a later section.

WAVE ENERGY DISSIPATION

Using equations (1) and (2) and assuming no energy loss, we can write:

$$(nEc)_i = (nEc)_o \tag{7}$$

and derive an expression for the wave height at point i on a wave ray:

$$H_i^2 = H_o^2\ (\cosh^2 kh / (0.5 \sinh 2kh + kh)) \tag{8}$$

To correct for refraction, assume b_o is the distance between two adjacent wave rays in deep water and that b_i is the distance (along the wave crest) at point i. The energy density varies with the ratio b_o/b_i, hence (8) can be written as :

$$H_i^2 = H_o^2\ (b_o/b_i)(\cosh^2 kh / (0.5 \sinh 2\ kh + kh)) \tag{9}$$

The stress exerted on the bottom by a moving fluid is expressed by:

$$\tau = c_f \rho u^2 \tag{10}$$

where c_f is a dimensionless friction coefficient, ρ is the mass density of water, and u is fluid velocity. The rate of energy dissipation due to bottom friction is (Taylor, 1919):

$$\begin{aligned} dE/dt &= \tau\, u \\ &= c_f\, \rho u^3 \end{aligned} \tag{11}$$

The change in wave height due to bottom friction can be written (using (1)):

$$dH/dt = 4\ c_f u^3 / \gamma H \qquad (12)$$

The horizontal component of wave orbital velocity (u) is (Airy, 1842):

$$u = \sigma H \cos (kx - \sigma t) / (2 \sinh kh) \qquad (13)$$

Inasmuch as u is periodic, equation (12) must be changed from the instantaneous rate to the mean rate. This is accomplished by integrating (12) over one wave period and dividing by the wave period:

$$\overline{dH/dt} = \frac{1}{T} \int_0^T (4\ c_f u^3 / \gamma H)\, dt \qquad (14)$$

Furthermore, u is oscillatory, and we wish to ignore the algebraic sign of the motion (i.e., energy is dissipated by both the to and fro motions); therefore (14) can be written:

$$\overline{dH/dt} = 2 c_f \sigma^3 H^2 / (3\pi\ \gamma \sinh^3 kh) \qquad (15)$$

The mean decrease in height per unit time (sec.) can be changed to decrease in height per unit distance traveled, a more useful expression, by multiplying by the wave period and dividing by the wavelength:

$$\overline{dH/dx} = 2T\ c_f\ \sigma^3 H^2 / (3\ L\ \pi\ \gamma \sinh^3 kh) \qquad (16)$$

where x denotes distance along the ray path.
The wave height at point x_i is determined by combining (9) and (16) to give:

$$H_i = H_o[(b_o/b_i)(\cosh^2 kh/(0.5 \sinh 2kh + kh))]^{1/2} - 0.67 \int_{x_o}^{x_i} T\ c_f\ \sigma^3 H^2 / (L\ \pi\ \gamma \sinh^3 kh)\, dx \qquad (17)$$

Equation (17) is numerically approximated by:

$$H_i = H_o[(b_o/b_i)(\cosh^2 kh/(0.5 \sinh 2\ kh + kh))]^{1/2} - 0.67 \sum_{j=1}^{i} (x_j - x_{j-1})\ (T\ c_f\ \sigma^3\ H_j^2 / (L_j \pi\gamma \sinh^3(k_j h_j))) \qquad (18)$$

where j is the jth increment along the path of the wave ray.

Because of the non-ideal character of the boundary, the coefficient of bottom friction c_f must be determined empirically. Though this is a non-dimensional number, it incorporates several variables and is probably inversely proportional to the Reynold's number (Rouse, 1961, p. 210):

$$c_f = fn\ (us\rho/\mu)^{-1} \tag{19}$$

where u is a velocity, s is a significant length, p is mass density, and μ is the dynamic viscosity of the fluid. The variable u is the wave orbital velocity and, as u is unsteady, it must again be considered in terms of its (absolute) mean value. The length variable s is less well defined. Longuet-Higgins (1970) considered it to be equal to the length of the semi-major axis (horizontal) of the wave orbit ellipse near the bottom. As we are attempting to describe the flow in a layer near the bottom, perhaps s should be determined by the thickness of this layer. That is, s may be the thickness of the layer of water that is significantly affected by the presence of the bottom. Observation over ripple-marked, sandy bottoms indicates that this is approximately equal to 2 to 3 times the height of the ripples. For medium-to-fine sand, intermediate water depths, and wave period of about 5 sec, s is about 10 cm.

The viscosity μ varies with temperature and salinity of the water, but it can be considered to be constant at 10^{-2} dyne-sec/cm^2 for the purposes of this study. If the mean value of u is assumed to be about 10 cm/sec, the Reynold's number is 10^4. According to the experimental work of Blasius (see Rouse, 1961, p. 246-247) with smooth pipes, this Reynold's number corresponds to a coefficient of friction of $c_f = 0.03$. Based on the work of Nikuradse (see Rouse, 1961, p. 250) with artificially roughened pipes, the corresponding coefficient of friction is also $c_f = 0.03$. This value is assumed to be a valid approximation for the purposes of the present study.

THE COMPUTER PROGRAM

A computer program was written that determines the changes in wave characteristics as the ray is tracked across the shoaling bottom to the breaker line. At the breaker line, the longshore component of wave power per unit distance along the shoreline (P_L) is computed.

The computer program requires the following input:

1. Bathymetry: A grid of depth values (XY matrix) taken from bathymetric charts of the study area. Choice of scale and grid-point spacing depends on the size of the area and the level of detail desired.

2. Wave characteristics: the deep water wave height, period, and propagation direction.

With this information, the computer program tracks a wave ray from a deep water starting point along its refracted path to the shoreline (this part of the program was modified from Wilson, 1966). The wave direction and phase velocity are recomputed at short intervals along the ray. The ray spacing is monitored by reference to equal points in time on adjacent rays constructed specifically for this purpose. The wave height is computed using (18). At a point where the wave height/water depth ratio exceeds 0.78, it is assumed that the wave breaks. The wave height at the breaker point is a function of the wave power that remains after dissipation by shoaling, refraction, and friction. Finally, the longshore component of wave power is computed using (4).

HYPOTHETICAL COASTLINE

To evaluate both the model and the approximation technique, several hypothetical coastal geometries were tested prior to the consideration of natural areas. One of these hypothetical cases (Fig. 3) is discussed in order to elucidate some of the characteristics and limitations of the technique. The shoreline geometry consists of a headland-bay head complex formed by arcs of equal radii. Offshore isobaths are geometrically similar to the shoreline and represent an offshore slope of .005.

If the values of longshore component of wave power (P_L) are plotted against distance along the shoreline (Fig. 3) it can be seen that the overall trend is complicated by considerable noise. The reasons for the local fluctuation of P_L is that a continuous surface (the bottom) is being approximated as a series of intersecting planes (the depth grid matrix). As a ray crosses from one depth grid segment to another, it must readjust to the new slope orientation. At times this occurs just prior to breaking, in which case there is insufficient time for proper readjustment of the ray. This may result in unusually large or small values, and may even reverse algebraic sign. If the value appears to be bad (this can be confirmed by checking the immediate history of the ray just before breaking) the value can be rejected.

The noise occurring in the raw data can be reduced by a smoothing technique. The first step is to reject obviously bad values as indicated above. The second step involves construction of a curve (shown in Fig. 3) by using the value for P_L occurring at points where the raw data curve (not shown in Fig. 3, for clarity) crosses the mid-points of grid segments. The mid-point values of P_L are used to calculate the longshore gradient (ΔP_L) and, thus, the potential for erosion or deposition.

Figure 3 also shows the effect of varying the initial wave ray orientation on longshore wave power and power gradient. Below the plan map of the shoreline are given three pairs of graphs corresponding to initial wave azimuth directions of 200°, 180°, and 160° (the compass direction from which the waves come). All points on the graphs project vertically onto the plan map at the shoreline.

The upper graph of each of the three pairs gives the relation of P_L to lateral position along the shoreline. Both the raw data points and the smoothed curve are shown. In this example, the longshore drift direction is from left to right (eastward). The points (a) and (e) indicate the endpoints of the system.

The lower graph of each of the three pairs gives the longshore gradient of P_L. This is computed by a differencing of adjacent P_L values to give ΔP_L (an index to dq/dx).

From this information the positions of the five critical points, (a) through (e), can be located. The varying of initial wave azimuth has little effect on the location of point (c), the divide between erosion and deposition, with this shoreline geometry. Similarly, the locations of points (b) (maximum erosion) and (c) (maximum deposition) are little changed by initial azimuth variation. The locations of the drift system endpoints, (a) and (e), are greatly changed by varying initial wave azimuth. The net results, however, are not as significant as would seem. The drift cell boundaries occur at sites of erosional divergence (headland) or depositional convergence (bay head). The lateral shift of these boundaries does not change the basic depositional or erosional character of the regime at a given place. A location on the headland may at different times belong to either of the two adjacent systems but remain under an erosional regime. Similarly, a point in the bay head may at different times belong to either of the two adjacent systems, but remain under a depositional regime.

Changes in shoreline (and isobath) geometry to less symmetric shapes causes the longshore distribution of P_L to become skewed. That is, the point (c) is not located midway between (a) and (e), but is closer to one or the other. The basic relationships among the points, in terms of erosion, transportation, and deposition, remain consistant. An indication of the effects of complicating shoreline geometry on the model can be obtained from Figure 4.

NATURAL COASTLINE

The model was applied to a natural coast to determine its usefulness. The coast between Cape San Blas and Cape St. George, Florida, was selected (Fig. 2). The choice was based on (1) suitable shoreline geometry (2) first-hand familiarity with the area by the authors, and (3) availability of detailed history of sediment drift and shoreline changes (Stapor, 1971).

The gross geomorphic characteristics of this coast include two shoal-based capes, San Blas on the west and St. George on the east, separated by a long (~35 km), concave-seaward beach. This general form is complicated by the presence of two inlets, Indian Pass and West Pass. West Pass (the more easterly of the two) is the primary outlet for the Apalachicola River and is far more significant than Indian Pass in terms of its influence on coastal form.

The general sense of littoral drift in the area has been to the west (Stapor, 1971). This has resulted in the maintenance and westerly extension of lunate sand bars from the east sides of the inlets (Fig. 2). Immediately behind the bars tidal currents maintain deep channels (up to 15 m in West Pass). Behind the tidal channels the shorelines are convex seaward, the only two points along the stretch of coast between the capes where such shapes occur, and appear to have prograded as a result of being sheltered by the breakwater effect of the lunate bars.

Surface wind data for the northeast Gulf of Mexico (U.S. Navy, 1955) show that easterly winds dominate (80 percent of the time). Due to the configuration of the coast, only winds from the southeast, south, and southwest can have significant effect on coastal processes. Of these winds (which occur 30 percent of the time) 80 percent have an easterly component. The velocities of the easterly and westerly winds are approximately the same, thus one would expect a strong tendency for net westward drift of sediment along the coast. As indicated above, this appears to have been the case.

Computer runs with initial wave approach directions from 135°, 180°, and 225° were performed on this coast. The results are summarized in Figure 4. The curves were drawn by visual approximation.

Note that the values of P_L are both positive and negative. The sign is meant only to designate the sense of drift direction, with eastward being positive and westward being negative. Values of P_L are considered in terms of their absolute magnitude when computing ΔP_L as an index to dq/dx (always computed in the downdrift direction).

The wind blows from the southeast octant (centered on 135°) approximately 58.7 percent of the time (considering only winds from the three south octants). Figure 4 shows the drift direction to be westward (negative values of P_L) along the entire stretch of coast between Cape St. George and Cape San Blas, except for less than one km of shoreline immediately east of San Blas (where P_L is positive indicating eastward drift). There are three zones of erosion (where P_L is absolutely increasing) located just west of (1) Cape St. George, (2) West Pass, and (3) Indian Pass. A fourth zone occurs just east of Cape San Blas resulting from eastward drift. There are, similarly, three zones of deposition located just east of (1) West Pass (2) Indian Pass, and (3) Cape San Blas. The last zone receives drift from the west and from the east.

Of the southerly winds, 26.7 percent come from the south octant (180°). The graph of P_L in Figure 4 shows that this wave direction subdivides the area into three discrete cells with headlands (points (a)) located at (1) Cape San Blas, (2) the convex-seaward shoreline just west of Indian Pass, (3) the convex-seaward shoreline just west of West Pass, and (4) Cape St. George. Values of P_L increase (absolutely) away from these points (erosion) to a maximum (points (c)) and back to zero at the bay heads (points (e)).

The remaining 14.6 percent of southerly winds come from the southwest quadrant (225°). The effect of Indian Pass on the system is greatly diminished compared to 180°. Drift is dominantly eastward, but the system is not integrated. That is, there are local zones of westward drift (negative P_L), as well as local zones of erosion and deposition.

SUMMARY

A model is presented that describes the littoral drift system in terms of the interaction between shoreline geometry and shoaling waves. From the model, a better understanding of present and past conditions may be obtained, as well as a basis for predicting future shoreline developments.

A computer program has been written that allows simulation of the shoreline model. It provides numerical data facilitating (1) evaluation of the model and (2) analysis of certain shoreline regions. Application of the technique to the Cape San Blas–Cape St. George area of northwest Florida supports the basic model and the computer simulation provides information that is meaningful in the interpretation of the geologic history of this coast. Additional studies involving other coastal areas are being made in order to further evaluate the technique and to solve specific coastal problems.

TABLE 1

Definitions of symbols.

Symbol	Definition	Dimension
H	wave height	L
I	immersed weight transport rate	MLT^{-3}
L	wave length	L
P	wave power per unit width of crest	MLT^{-3}
T	wave period	T
c	wave phase velocity	LT^{-1}
c_f	coefficient of bottom friction	-
h	still water depth	L
k	wave number ($2\pi/L$)	L^{-1}
n	ratio of group to phase velocities	-
q	volume transport rate	L^3T^{-1}
s	linear distance or length	L
t	time coordinate	T
u	horizontal wave orbital velocity	LT^{-1}
x	horizontal space coordinate	L
β	wave angle	
γ	acceleration of gravity	LT^{-2}
μ	dynamic viscosity	$ML^{-1}T^{-1}$
ρ	mass density of water	ML^{-3}
ρ_s	effective mass density of sediment (corrected for immersion and porosity)	ML^{-3}
σ	wave angular velocity ($2\pi/T$)	T^{-1}
τ	bottom stress due to fluid motion	$ML^{-1}T^{-2}$

Subscripts

o	refers to deep water
i	refers to point in shoal water
b	refers to point at breaker line
L	refers to shore-parallel component or vector

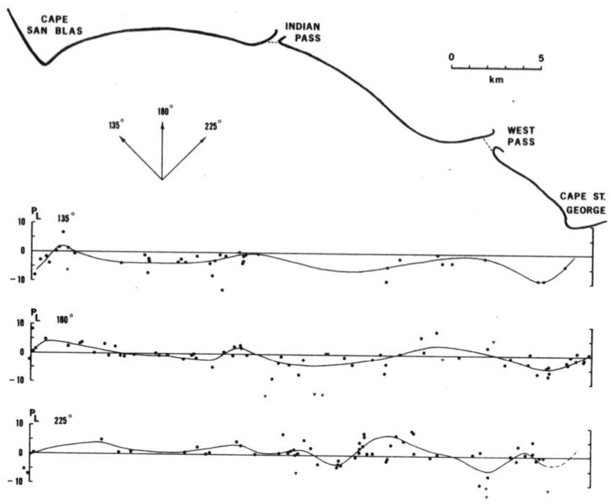

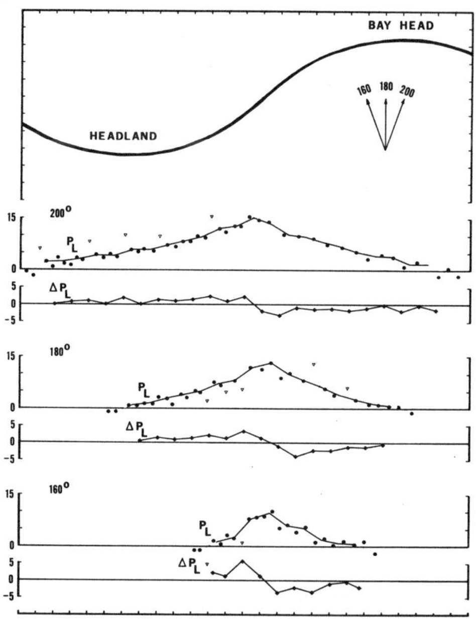

Figure 3. Map of hypothetical shoreline and graphs of longshore variation in shore-parallel component of wave power (P_L) and its longshore gradient (ΔP_L). Points on the graphs project vertically onto the plan map at the shoreline. The vertical axis of the graphs has the dimension joules/m-sec. The open triangles represent rejected data points (see text for rejection criterion).

Figure 4. Map of Cape San Blas-Cape St. George study area and graphs showing longshore variation in shore-parallel component of wave power. Notation and conventions same as in Figure 3.

REFERENCES

Airy, G. B. 1842. *On tides and waves: Encyclopaedia Metropolitana,* London, p. 289.

Bretschneider, C. L., and Reid, R. O. 1954. Modification of wave height due to bottom friction, percolation, and refraction: Beach Erosion Board, U.S. Army Corps Eng., Tech. Memo. 45, 36 p.

Caldwell, J. M. 1956. Wave action and sand movement near Anaheim Bay, Cal.: Beach Erosion Board, U.S. Army Corps Eng., Tech. Memo. 68, 21 p.

Eagleson, P. S., and Dean, R. C. 1966. Small amplitude wave theory: *in* Ippen, A. T., (Ed.), *Estuary and Coastline Hydrodynamics,* McGraw Hill Book Co., New York, p. 1-92.

Inman, D. L., and Bagnold, R. A. 1963. Littoral processes: *in* Hill, M. N., (Ed.), *The Sea,* Vol. 3, Interscience, New York, p. 529-553.

Keulegan, G. H. 1948. Gradual damping of solitary waves: Jour. Res. Natl. Bur. Stand., v. 40, p. 487-498.

Komar, P. D. 1971. The mechanics of sand transport on beaches: Jour. Geophys. Res., v. 76, p. 713-21.

Komar, P. D., and Inman, D. L. 1970. Longshore sand transport on beaches: Jour. Geophys. Res., v. 75, p. 5914-27.

Lamb, H. 1945. *Hydrodynamics:* Dover, New York, 6th Edition, 738 p.

Le Mehautě, B., Divoky, D., and Lin, A. 1968. Shallow water waves: a comparison of theories and experiments, Proc. 11th. Conf. Coastal Eng., v. 1, p. 86-107.

Longuet-Higgins, M. S. 1970. Longshore currents generated by obliquely incident sea waves (parts 1 and 2): Jour. Geophys. Res., v. 75, p. 6778-6801.

Putnam, J. A., and Johnson, J. W. 1949. The dissipation of wave energy by bottom friction: Trans. Am. Geophys. Union, v. 30, no. 1, p. 67-74.

Rayleigh, Lord. 1876. On waves: Phil. Mag., 5th Ser., v. 1, p. 257.

Rouse, H. 1961. *Fluid Mechanics for Hydraulic Engineers:* Dover, New York, 422 p.

Savage, R. P. 1959. Laboratory study of the effect of groins on the rate of littoral transport: Beach Erosion Board, U.S. Army Corps Eng., Tech. Memo. 114, 55 p.

Savage, R. P. 1953. Laboratory study of wave energy losses by bottom friction and percolation: Beach Erosion Board, U.S. Army Corps Eng., Tech. Memo. 31, 25 p.

Stapor, F. W. 1971. Sediment budgets on a compartmented low-to-moderate energy coast in northwest Florida: Marine Geol., v. 10, p. M1-M7.

Taylor, G. I. 1919. Tidal friction in the Irish Sea: Phil. Trans. Royal Sec., A, v. CCXX, p. 1-93.

U.S. Navy. 1955. *Marine Climatic Atlas of the World:* Vol. I, The North Atlantic, Washington, D. C.

Watts, G. M. 1953. Study of sand movement at South Lake Worth Inlet, Fla.: Beach Erosion Board, U.S. Army Corps Eng., Tech. Memo 42, 24 p.

Wiegel, R. L. 1960. A presentation of cnoidal wave theory for practical application: Jour. Fluid Mech., v. 7, pt. 2, p. 273-286.

Wiegel, R. L., and Johnson, J. W. 1951. Elements of wave theory: Proc., First Conf. Coastal Eng., Long Beach.

Wilson, W. S. 1966. A method for calculating and plotting surface wave rays: Coastal Eng. Res. Center, U.S. Army Eng., Tech. Memo 17, 57 p.

Chapter 3

PROBABLE CAUSES OF SHORELINE RECESSION AND ADVANCE ON THE SOUTH SHORE OF EASTERN LONG ISLAND

C. Larry McCormick

ABSTRACT

The shoreline along the south shore of eastern Long Island changes regularly in character along an east-west traverse. A profile of the shore zone drawn at any particular point can be used to infer the recent history of that particular segment of shoreline. The coastal profile west of the till cliffs at Montauk indicate a slowly prograding shoreline from Napeague Harbor to East Hampton Village. West of East Hampton the profiles indicate a receding beach with a few segments exhibiting temporary stability. Further west between Shinnecock and Moriches Inlets the form of the shore zone indicates a rapidly receding barrier island.

Data on the migration of the high water line support the geomorphic evidence of shoreline movement. A close correlation of increasing recession rate with the time that Shinnecock and Moriches Inlets were established suggests a causal relationship. Calculations of amount of sand trapped by Shinnecock and the amount removed by increased recession of the shoreline further suggests that the inlets are responsible for accelerated recession on the intervening beaches.

The recession rate of the cliffs at Montauk govern to a large extent the rate of shoreline advance or retreat for beaches west to Shinnecock Inlet. Advance of the shore in the area from Napeague Harbor to East Hampton represents the infilling of a re-entrant along an otherwise straight shoreline. The rate of recession between Shinnecock and Moriches Inlets is explained as the net effect of natural island migration and the result of reduced sand supply. Increase in sea level is not considered to be a significant factor in recession of beaches along the south shore of eastern Long Island.

INTRODUCTION

The shoreline along the south side of eastern Long Island exhibits a wide variety in character from Montauk Point westward. From Montauk Point to

---

Southampton College of Long Island University, Southampton, New York 11968

Napeague Harbor it is characterized by cliffs up to 60 ft high overlooking narrow cobble beaches. A relatively short segment of the shore between Napeague Harbor and the village of East Hampton consists of a series of low dune ridges trending roughly parallel to the present water line. From East Hampton to Southampton Village is a sandy beach bordered landward by a single narrow dune ridge that rests on the truncated margin of a gently sloping outwash plain. Westward from Southampton the beach and dune ridge are separated from the rest of Long Island by a series of shallow bays. At one time this barrier island system was continuous to Fire Island Inlet, but is now interrupted by Shinnecock and Moriches Inlets.

Much of the shore zone from East Hampton westward is densely populated. The construction of numerous dwellings close to the water line has caused much attention to be focused on the continual recession of the beaches in this area. This concern has resulted in periodic rather piecemeal beach nourishment programs as well as construction of groin fields to trap sand from the prevailing westerly littoral drift. Shoreline recession results from both natural processes and from works of man. An understanding of the causes should be helpful in determining where beach stabilization programs are needed, and to some degree, guide the selection of the kind of remedial action necessary.

CHARACTER OF THE SOUTH SHORE

The Montauk part of the shoreline (Fig. 1) is more irregular than any other segment along the south shore. The high cliffs of Montauk Till and later Wisconsin deposits stand out as bold promontories separated by re-entrants that are backed by similarly high cliffs. The poor correlation between irregularities in the shoreline and the height of the cliffs suggests that topography exerts little control on the present configuration of the coast. Large boulders are more abundant in front of the promontories suggesting that local areas of boulder concentration within the till is responsible for the development of a natural riprap.

The re-entrants contain pocket beaches of coarse sand, gravel, and cobbles that represent the finer fraction of the lag concentrate generated by erosion of the cliffs. The more mobile medium and fine sand is swept longshore by the prevailing westerly drift and the fines are lost offshore. The abundant medium and fine sand available in the till is probably the primary source of material for the beaches to the west. However the subangular character of the large till fragments and the well rounded nature of the cobbles and boulders on the beach indicate that the grinding action of these particles in the surf zone is actively manufacturing finer more mobile constituents that may nourish beaches to the west.

Cliff recession rates vary according to the amount of natural riprap, degree of induration of material composing the cliff, dispersal of wave energy, and the rate at which debris can be removed longshore. Data from historic surveys of this segment of the coast are too scant to allow a precise estimate of the overall rate of cliff recession. However George Washington is said to have surveyed the location of the Montauk Lighthouse and using a recession rate of about 1 ft/yr located the Lighthouse far enough inland to assure its existence for about 200 yrs. Until comparative measurements are made from a series of aerial photographs this estimate seems a reasonable value.

Westward from the cliffs in the Montauk area to the village of East Hampton is a narrow neck of land consisting of several dune ridges sub-parallel to the present shoreline (Fig. 2). Following recession of the Wisconsin ice sheet and near to the time when the sea reached its present level this area was a shallow expanse of open water separating Hither Hills from the rest of Long Island. A spit built westward from Hither Hills toward a low island in the area of Promised Land. At the same time a spit built westward from Promised Land Island and eventually joined the cliffs that had developed at what was then the eastern tip of Long Island. At this stage the land connection between Montauk and the rest of Long Island was complete. As sand drifted westward the re-entrant formed by the joining of the two spits was filled in. The series of subparallel dune ridges indicate positions of temporary stability as the shore built seaward. This is the only area that appears to have exhibited a somewhat continuous progradation along the Atlantic side of eastern Long Island and in this sense is unique.

The presence of many profiles of progradation on the seaward side of the dune face in the Napeague Harbor area indicates that the advance of the shore has continued to the present. A profile of progradation is a convex upward slope on the seaward face of the dune ridge (Fig. 3). This convex slope is generated by accumulation of wind blown sand in front of the dune and is preserved only if the beach is widened to the point that the storm beach does not allow attack of the dune face.

Since a profile of progradation may develop in a few years it is indicative of only the recent past for a particular segment of beach and does not exclude the possibility that the beach may be receding if viewed over a long period. However, presence of many of these constructive profiles along a given segment of beach causes one to favor a long term condition of progradation. The encroachment of grass over the seaward convex slope indicates greater permanence of the feature and this in turn lends stronger support to the supposition of long term advance of the beach.

By contrast most of the dune faces westward of the village of East Hampton display a steep concave upward profile. This profile is generated by undercutting of waves that have recently attacked the dune face. Like a profile of progradation it is not definitive of long term change, but is symptomatic of such long term trends. It definitely indicates recent recession of the shoreline and suggests the same may be true over a long period of time.

All gradations exist between profiles of progradation and recession. Mixed profiles exhibit a nearly vertical upper seaward facing slope interrupted lower on the dune face by a more gentle convex upward slope that grades into the beach

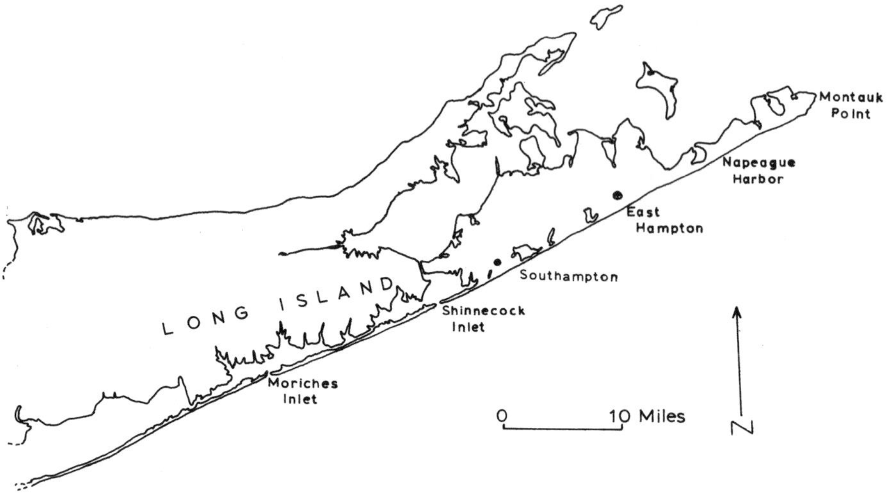

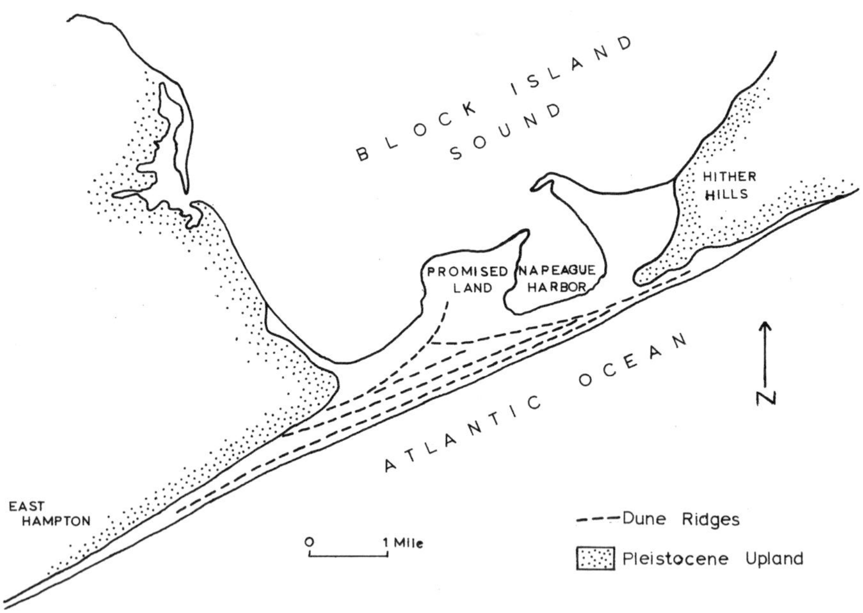

Figure 1. Location map eastern Long Island, New York.

Figure 2. Dune ridges in the Napeague Harbor area.

berm. They are indicative of recent erosional conditions followed by deposition, and probably develop most often along relatively stable shorelines. Additional details of recent beach history can be inferred from the degree of stabilization or size of the convexity on constructive profiles and the freshness of the scarp on profiles of recession.

From East Hampton to Southampton the beach truncates the margin of the gently sloping surface of the Ronkonkoma outwash plain. The normal recessional profiles along this segment of beach are frequently interrupted by a low terrace of glacial outwash that projects beneath the dune into the beach berm. These terraces are most evident immediately following a storm and are then quickly covered by slump from the dune face and wave generated swash bars.

West of Southampton the beach and dune ridge are separated from unmodified Pleistocene deposits by a series of shallow bays. The dune profiles from Southampton to Moriches Inlet exhibit mostly recessional profiles with notable exceptions extending approximately 2 mi eastward from Shinnecock and Moriches Inlets. At these localities the east jetties have resulted in trapping sand from the dominant west drift and the shoreline has advanced. This advance is reflected in the dune profiles.

CAUSE OF SHORELINE CHANGE ATTRIBUTED TO WORKS OF MAN

Five surveys of the high water line from Montauk to Moriches Inlet were conducted by the United States Coast and Geodetic Survey and the Army Corps of Engineers in 1838, 1891, 1933, 1940 and 1956. The data pertinent to these areas are listed in Table 1. These figures represent changes in the shoreline for 41 points spaced approximately 1 mi apart between Napeague Harbor and a location 6 mi west of Moriches Inlet.

Due to the short time interval the data for the 1933 to 1940 interval and the 1940 to 1956 interval were combined into one time period from 1933 to 1956. The variations in character of the shoreline and the locations of the two artificially maintined inlets suggested a natural grouping of the data into three segments: (1) Napeague Harbor to East Hampton (2) East Hampton to Shinnecock Inlet (3) Shinnecock Inlet to a location 6 mi west of Moriches Inlet.

The amount of advance or retreat for all points in each of the three areas was averaged and the yearly rate of advance or retreat was calculated for each interval between surveys. These rates and time intervals are given in Figure 4.

The shoreline westward from East Hampton has continuously retreated for the 118 year period between 1838 and 1956. During this period the rate of retreat between East Hampton and Shinnecock Inlet varied within the relatively narrow limits of 0.9 ft/yr to 1.4 ft/yr. During the first 95 yrs of this period the erosion rate for the Shinnecock to Moriches segment varied from 1.6 ft/yr to 0.7ft/yr. However, during the period from 1933 until 1956 the Shinnecock to

Table 1. HIGH WATER LINE CHANGES

Interval																	Mean
	Napeague Harbor						East Hampton										
1838-1891	+ 50	+ 90	+110	+ 20	+100	+150	- 50	0	-160								+ 34.4
1891-1933	0	0	+150	+150	+100	0	+150	+100	0								+ 72.2
1933-1956	+230	+180	-110	-140	+ 30	+ 60	+ 80	0	+200								+ 83.3
Net Change	+280	+270	+370	+ 30	+230	+210	+180	+100	+ 40								+190.0
	East Hampton													Shinnecock Inlet			
1838-1891	- 10	+100	- 20	-180	- 10	+ 50	- 50	- 20	- 50	-200	-220	-100	- 40	- 50	-110		- 60.6
1891-1933	-100	-120	-200	- 60	-150	-110	0	-100	0	+ 50	+ 10	-100	-150	0	+120		- 60.6
1933-1956	+100	0	+ 60	0	+ 50	-160	-160	-100	- 10	+160	-150	0	-100	0	- 10		- 21.3
Net Change	- 10	- 20	-160	-240	-110	-220	-210	-220	- 60	+ 10	-360	-200	-290	- 50	0		-142.6
	Shinnecock Inlet										Location 6 mi. West Moriches Inlet						
1838-1891	0	+200	+ 80	-300	-260	-120	-180	-200	-120	-160	-180	+ 80	-100*	-140*	0*	+ 80*	- 82.5
1891-1933	0	-120	0	+ 20	-140	0	+100	+100	0	0	0	-140	0*	-140*	- 40*	-160*	-32.5
1933-1956	-280	-140	-140	+ 20	0	-100	-140	-140	-180	-380	0	-160	-420	-140	-160	-160	-157.5
Net Change	-280	- 60	- 60	-260	-400	-220	-220	-240	-300	-540	-180	-220	-520	-420	-200	-240	-272.5

\+ Advancing Shore Line
\- Retreating Shore Line
* Data from 1888 Survey used where 1891 Data unavailable.

All Data in Table From Taney (1961)

Moriches portion of the beach exhibited a dramatically increased rate of 6.8 ft/yr.

By comparison of aerial photographs from 1933 to 1960, the Soil Conservation Service reported a recession rate of 2.5 to 4.5 ft/yr one mi east of Shinnecock Inlet and 8 to 12 ft/yr at Tiana Beach west of the inlet. This unpublished study corroborates the marked difference in erosional rates in the areas east and west of Shinnecock Inlet since 1933.

The first permanent natural break in the barrier island beach system west of Montauk Point is Fire Island Inlet and this opening has constantly migrated to the west in historic times. During the period since 1838 storms have caused breaks through the island at several points but none were stabilized by works of man and all have long since drifted closed. Moriches Inlet opened in 1931, and except for a one year period, has remained open to the present. If it were not for attempts to stabilize this inlet in 1947 and 1952 it would probably have closed. Shinnecock Inlet has a similar history, it was opened in the hurricane of 1938 and by 1952 was stabilized by the construction of stone jetties. The close correlation between the time the inlets were opened and the abrupt change in the erosional rate of the beach between the inlets strongly suggests a causal relationship.

The artificially maintained inlets appear to have acted as barriers to the normal movement of sediment in the littoral drift. While the net littoral drift is to the west, easterly drift is experienced as well. As sediment is shifted past the mouths of the inlets ebbing and flooding tides distribute it over ebb and flood tidal deltas. The sediment trapped by this process represents a net loss to the littoral drift and may explain the greatly accelerated erosion rate between the inlets.

To find if this process does remove a significant amount of sand from the littoral drift, the volume of sand being trapped was calculated from aerial photographs of the flood tidal delta taken in 1955 and 1969. The total volume added was found to be 22.3 million ft^3 or 1.6 million ft^3/yr. The calculation of this volume was made in the area where the delta showed a significant advance into the bay. It does not include any vertical accretion over parts of the delta already present in 1955, and must represent a minimal figure for total accretion.

The volume of sand being removed from the beach face each year was calculated by the equation: $V = a\ b\ c\ (\sin \theta)$ where a is the length of a particular beach, b is the erosional rate per year, c is the length of the beach face normal to the water line within the vertical interval of maximum change, and θ is the average beach slope. This equation is the formula for the area of a parallelogram times the beach length. One side of the parallelogram is given by the erosional rate and the other side by the length of the beach face normal to the shoreline within 8 ft vertical limit. The 8 ft vertical limit was determined empirically by surveying 50 beach profiles to a level below the low tide line. These profiles show marked changes in the beach face in the 8 ft interval below the beach crest,

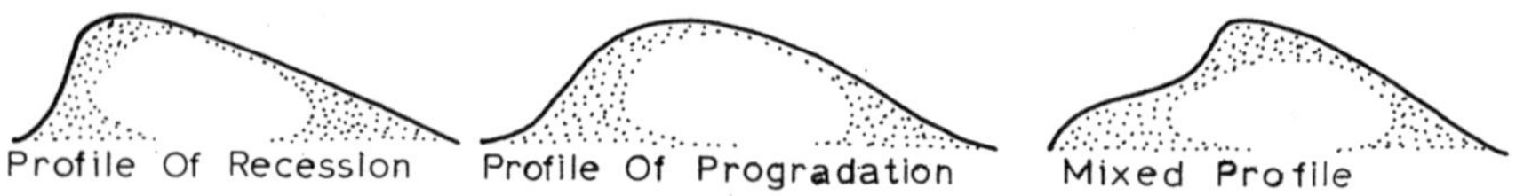

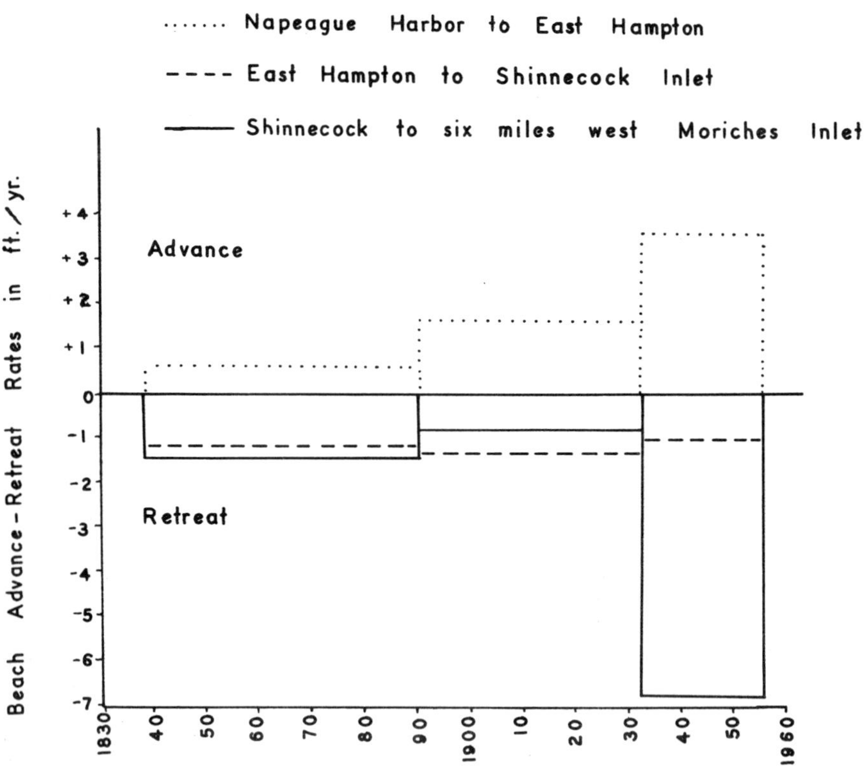

Figure 3. Dune profile types.

Figure 4. Erosion rates for three beach segments on the south shore of eastern Long Island.

but they tend to converge below the 8 ft line. In other words the envelope of change exhibits maximum variation within an 8 ft vertical limit. The length of the beach face normal to the water line was calculated from the average slope of the beach face and the vertical interval of maximum change.

The volume of sand lost annually from beaches between Moriches and Shinnecock Inlets was 0.8 million ft^3 during the interval from 1838 to 1933. From 1933 until 1956 the annual sand loss from the beaches between Moriches and Shinnecock Inlet increased to 4.6 million ft^3, a difference of 3.8 million ft^3/yr. A minimum loss of 1.6 million ft^3/yr is accounted for by trapping of sand on the flood tidal delta at Shinnecock Inlet. If it is assumed that Moriches Inlet, which is similar in size and geometry to Shinnecock Inlet, has the same annual trapping effect, nearly all of the 3.8 million ft^3/yr can be accounted for by loss to the inlets.

It is clear that the figures cited for loss of sand are of the right order of magnitude to account for the rapid acceleration of the erosion rates between the inlets. The loss of sand at Moriches Inlet probably plays less of a role in acceleration of the erosional rate than the loss at Shinnecock Inlet because the net drift is west. However, an erosional shadow does stretch several miles westward of Moriches Inlet.

NATURAL CAUSES OF SHORELINE CHANGE

Perhaps the most apparent natural cause of shoreline recession is that material is being removed from a particular beach more rapidly than it is being supplied. This is certainly the case for the cliffs at Montauk where the prevailing westerly drift sweeps the finer detritus away and the only source for replacement is the cliff section backing the beach.

It is not possible to apply this explanation to the barrier island portion of the shoreline west of Southampton. The age of a peat layer at a depth of minus 7.5 ft mean sea level in Shinnecock Bay indicates that the bay and consequently the barrier island have been in existence for at least 2300 yrs (personal communication Edward Belt, 1972). The barrier island would have been destroyed some time in the past if the recession rate prior to establishment of the inlets is extrapolated backward, and if the recession is assumed to be due to an imbalance between erosion and deposition. The only explanation that seems consistent with the long continued existence of the barrier island, as indicated by the peat, is that erosion and deposition are in balance and the island is a migrating topographic form.

Onshore winds remove sand from the berm and carry it into the dunes. Waves replenish this material and during storms may carry large quantities of sand through the dunes to form washover fans. The movement of material by these processes from the ocean side to the landward side of the island can be considered as a constant pressure applied to the island that results in the establishment of a natural migration rate. Viewed in this way a profile of the island represents a moving wave form, the details of which are in equilibrium with local climatic conditions and incident wave energy.

A prerequisite for barrier island migration is that arrival and removal of sand on the beach face by longshore transport be in balance. An imbalance in either direction for a prolonged period will result in either advance of the shoreline or destruction of the barrier island. Since the source of sand for the barrier island is supplied by erosion of the cliffs to the east and the shoreline must be relatively straight to permit balance between arrival and removal of sand be longshore drift the landward migration rate of the island must be dependent in large part upon the erosional rate of the Pleistocene deposits to the east. Fairbridge and Newman (1968) have suggested relative sea level is rising at a rate of 2 to 4 mm/yr in the New York area and infer that about the same rate of rise applies to Long Island. It is appealing to use this rise in sea level as the cause of barrier island migration. Like water rising and spreading over the inclined bottom of a swimming pool some individuals see an analogy between the migrating water line and recession of the barrier island. The analogy is applicable along sloping shorelines that are not easily eroded, but the barrier island is composed of particulate material that is capable of adjusting to changing conditions of wave energy and tidal height. The beach crest and berm adjust to changes in tidal height that may occur during a single storm or to unusually high or low astromonical tides. Since they possess the ability for such rapid vertical adjustment to change in water level it seems correct to assume that a rise of 4 mm/yr could be accommodated by increase in height of the berm and beach crest. Consequently it would appear that rise in sea level does not play a significant role in controlling recession of the barrier island.

CONCLUSIONS

The rate at which the cliffs at Montauk recede is dependent upon the degree of induration of the material composing the cliffs, the incident wave energy, and degree of natural protection. Immediately west of the cliffs in the area from Napeague Harbor to East Hampton the beach has built oceanward. The advance of the beach in this area is explained by its unique position as a re-entrant along an otherwise straight shoreline. The existence of subparallel dune ridges, numerous constructive dune profiles, and historic surveys of the high water line are consistent with this interpretation. However as the cliffs at Montauk and the outwash deposits west of East Hampton continue to recede this trend will be forced to reverse itself.

From East Hampton to Shinnecock Inlet the shoreline is retreating at a rate governed largely by the recession rate of the till cliffs to the east. Retreat along this segment of the beach is indicated by historic surveys of the water line, and numerous profiles of recession.

In the area between Shinnecock and Moriches Inlets the shoreline has receded at an accelerated rate in response to the loss of sand from the littoral system largely attributable to the tidal action at the inlets. The rate of recession between the inlets appears to be the net effect of the natural migration rate of the island and the effect of a reduced sand supply. The former allows for shoreline recession while the profile of the barrier island is maintained, but the latter has resulted in partial destruction of the barrier island. The weakened condition of this segment of the south shore barrier system may prove hazardous to man and the entire ecologic setting of the associated bays during times of extreme storms.

It is suggested that the distinction made between areas that are suffering island migration as opposed to island segments that are undergoing destruction is a significant criterion in consideration of localities that should be artifically maintained.

ACKNOWLEDGMENTS

The financial support of Southampton Town made this report possible.

REFERENCES

Fairbridge, R.W., and Newman, W.S. 1968. Postglacial subsidence of the New York area: Z.Geomorph.. v. 3, p. 296-317.

Taney, N.W. 1961. Geomorphology of the South Shore of Long Island, New York: Corps of Engineers, Tech. Memo. No. 128, 50 p.

Chapter 4

DYNAMICS OF BEACH ACCRETION IN SOUTH LINCOLNSHIRE, ENGLAND

Cuchlaine A. M. King

ABSTRACT

Beach accretion is taking place rapidly in the area between Skegness and Gibraltar Point in south Lincolnshire, England. The processes involved are generating dune ridges, saltings, new salt marsh in the shelter of a spit, foredunes with intervening marsh strips, and ridges and runnels on the wide foreshore. The formation of these features is related to the pattern of offshore tidal channels and banks. The source of material lies to the north and it is brought into the area by tidal streams and wave drifting. A source in mixed glacial and fluvioglacial deposits, which crop out on the foreshore and cliffs to the north, is supported by a study of the sediment size distribution and electron microscopy. The reason for accretion at this point is probably related to the offshore morphology. Sediment reaches this particular area on account of the nature of the tidal streams and their residual flow, which allow sediment to move with the south-flowing flood stream onto the lower foreshore in the zones of maximum accretion. Trend analysis reveals the marked regularity and magnitude of the accretion in terms of beach ridge size and movement, accretion of sediment on the foreshore and in trends of spit growth. The sediment builds up convex-seaward elongated foredunes on a foundation of stabilized beach ridges on the upper foreshore, while salt marsh accumulates in the sheltered runnels to landward. The relationship between ridge height and accretion confirms the hypothesis that ridges develop where there is a surfeit of sand on the beach, producing a gradient lower than the equilibrium swash slope gradient. In this process waves build the coast out seawards on a broad front, by the creation of "nesses" of accumulation, the position of which is related to the pattern of offshore banks and channels. Thus tidal processes are of fundamental significance on this coast in determining the location of maximum accretion, which changes with time as the offshore morphology is modified by the tidal streams.

Department of Geography, University of Nottingham, Nottingham NG7/2RD, England

INTRODUCTION

Beaches are among the most dynamic of landforms, and for this reason provide excellent material for the study of processes. This contribution aims to analyse the processes by which accretion takes place along a short stretch of the coast of south Lincolnshire in the neighborhood of Gibraltar Point (Fig. 1). A number of questions must be posed, and answers attempted, in such an analysis. They provide examples of different techniques of field observation and methods of analysis.

The main questions that will be considered are the following: (1) where does the material come from? (2) why does it come? (3) how is it brought? (4) how fast and regularly does it come? and (5) in what form is it deposited? The first question necessitates a study of the character of the sediment in the area, so that it can be matched with possible source areas. Modifications en route must also be considered, and these can be studied by several techniques, such as stone shape analysis and sand grain electron microscope studies. In order to answer the second question it is necessary to examine the geomorphic setting of the area in relation to the conditions required for accretion to take place. This leads to the third question, which necessitates a study of the processes that operate in and around the area. The processes provide the dynamic aspects of the study; they are instrumental in bringing the material into the area. They include the waves, tides and the wind, and their associated currents. The fourth question is concerned with the time element, itself an integral part of the dynamic approach. It provides an opportunity to apply time series analyses of various types in order to assess the regularity and other temporal characteristics of the operation of the processes. The final question seeks to elucidate the form of the features built up, and hence is an essentially morphologic aspect of the enquiry. Relationships, however, can be established between form and process, in terms of the amount of material added to the beach and the resultant beach morphology.

The stretch of coast under consideration lies at the northwest corner of the Wash (Fig. 1). Between Skegness at the north end and Gibraltar Point to the south the coast runs approximately north-south, while to the south of Gibraltar Point the coast turns abruptly inland to the west. The 3 mi coastal stretch between Skegness and Gibraltar Point consists of a number of distinct geomorphic units, each of which plays a distinct part in the processes of accretion in the area. As a whole the zone of accretion is sharply differentiated from a zone of erosion extending northwards from Skegness to Mablethorpe (Plate 1). Erosion has taken place intermittently in this area over the last 750 yr, while the zone of accretion to the south suffered from erosion, at least at its northern edge until about 150 yr ago.

The zone of accretion is characterised by features that owe their form to a variety of different processes. It is divisible into (1) dune systems, (2) salt

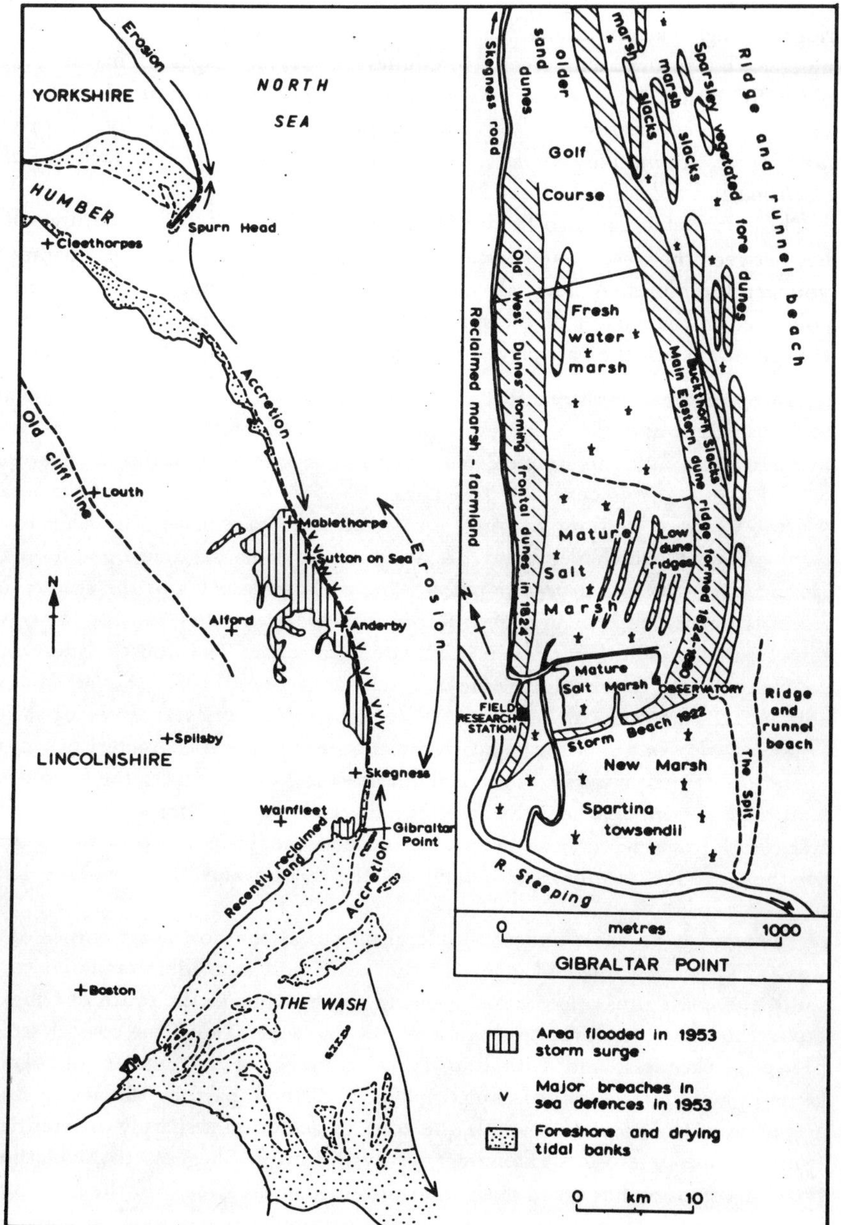

Figure 1. Map to show the setting of the Gibraltar Point area of accretion. The different environments within the area are shown in the inset.

marshes, (3) a spit, (4) a ridge and runnel beach, and (5) offshore banks.

The dune system can be furthur divided into an older, mature western dune ridge, a high eastern dune ridge, which formed during the latter part of the nineteenth century, and recent foredunes of varying stages of maturity. The older foredunes carry a vigorous growth of *Hippophae rhamnoides* while the youngest ones nearest to the sea are just being colonised by *Agropyron junciforme, Ammophila arenaria* and annual sand-loving plants, such as *Cakile maritima* and *Salsola kali* (Plate 2).

The salt marsh can also be divided into a number of types; a mature salting lies between the western and eastern dune ridges, while strip saltings separate the younger foredunes. A new salt marsh, in the slobland stage, is being actively colonized in the shelter of the spit. Both the dunes and the salt marsh illustrate the importance of the part played by vegetation in this zone of coastal accretion.

Plate 1. Oblique air photograph to show the start of the zone of accretion on the northern edge of Skegness. The beginning of an upper foreshore runnel and embryo dunes can be seen in the foreground.

One of the most important distinctions between the mature marsh and the recent marsh development is the presence of *Spartina townsendii* in the newer marshes. This vigorous plant leads to a rapid rate of accretion where it grows profusely in the new marsh and some of the strip saltings.

The spit prolongs the alignment of the coast south of Skegness thus forming an embayment in its shelter where the coast swings westwards. The spit was initiated after a severe storm in 1922, when the southern end of the eastern dune ridge was truncated and driven over the mature marsh. A reversal of the drainage pattern in the mature marsh resulted. The development of the spit during the last two decades has been recorded by repeated plane table surveys, mainly at annual intervals. The spit is a wave-formed feature that prolongs the upper beach to the north.

The most significant feature of the foreshore is the presence of well developed

Plate 2. Oblique air photograph of the area of accretion south of Skegness. The eastern dune ridge runs down the centre of the photograph. The convex curvature of the foredunes seawards of the eastern dune ridge is apparent. Their increasing maturity towards the eastern dune ridge can also be seen. The mature marsh lies to the west (left) of the eastern dune ridge, with the western dune ridge on its western margin. Two nesses of accumulation can be seen in the ridge and runnel pattern on the foreshore. Profile 4 lies across the northern one and profile 2 across the southern one.

ridges and runnels. They represent an imbalance between the overall beach gradient, which is low on this actively accreting shore, and the natural equilibrium beach gradient, which is determined by the type of material and the wave dimensions. The ridges disappear on the lower foreshore, giving way to a muddy, low-gradient zone, which is protected from wave action by the presence of banks offshore.

The offshore banks are formed by the tidal currents. In this area tidal processes are particularly important in connection with coastal accretion, and the offshore zone must be considered as an integral part of the zone of coastal accumulation. The tidal streams are rectilinear and reach considerable velocities so that they are effective transporting agents. The streams are concentrated in separate channels on either side of the banks, some channels being dominated by the south-flowing flood currents and others by the north-flowing ebb currents. In each type of channel the residual of the tidal streams ensures that material moves in a dominant direction along the channel. The pattern of banks and channels is, therefore, of considerable importance in connection with the major zones of accretion.

SOURCE OF THE MATERIAL

The wide range of different environments present in the Gibraltar Point area means that a considerable range of type of material is involved in the accretional processes. The dunes are formed exclusively of sand, while the marsh deposits are much finer. Silt sized particles predominate over much of the salt marsh area, although in some parts there is a considerable admixture of sand. The spit is mainly composed of sand, although pebbles are also present on the spit crest. Pebbles also occur in the foundations of the old beach ridges that are exposed in parts of the mature and new salt marshes. The foreshore is basically sandy, although in this environment also other materials occur. There are pebbles on the upper foreshore and on some of the upper ridge crests, while on the lower foreshore the shelter of the offshore banks allows finer silty sediment to accumulate. Fine material can also settle in the runnels particularly near the top of the beach. The offshore banks are formed of sand, but the vigorous tidal streams in the channels do not allow fine sediment to settle in them. Thus it is apparent that material of a wide range of size must be available to allow accretion to take place in the variety of environments involved in this area.

There are two possible major sources of material. It could come from inland via the rivers, or it could come from the sea. The latter source involves transport of material either along the coast by longshore drift, or from offshore. The character of the sediment in the differet environments in the area must be examined in assessing the relative importance of the possible sources and methods of transport.

The sediment ranges form clay, through silt and sand to pebbles. The most likely source for such mixed sediment is the glacial material that crops out on the Lincolnshire foreshore in places to the north of Skegness, and in the rapidly eroding cliffs of Holderness, Yorkshire, north of the Humber. Post-glacial silts also suffer erosion on the foreshore to the north of Skegness, when the beach is low or lacking. The glacial deposits could provide all the necessary grades of sediment. The rivers, on the other hand, that now flow into the sea near Gibraltar Point are very sluggish and could only supply the finer sediment. It thus seems reasonable to consider that the bulk of the sediment reaching the area comes from the coastal exposures of till and glacial sands and gravels to the north.

The variety of material in the area is related to the range of environments present, each of which can be characterized by its energy level in terms of the sediment character. The area as a whole can also be considered from the point of view of the energy generated by the processes. The area is situated on the east coast of England and is thus sheltered from the main Atlantic waves, and the strongest winds, which are from the west, blow offshore. Furthur protection from wave action is provided by the offshore banks, on which the waves break at high tide and which dry out at low spring tide. The complex offshore relief and shallow nature of the North Sea mean that wave action is rarely violent in this area.

The low energy conditions are reflected in the nature of the quartz sand grains as studied through an electron microscope. Five samples from the foreshore and offshore banks have been examined and compared with sand from an exposed west coast Irish beach and with fluvioglacially modified glacial sand. The east coast sand can be readily differentiated by its extremely weathered appearance, intense surface weathering is apparent in the form of linear peeling of the surface of the grains, cracks and a cauliflower type appearance. Despite the high degree of chemical weathering the grains also show evidence of previous glacial action in the form of parallel lines on the grains. This feature confirms their original source. A few marine features are also evident. The present character of the sand indicates a low energy marine environment in which chemical activity is effective in modifying the surface of the quartz sand grains. Both accretion and erosion by peeling are evidence of chemical activity rather than mechanical action. The wetting and drying to which the particles are subject in this macrotidal environment probably play an active part in the chemical processes. The low energy environment is also apparent in the large amount of fine sediment present in the area. Thus both the detailed electron microscope study and the gross distribution confirm the area to be one of low energy.

Despite the general low energy of the environment, there are considerable variations within the area. These variations are related to the nature of the sediment. Some examples of the range of sediment size distribution are given in Figure 2, in which mean size is related to skewness and sorting. The 20 samples are derived from two profiles surveyed across the beach and from the channels and banks offshore (Figs. 3 and 4). The mean size, M_z in ϕ units ranges between -1.29 and 5.07. The former is found in the tidal stream channel offshore at a depth of 33 ft and the latter is a low tide sample, at a point where the beach is

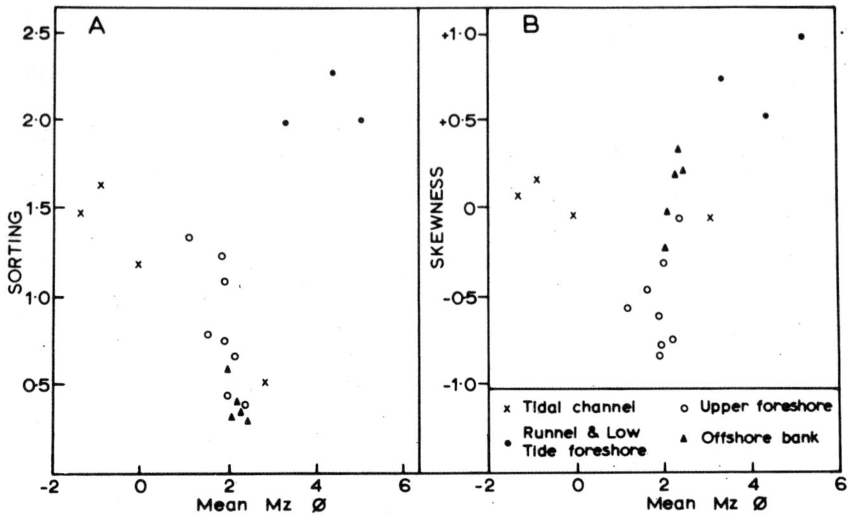

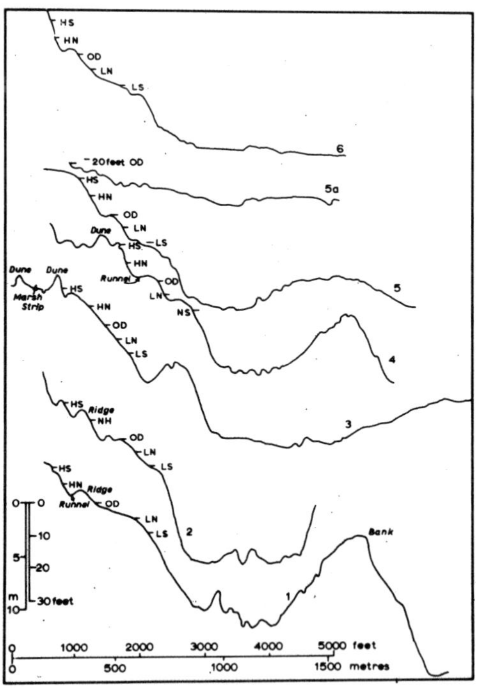

Figure 2A. The relationship between material size and sorting for sediment from four different environments. The offshore sampling sites are shown in Figure 4

Figure 2B. The relationship between material size and skewness for sediment from four different environments.

Figure 3. Profiles surveyed in the offshore zone between Skegness and Gibraltar Point. The position of the profiles is shown in Figure 4.

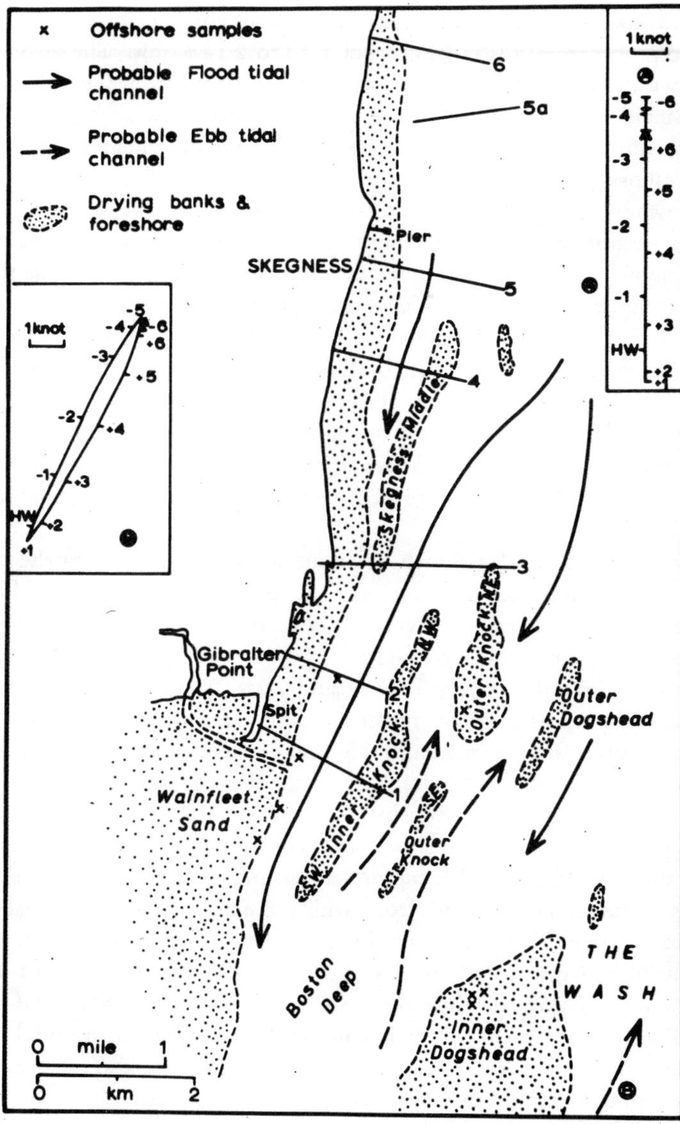

Figure 4. Map of the offshore zones between Skegness and Gilbraltar Point. The inset shows the tidal residual vectors at points A and B. H.W. refers to High Water at Immingham in the Humber to the North.

protected by offshore banks from wave action. The 11 samples from the two profiles fall into two major groups. In the first are the fine sediments of the runnels and lower foreshore, and these fall into silt and very fine sand grades. In the second group are sands ranging from 1.13 to 2.14 ϕ units. The second group includes two dune samples, which can be differentiated from the beach samples by their smaller mean size. The differences, however, are slight, owing to the limited amount of wind action on the dunes that are only in an embrionic form on the foreshore.

The skewness (Figure 2B) also strongly differentiates the sediments of the quieter environments from those in the more vigorous ones. The fine samples from the runnels and lower foreshore all show positive skewness, while the ridge and dune samples show negative skewness. This is a reflection of the more active wave action that builds up the ridges, and also indicates that the wind has played little part in changing the size distribution in the low embryo foredunes on the foreshore. The negative skewness indicates an absence of fines, a quality characteristic of a more energetic environment. The fine material can, however, settle in the runnels and the lower foreshore from which wave action is excluded. A tail of fines, therefore, gives the positive skewness in these situations.

Negative skewness is also shown by some samples derived from the offshore channels. The channel samples can also be distinguished from the foreshore samples by their much coarser nature. Between 25 and 33 ft below O.D. (7.6 and 10 m) the mean sizes in ϕ units of 3 samples were -0.90, -0.06 and -1.29, falling into the granule or very coarse sand class. The deeper samples can be differentiated in terms of their greater size and some show a positive skewness characteristic of a flowing medium. The banks were mostly composed of fine sand between 2.06 and 2.48 ϕ units. Most of these samples also showed a positive skewness suggesting sedimentation by tidal streams rather than by wave action. The median diameter of the offshore bank samples closely resembles that of the ridges on the foreshore, and this suggests a possible connection between the two, the banks forming the reservoir from which the foreshore is replenished. The differences of size distribution could be accounted for by the variation in sedimentation processes in these two different environments, one of which is basically wave dominated while the other is dominated by the tidal streams. Sorting (Figure 2A) on the whole is considerably better on the tidal offshore banks than it is on the foreshore, where stones and silt are also being carried by the waves for deposition in suitable locations within the area.

The main conclusions to be drawn from the study of the sediments are: (1) they are derived from a mixed source; the glacial deposits exposed on the foreshore and in the cliffs to the north are the most likely source. (2) the general energy level of process activity in the area is low. (3) within the area of accretion there are several distinct environmental units each characterised by its own distinctive sediment characteristics. Further points concerning the significance of sediment character will be mentioned in connection with beach ridge morphology in a later section.

CAUSE OF ACCRETION

One of the main problems to be considered in connection with the zone of accretion is its very abrupt termination northwards where it abuts against a stretch of coast that has been and is still liable to erosion (Plate 1). Another point that requires elucidation is the reason for the very localized and varying positions of maximum accretion. At these positions "nesses" of accumulation build out and their pattern can be studied by examining the pattern of beach ridges as revealed on air photographs (Plate 2).

A significant point in connection with the abrupt termination of the zone of accretion is its coincidence with a change in the offshore morphology. The two are very likely to be causally connected. To the north of Skegness the offshore profile, surveyed in 1964, shows a sharp drop below low spring tide level to a depth of 5 m about 300 m offshore, beyond which point the offshore gradient is very flat and level (Figure 3, profiles 5a and 6). The sea bed is only thinly covered by sediment in this zone according to A. H. W. Robinson (1964), whose work in the offshore zone is of interest in this connection. The smooth, level offshore relief continues to the point where accretion starts near the pier at Skegness. The profiles offshore to the south of this site show increasingly well developed offshore banks and intervening channels (Figure 3, profiles 1 to 5). These features are formed by the reversing tidal streams of this area. The pattern of banks and the position of the profiles are shown on the map (Figure 4). The most northerly bank is the Skegness Middle Bank, which is separated from the shore by a flood tidal stream channel. This channel is dominated by the south-flowing flood stream and as a result it shallows southwards running up onto the lower foreshore just south of profile 4 at present. Profile 4 is situated at the southern edge of Skegness in the zone of maximum accretion. A study of earlier Admiralty charts shows that this bank has been developing steadily in size. Its associated flood channel has become more effective between 1918 and 1956, the dates of the last two available surveys, and it facilitates the southward movement of material up onto the foreshore in the area of maximum accretion. The feature was well developed about 1828, but was not conspicuous in 1871, dates when surveys were made. Another flood channel lies seawards of the Skegness Middle Bank, but it does not run up onto the foreshore. Instead it merges southwards into the Boston Deep, which is an ebb channel, in which the residual of the tidal streams is directed northwards. This ebb channel then divides around the Southeast Outer Knock Bank and the Northeast Outer Knock Bank, while flood channels occur on either side of the Outer Dogshead Bank, which lies further offshore. The Inner Knock Bank has a flood channel on its landward side, and an ebb channel on its seaward side. The offshore morphology, therefore, is very complex in the zone off the accreting coast. The pattern of banks and channels also changes with time. The channels provide the main routes of sediment transfer, while the banks provide the reservoir of sediment that builds up the beaches, as suggested by the sediment size analysis.

The variability of the pattern of banks and channels offshore is associated with the variability of the zones of maximum accretion on the foreshore. The zones of maximum accretion on the foreshore at any one time will be those where flood channels tend to allow material to come ashore from the north, as the flood tidal stream residual is directed southwards. This is now occurring between profiles 3 and 4. The pattern of dune ridges in the zone of accretion (Plate 2) shows that different stretches of the coast of accretion have received most sediment at different periods of time. There is not enough detailed evidence of former movement of the offshore banks and channels to relate these in great detail to the changes on the foreshore, nevertheless the relationship between the two sets of features is probably close. More detailed studies of the offshore morphology are about to be undertaken to enable the relationship to be studied more closely.

Tidal processes are of major importance in accounting for the accretion and its variation over time on this stretch of coast. They are fundamentally responsible for bringing material into the zone of accretion, where it is built into the characteristic features by the waves. The pattern of the offshore banks suggests a complex route of sediment supply, under the influence of tidal residual streams. The movement of the sediment is predominantly alongshore in the zone in which these streams are dominant, but at certain localities the morphology is such that material is directed up onto the foreshore. In these positions the characteristic nesses of accumulation build up and out, thus causing accretion on a relatively broad front.

PROCESSES OF ACCRETION

The discussion of the offshore morphology pointed to the importance of tidal processes in accounting for the accretion south of Skegness. Thus both tides and waves must be taken into account in considering processes of accretion, while in the formation of dunes the wind is also important, and vegetation plays a major part in the development of both dunes and salt marshes.

The tide is important both in bringing sand to the foreshore and in the accretion of finer sediments in the salt marsh environment. The tidal range in the area is considerable, exceeding 6.2 m at extreme spring tides, while the mean neap tide range is 3.0 m. The maximum range is 6.6 m at the spring equinoctial tides. Tidal streams are also strong, exceeding 4 knots (200 cm/sec) in the tidal channels between the banks and the shore. Even in the open water off Skegness the tidal streams reach 2 knots (100 cm/sec). Both the tidal range and the tidal streams play a part in accounting for the accretion in the area, but the streams are the most important influence in that they are responsible for the bringing of sediment to the area, while the tidal range affects the resultant morphology.

Some evidence on the transporting capacity of the tidal streams was obtained by Robinson (1964) in his work with sea bed drifters. These are carried from their release point by the tidal streams and eventually those that become stranded on the beach and are returned can be used to establish the pattern of tidal residual movement. The returns from a set of drifters released off Spurn Point at the mouth of the Humber showed that many came ashore on the south side of the Humber in the area of coastal accretion in North Lincolnshire. None were recovered from the stretch between Mablethorpe in the north and Huttoft in the south. This is in the zone of erosion and it is significant that no drifters came ashore in this area, which is heavily frequented and has a relatively narrow beach, so that any drifters coming ashore in this area would almost certainly have been found. Furthur drifters, however, were recovered from the beach to the south of Skegness in the zone of accretion under discussion. These drifters took a considerable time to reach the Gibraltar Point area, some not reaching the foreshore until after the 91 days or even 121 days. The slow time of arrival is probably the result of the complex travel route of the drifters along the tidal channels off Gibraltar Point. Further drifter studies are being planned to study the pattern of sediment transport in more detail in this complex offshore area. This method of study provides useful information, but only long term movements are indicated and the precise route and rate of progress cannot be established by this means.

Other methods of studying tidal processes include current observations at one point over time. These observations provide evidence that can be used to establish the tidal stream residual direction and strength by the construction of vector diagrams. The results of such diagrams confirm that the tidal residuals are important in influencing the transport of material along the channels. Observations of sediment also support these findings. In the flood channel off profile 1, off the spit at Gibraltar Point, samples of sediment became progressively finer along the length of the channel as the speed of the tidal streams was reduced. The coarseness of the material where the channel is best developed is evidence of the strength of the tidal stream within it. The morphology of the offshore banks also indicates the effectiveness of the tidal streams. On the Inner Dog's Head Bank a series of well-developed ebb-tide sand waves had a strong asymmetry, indicating their formation by the ebb tide which is dominant in this area. The sand waves were aligned, normal to the ebb tide flow and had a mean length of 27.5 ft (8.4 m) and a mean height of 1.2 ft (36.6 cm).

The waves are, however, the dominant process on the foreshore. They build up the ridges on the foreshore in those areas where the general overall gradient of the foreshore is less than the equilibrium gradient they require. Most of the waves are short, having been generated within the North Sea, and high waves are rare. L. Draper (1968) gives the most common wave size as recorded at the Smith's Knoll Light Vessel in the North Sea 22 mi (35.2 km) east-northeast of Great Yarmouth as 2 to 3 ft (60 to 90cm) high and with a period of about 6 sec. The highest wave recorded was 24 ft (7.3 m) with a period of 8.4 sec. Waves were considerably higher in winter than summer.

The short waves, with a relatively high steepness value of about 1:40, would need a fairly steep equilibrium swash slope gradient, which would exceed that of the gentle overall gradient which is lowest where accretion is greatest. The ridges, therefore, represent the attempt of the waves to build up their equilibrium gradient on the too flat overall gradient. The large tidal range results in ridge building taking place over a wide stretch of foreshore. The ridges are formed by the swash of constructive waves, which move sand up the beach. The ebbing tide is concentrated landward of the ridge and the runnel develops by this means. The action of the swash of the waves in building the ridges is seen in the texture of the foreshore. The steep swash slope is formed of firm, smooth sand. The developing runnels, the flatter areas between the ridges, and the ridge crests are of soft, rippled sand, because in these low gradient parts of the foreshore the swash is ineffective. The exposure of the offshore banks towards low water furthur protects the lower foreshore and prevents ridge building in this zone, but allows the deposition of fine sediment.

The maximum fetch from which the flatter, ridge-building waves come is to the north and northeast. Waves from this direction reach the coast at a slight angle, which is reflected in the alignment of the ridges. These diverge slightly southwards away from the foreshore so that they move bodily southwards along the coast under the influence of the waves from the north. On any single profile this reults in a landward movement. The pattern of the ridges on the foreshore confirms the fact that the dominant direction of sediment movement under wave action on the foreshore is to the south. Thus along this stretch of coast transport of sediment by both flood tidal stream residuals and wave movement is predominantly southwards. Thus further confirming that the source of material lies to the north. From the point of view of wave action, as a whole, the coast is in a zone of low energy, a point that is confirmed by the electron microscope examination of the character of the quartz sand grains.

RATE OF ACCRETION

Rates of accretion vary in the several environments within the area and differing methods of measurement are needed in the different zones. Observations have been made on the rate of marsh accretion in the new marsh behind the spit, the movement and size of the beach ridges have been recorded, and the growth and movement of the spit has been assessed over a period of about two decades. The growth of the ridges and the spit are intimately linked with the accretion on the foreshore, and this has also been measured. Time series analysis provides techniques that can be used to analyze these data.

Salt marsh growth has been recorded by means of several detailed surveys of the new marsh by plane-tabling and contouring. Three surveys have been carried out at roughly 10 yr intervals in 1951, 1959 and 1969. They reveal rates of silt accumulation of up to 9 in (23 cm) in 10 yr over much of the marsh. The rates of growth are rather slower during the second interval than the first. The slowing down of accretion in the marsh probably reflects the stage of development, and the nature of the vegetation, *Spartina townsendii* was spreading very rapidly over the marsh during the first interval, but was becoming less vigorous in parts of the marsh during the second interval, owing to build up to levels above its optimum. In the second period growth has exceeded 1 ft (30 cm) alongside the spit, where waves wash sand onto the marsh across the spit at the highest tides. Accretion was also rapid in the area of the marsh nearest the mature marsh where the vegetation is mixed and in the most advanced state and is most profuse. The axis of the marsh has built up least rapidly, probably owing to the greatest rate of tidal drainage from this area, allowing less material to settle here. The accretion in the salt marsh is dependent on material brought in by the tide, which regularly floods the marsh twice each day except in the higher part of the marsh at neap tide.

Accretion on the beach is dependent on sand brought onto the beach and built up by the waves, although the part played by the tidal streams is also very important as already considered. The accretion on the beach is in the form of ridges and an estimate of the movement and development of the ridges over time can be obtained from the long series of profiles that have been surveyed between 1951 and 1971. Measurements of the position of the ridges on the profile, as well as changes in their height over time, have been analyzed. The fact that the ridges move landwards on any one profile is due to their transference south, as sand is pushed over their crests by the waves.

The rate of the landward shift can be analyzed by trend analysis. As long as no ridge grows up to seaward to shelter a high ridge on the beach, the upper ridge will move rapidly landwards at a rate determined by the angle it makes with the line of the coast, its position on the foreshore and the rate of sand supply. Movement is most rapid in the central part of the foreshore where the waves can operate for a long period, and where the level is above that to which the offshore banks protect the beach from waves. The maximum recorded rate was 340 ft (104 m)/yr during 1964 for a ridge on profile 4. The trend pattern of ridge movement is shown in Figure 5. The ridges move rapidly up the foreshore until they become stabilized by the growth of ridges to seaward. The upper ridge then becomes static, and develops into a foredune, as windblown sand accumulates around debris and vegetation becomes established.

The trend of ridge height is closely connected with the amount of accretion on the beach because the more the accretion the gentler the overall beach gradient becomes, and as a result the higher the ridge must be to reach the equilibrium gradient. This relationship will be considered further in the next section, as it is concerned with the form of the beach.

The amount of accretion on the beach can also be analyzed by various time series methods. An analysis of the trend of accretion on profiles 1, 2 and 4 has been carried out over the period from 1953 to 1971, the resulting trend equations being $Y = -5 + 13.35\ X$ for profile 1, $Y = 976 + 124.4\ X$ for profile 2 and $Y = 1690 + 190\ X$ for profile 4 (units in ft) for the 19 yr period. The equations illustrate important differences in the amount of beach accretion over time on the three profiles. The maximum amount is accumulating at profile 4, while profile 2 is also gaining a considerable amount of material. Profile 1, which lies across the spit, has made little consistent gain, and in the last two yr, as well as in the earlier period, it lost a considerable amount of sand. There was a considerable addition of material during the period 1963 to 1969.

In order to assess the regularity of the trend and the relationship between the three profiles cross spectral analysis has been used for 54 observations of beach accretion over the period 1953 to 1971. The number of readings, some of which are interpolated, is too small to reveal any annual cycle of accretion, but some indication of the steadiness of the accretion can be obtained. Profile 4 shows the most regular fluctuations around the trend (Figure 6), while accretion on profile 2 lies above the trend line for the early and late part of the period and below in the middle, due to rapid increase in accumulation in the latter half of the period. Profile 1 shows a similar oscillation around the trend.

The only one of the three observations to show any cyclic form of accretion was profile 4, which shows some indication of a 6 yr cycle, the spectral density reaching a peak of 0.6736 for the 6 yr period (Figure 6, inset). The reason for this minor cyclic rate of accretion is not easy to determine and it is not apparent in either of the other two profiles, which show no cyclic pattern at all. Owing to the presence of a weak cycle in only one of the profiles the coherence between the pairs of profiles is low, apart from a weak coherence between profiles 1 and 2 at a period of 3.60 or frequency of 10. The coherence is 0.3892 at this period, which is just a little over 1 yr. This could suggest some weak annual effect that is felt on these two profiles. The main result of the spectral analysis is to confirm the regular pattern of accretion on these profiles, because the trend element is by far the strongest in the time series.

The growth of the spit has also been analyzed by trend analysis. Various dimensions have been recorded, including the length from a fixed peg to the distal end at three elevations, the horizontal position relative to the fixed point, the volume above 10 ft O.D. Liverpool, and the area above each contour in feet. These trends are shown in Figure 7 and the equations are listed in Table 1. On the whole the trends are positive showing that the spit has been growing in most of its dimensions over the two decades of observations. The positive trend appears, however, to be slowing down and reversing at present. The volume in 1971 was down to the value of 1964 owing to a loss of material from the higher part of the spit. The area above 10 ft has stayed very static on the whole, as indicated by its low trend coefficient. The area above 12 ft was almost identical in 1971 as it was in 1962. The same applies to the area above 11 ft, a peak having been reached in 1965. One very consistent trend recorded over the period of observation is the steady landward shift of the whole feature. A peg originally fixed on the landward side of the spit reappeared later on the seaward side, while another originally at the margin of the spit is now buried nearly under its highest point. The spit is thus encroaching onto the new marsh, which is developing in

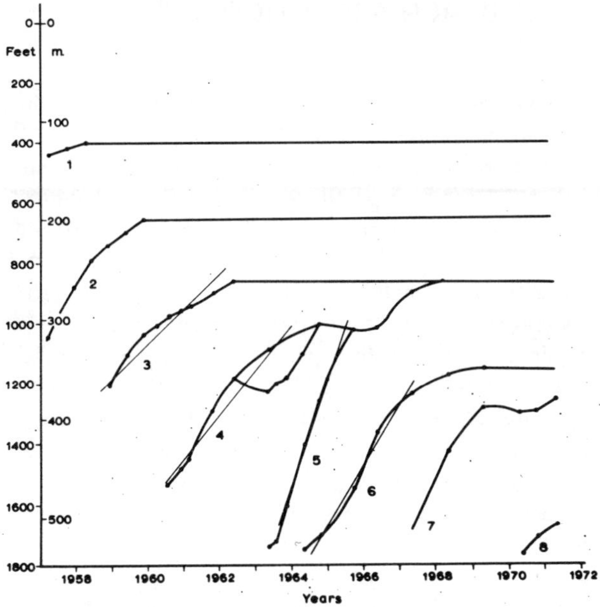

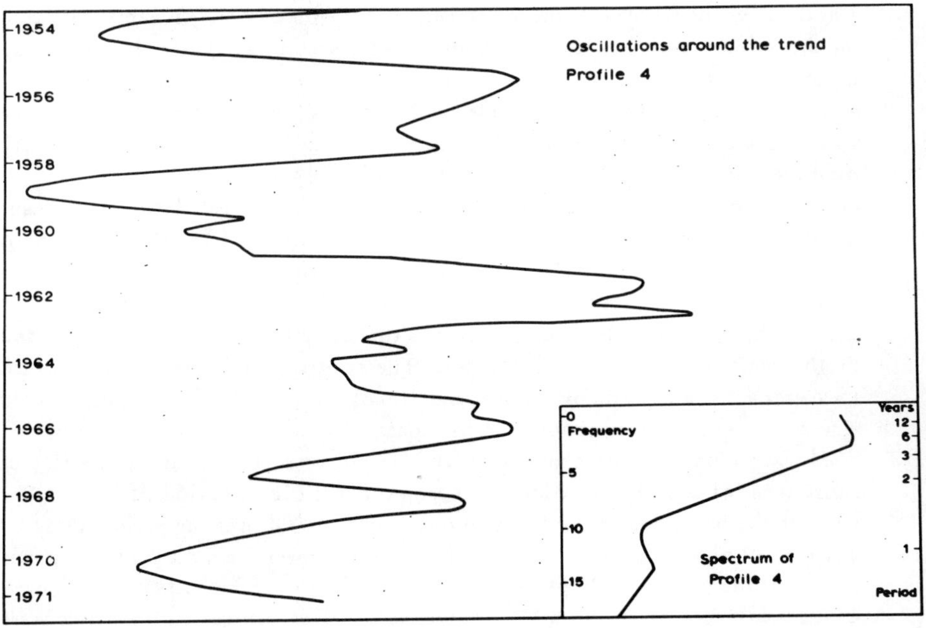

Figure 5. The pattern of ridge movement on profile 4, south Skegness, is shown.

Figure 6. Oscillations about the trend for accretion on profile 4. The inset shows the spectrum for the time series, indicating a minor 6-yr. cycle.

its shelter. This movement accounts for the rapid gain in height in the parts of the marsh adjacent to the spit.

A close correlation has been established between the volume of accretion on the spit and that recorded on profile 2, which is situated about 0.5 mi (0.8 km) north of the spit. The correlation coefficient is 0.90, explaining 80.8% of the variation in the spit volume, the regression equation being $Y = 174.1 + 0.20\,X$, where Y is the spit volume and X is the accretion on profile 2. The correlation is high owing to auto-correlation of both sets of values, nevertheless it does indicate a significant relationship, which is lacking between the accretion on profile 1 and the volume of the spit. The close relationship suggests that the spit is built up by material moving along the upper foreshore from profile 2 and the downward turn of the two variables in the last two years may be significant, although the change at profile 2 has been up and down in the period. It remains to be seen whether these changes will be prolonged. The relation between accretion on profile 2 and the spit volume does, however, furthur confirm the movement of material southwards along the coast by wave action on the upper foreshore. The spit is built up by waves drifting material along the upper beach at high tide. The lack of correlation between the accretion of material on profile 1 and 2 over the whole foreshore suggests that the bulk of the movement of sand on the foreshore under wave action must take place at high water. The waves are more vigorous at this time owing to the deeper water offshore. The tidal streams are also flowing south at this time, and thus add to the effectiveness of the waves. Wave refraction will also be less effective at high water when the depth offshore is greatest and the offshore banks will affect the waves less.

The study of the changes in this area over time has revealed some significant points. The changes can be shown to be steady on the whole, in that the trend is the major element in the time series that have been analyzed by cross spectral analysis. Regular movements are also shown in the movement of the ridges up the beach, in the accretion on the profiles that are building up most rapidly, and in the various dimensions of the spit. The relatively long period over which measurements are available make the trends of considerable significance in assessing possible future changes on the coast.

The steadiness of the changes on the beach were confirmed by a series of short term observations carried out near profile 2 over a period of 1.5 yr. The level of the beach was recorded at intervals of 2 to 3 d over the period. Only 22 out of 582 observations indicated changes in level over 5 cm during the summer period from March to September, while 42 out of 480 observations showed changes in excess of 5 cm in the winter period. The changes on the whole were slow and regular and were associated with slow ridge growth and movement. The larger changes were caused by runnel formation. Total changes in level did not exceed 4 in (10 cm) in the summer period. The results confirmed the stability of this fine sand beach and hence the value of relatively widely spaced profiles in time.

The reason for the regularity of the beach changes is probably due to the

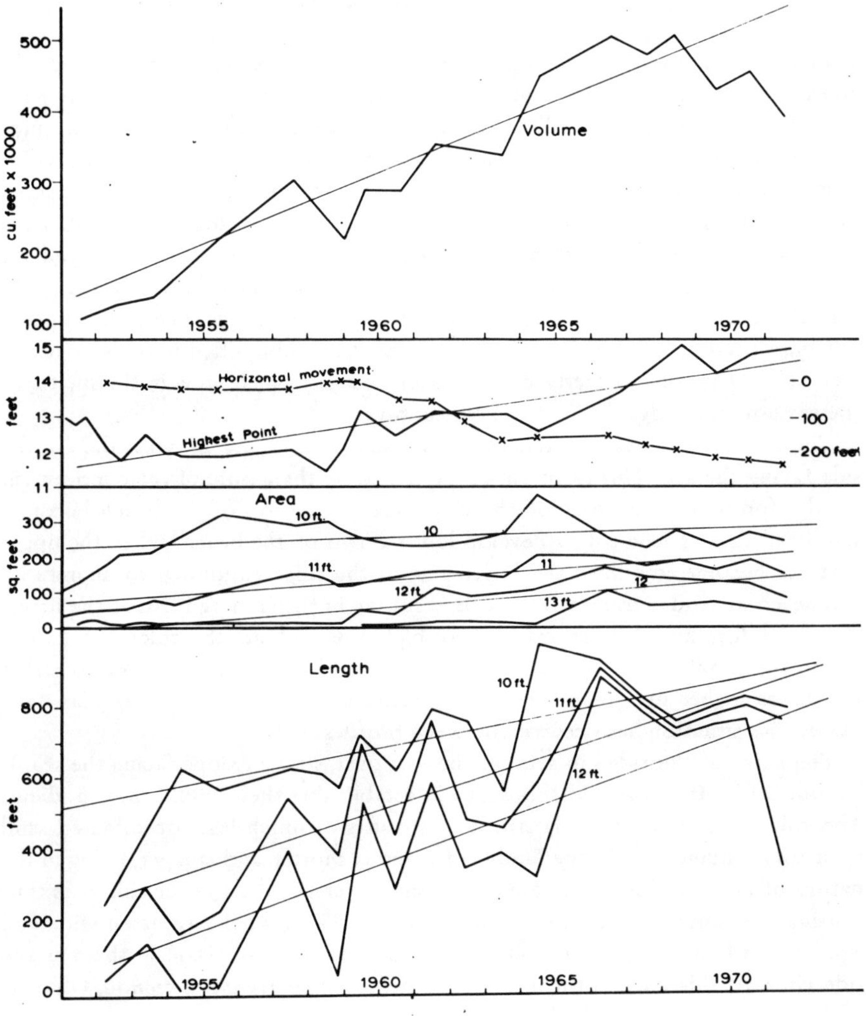

Figure 7. Trends of spit growth and movement at Gibraltar Point. The calculated trend lines are entered.

regularity of the tidal streams, which are the major transporting agent of the material into the area, and the relatively low energy of the waves, which build up the material supplied by the tidal streams into the ridges. These features reflect the amount of accretion of the foredunes, by means of which the coast builds out seawards.

FORM OF ACCRETION

In this section the morphological results of the processes of accretion will be considered briefly. The range of environments within the area has already been mentioned and emphasis will be placed on the beach foreshore zone, although the dunes and intervening salt marsh strips are the ultimate forms in which accretion takes place. The dunes cannot form until the ridges on the foreshore have grown up high enough to become stabilized above high tide level. Then vegetation and wind blown sand can accumulate. Shelter behind a stabilized ridge is required before mud can settle and plants can become established in the runnels to landward of the ridges to initiate the strip saltings. The wider areas of salt marsh require shelter in the form of a spit, as exemplified in the new marsh area at present, and formerly in the mature marsh, which grew in the shelter of the eastern dune ridge during the 19th century.

The foredune ridges are arcuate in plan shape (see Plate 2), with the convex side facing the sea. This form can be explained by the nature of ridge movement on the foreshore. The part of the ridge nearest the top of the beach becomes stabilized first as new ridges develop lower down on the beach. When the upper part has become stabilized the lower part of the ridge continued to migrate up the beach as sand is transferred south, resulting in the arcuate form of the newly stabilized foredune. Water trapped at high tide behind the ridge breaks out where the distal end is lowest and when this end has been driven close to earlier foredunes higher up the beach, local erosion of these earlier dunes may take place. This situation has occurred between profiles 3 and 4.

The form of the ridge in section shows a firm, straight slope facing the sea. It is built up by the swash of the waves in establishing their equilibrium gradient. The ridge crest is usually flattish, and consists of much less consolidated sand with some shingle. while the landward slope is shorter and steeper, being in the nature of an accretion slope. Sand is pushed over the crest as the tide rises thus causing the landward movement of the ridge. The runnels are often wide and rippled until the ridges are high enough to allow mud to settle in them as the tide ebbs. At this stage salt marsh plants, including *Spartina townsendii, Glyceria maritima* and *Salicornia herbacea* can take root, and help trap more mud. Successive profiles illustrate clearly the landward movement of the ridges and their form, as well as the filling of the runnels to form strip saltings or marsh slacks.

Successive profiles plotted on the same diagram indicate the variability of the beach, and the envelope within which the profiles lie is called the sweep zone. It is exceptionally wide on the Lincolnshire coast owing to the presence of mobile ridges on the foreshore. Long term changes in the beach morphology can be assessed by means of a series of superimposed sweep zones, as shown in Figure 8. This method eliminates the short term variations, which on many beaches are due to changes in the wave type. On the beaches around Gibraltar Point, however, they are mainly due to the position of the ridges and the channels cut across them by the tide in draining off the beach. The pattern of superimposed sweep zones shows clearly the rapid accretion that has taken place on profiles 4 and 2. The narrower sweep zone of profile 1 reflects the smaller dimensions of the ridges on this profile, where the surfeit of sand is much less marked than on the other profiles.

The width of the respective sweep zones of profiles 1, 2 and 4 indicates a relationship between the ridge size and the amount of accretion on the beach. This relationship has been studied by means of the analysis of co-variance model, which can be used to relate two variables while holding another constant. In this instance the accretion and ridge size were related, while the three profiles were standardized by combining their means. Under the circumstances existing in this instance a marked improvement of the correlation between the two variables is achieved by this means. The plot of the data for accretion and ridge height shows that there is a close relationship between the two variables for profiles 2 and 4 as shown in Figure 9. A warning must be given again, however, that autocorrelation may make the relationship better than it should be. The relationship is not, on the other hand, apparent for the profile 1 values.

The curves for profiles 2 and 4 lie approximately parallel, but the values for profile 4 are consistently greater than those for profile 2. Thus by merging the means into a grand mean, through which the new regression line is drawn for the combined data, the level of explanation is raised from about 50 percent to just over 80 percent. The result of the analysis shows that the ridge size is closely associated with the amount of accretion taking place on the beach, the higher ridges being associated with the greater accretion. The large amount of accretion leads to a flatter overall gradient and hence the necessity for higher ridges. Profile 1, which has not received nearly so much sand, does not need high ridges to establish the equilibrium gradient.

The difference in ridge size between profiles 2 and 4 cannot be wholly accounted for in terms of the amount of accretion taking place because the two regression lines are parallel, indicating that for any given amount of accretion the ridges need to be higher to reach equilibrium on profile 4 than on profile 2. This situation can be accounted for in terms of the size of material on the two profiles. The sand on profile 4 is coarser than that on profile 2, samples gave the respective mean diameters of 1.35ϕ and 2.27ϕ on profiles 4 and 2 respectively. Coarser sand requires a steeper equilibrium gradient than finer sand, and hence the reason for the higher ridges on profile 4 compared with profile 2.

The growth of ridges on the foreshore gives rise to shallow nesses of accumulation, with a general curvature convex to the sea. The ness that is

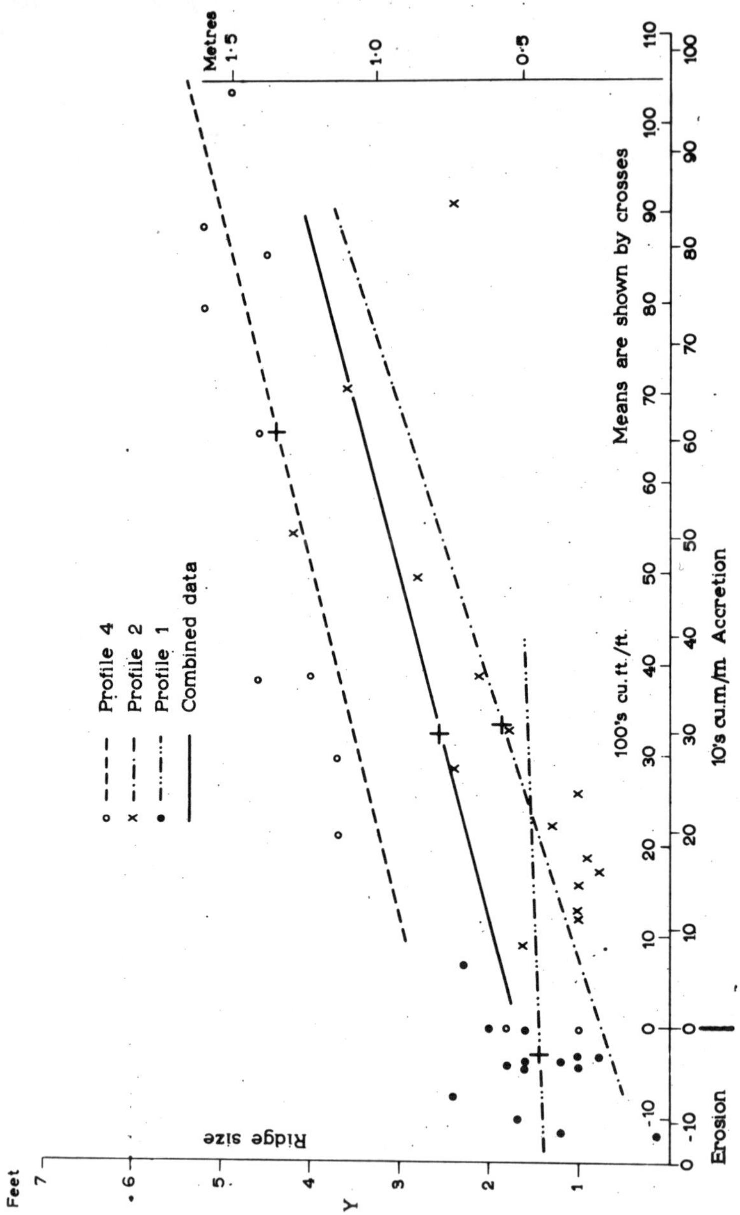

Figure 8. Sweep zone profiles for three periods for profiles 1, 2, and 4 south of Skegness. The position of the profiles is shown in Figure 4.

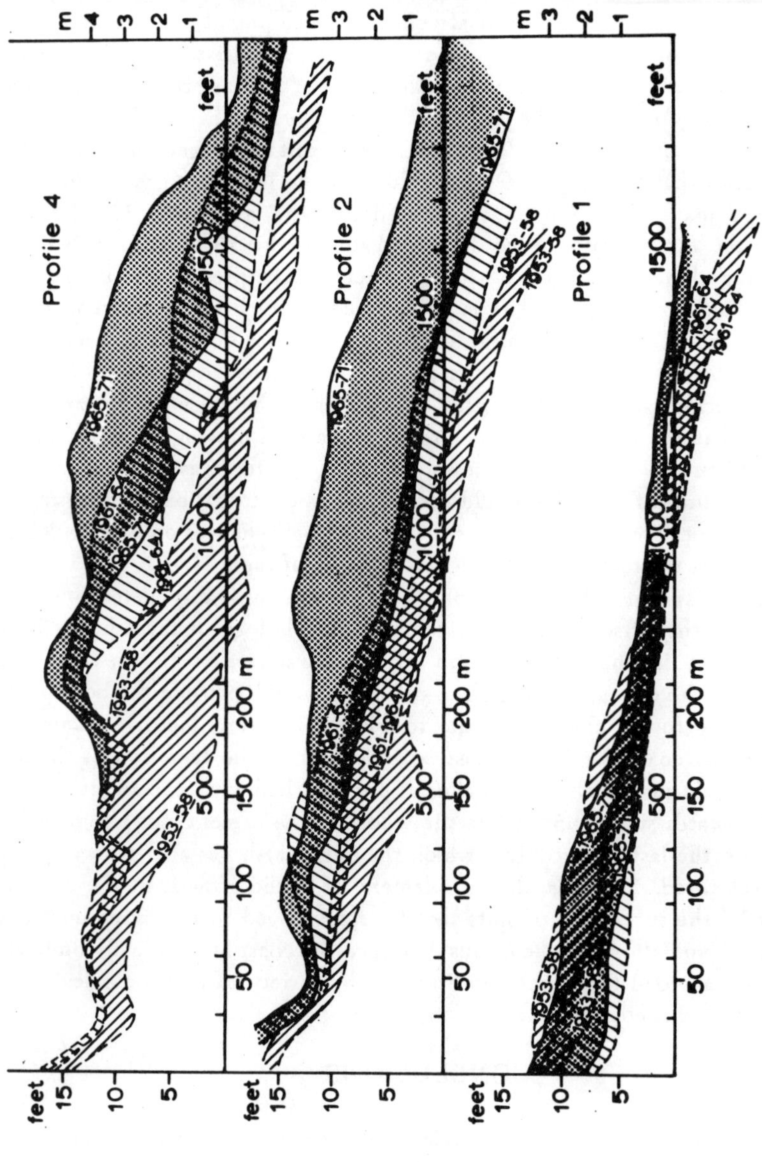

Figure 9. Analysis of co-variance of ridge size and accretion on profiles 1, 2, and 4.

currently developing along this stretch of coast reaches its apex just to the south of profile 4. Profile 2 lies near the southern margin of a ness of accumulation, while profile 1 lies beyond it. The positions of the nesses of accumulation change with time, as indicated in the pattern of stabilized beach ridges and the dune system. This pattern in turn affects the development of spits along this coast. The present spit is one of a series, and it was initiated after the severe storm of 1922, which truncated the southern end of the eastern dune ridge. The spit subsequently prolonged the line of the coast rather further offshore than the former position of the dune ridge. A still earlier spit prolonged the line of the western dunes that formed the frontal system in the 18th century, as shown on Armstrong's map of 1779. This spit ceased to grow, and broke into isolated islands as the eastern dune ridge developed during the 19th century, because sand moving alongshore could no longer reach it. The present spit is likely to undergo a similar change, as its supply is reduced by the accumulation of material to the north and the outgrowth of the ness around profiles 3 and 4. As the ness grows further seawards sand may be diverted further offshore, eventually forming a new spit further seaward than the present one. Evidence has shown that the spit is now replenished by material moving along the upper foreshore from the north. If the supply reaching profile 2 is reduced further, then the volume of the spit could also show a continuation in the present pattern of reduction in volume. Thus the morphologic changes on one stretch of this coast affect the features on the down-drift side of them.

The whole system of changes on this coast is probably determined by movement of the offshore banks so that the morphology of these banks is an essential aspect in the interpretation of coastal accretion in the area. An assessment of the changes taking place offshore would provide the best means of determining future changes on the foreshore. The time factor or tempo of change on this coast is relatively fast, as indicated by the accretion of the last two decades, during which the shoreline at profile 4 had advanced about 1200 ft (366 m) seawards since 1951, while the whole of the present spit has probably developed in the last 50 yr during which the new marsh has also grown up and become vegetated. There is also an element of cyclic growth in the area, as indicated by the succession of spits that have prolonged former dune lines and nesses of accumulation. By the means described the coast has gradually built out seawards on a broad front, thereby creating the variety of morphological units that constitute this coastal zone.

CONCLUSIONS

The discussion in the previous sections allows some tentative answers to be given to the questions posed in the introduction. 1. The material comes from the exposures of glacial deposits on the foreshore further north in Lincolnshire and from the exposed glacial cliffs of Holderness, where erosion is very rapid. The evidence for this is seen: firstly, in the nature of the sediments in the area at present, secondly in the direction of coastal drift, thirdly in the results of studies using floats, and fourthly in the pattern of tidal stream residuals in the offshore zone. Little material comes from inland at present.

2. The material comes to and accumulates in this area because at this point along the coast there is a mechanism operating that allows more material to reach the foreshore than is removed. This mechanism is related to the pattern of offshore banks in the vicinity. They reflect the way in which the tidal streams transport sediment in the offshore zone. Thus the presence of abundant sediment that can be shaped into tidal offshore banks allows a pattern of transport to develop that can feed the foreshore in this vicinity.

3. The material is brought by the tidal streams. The high tidal range and the residual pattern ensures effective transport of material by strong tidal currents. The pattern of movement is probably complex, but the fact that a flood stream channel runs up onto the lower foreshore in the zone of maximum accretion allows material to come onto the foreshore here. The waves then move the material up the beach and drift it alongshore southwards, particularly on the upper foreshore where the waves are acting in conjunction with the southerly-directed flood tidal stream. The growing ness will also itself cause some sand to be trapped as the sand passes south down the coast.

4. The material comes very regularly into the area, but the rate varies along the coast and in the different environments within the area. The fastest recorded rate of accretion is along profile 4 near the apex of the ness where the mean rate of material arrival on a profile of unit width of beach is 953 ft^3 (89 m^3)/year for the period 1953 to 1971, providing a total buildup of 17,100 ft^3 (1,590m^3) in 18 yr. There is a very small 6-yr period cyclic element in this growth rate, but the regular trend is by far the most important element of the time series. No cycles were identified in the time series for the profiles further south. The trend was still fairly regular at profile 2, where it has been consistently positive. Profile 1 has not gained nearly as much material as those further north, which are situated within the ness of accretion. Another regular trend recorded is the movement of the ridges up the foreshore until they become stabilized by the growth of others to seaward. The regularity of the accretion suggests that the tides are largely responsible for the arrival of the material, as they are the most regular phenomenon.

5. The material is deposited in the form of beach ridges on the foreshore. These ridges represent the response of the waves, which are short, to an overall gradient that is too low. They build up their equilibrium gradient by constructive action, thereby forming a ridge. The ridges are highest where the overall gradient is lowest in the zones of greatest accumulation. The ridges are also higher where the material is coarser, as this also requires a steeper equilibrium gradient. The ridges move up the beach until they become stabilized, the distal end continuing to move up so that in plan they are convex-seaward arcuate features. The ridges are colonized by sand-loving plants as they become stabilized and marsh plants grow in the runnel to landward. The vegetation helps

to stabilize the features further, by assisting the build up of wind-blown sand and tidal silt, respectively, on the dunes and in the marsh strips. By this mechanism the coast builds out seawards on a broad front. The spit prolongs the coast and provides shelter for a wider salt marsh where the land turns abruptly westwards at Gibraltar Point.

The study of this coast of accretion suggests that, in this area at least, the tide is more important than the waves in bringing in the material that is essential for the accretion to take place. The offshore tidal morphology is an essential part of the coastal system, which includes the foreshore and backshore regions that cover the beach, spit, dunes and salt marshes. Thus many processes operate to account for the wide variety of coastal type within this small zone of accretion. The rapidity with which changes occur illustrates the effectiveness of the dynamic processes that are operating along the coast. They may be expected to continue to build the coast seawards, as long as suitable offshore morphology enables material to come ashore in the area. The precise areas of accretion are, however, likely to vary with the migration of the offshore banks, in response to the activity of the tidal streams in the offshore zone. The present spit may also decay if a new one prolongs the growing ness to the north and cuts off the supply of wave-drifted material to the present spit. A new spit would in turn initiate the growth of another salt marsh in its shelter. Thus tides and waves work together to give the area its character.

Table 1 Trend equations for the changes in spit size and position

Volume above 10 ft	–	Y = 346 + 24.9 X	for 1951-1971
Area above 10 ft	–	Y = 260 + 3.21 X	"
Area above 11 ft	–	Y = 142 + 6.78 X	"
Area above 12 ft	–	Y = 67 + 7.86 X	"
Length at 10 ft	–	Y = 686 + 30.0 X	for 1953-1971
Length at 11 ft	–	Y = 525 + 40.0 X	"
Length at 12 ft	–	Y = 383 + 47.5 X	"
Height in ft	–	Y = 13.04 + 0.149 X	For 1951-1971

REFERENCES

Draper, L. 1968. Waves at Smith's Knoll Light Vessel, North Sea: Nat. Inst. Oceanography Internal Report, No. A33, 5 p.

Robinson, A.H.W. 1964. The inshore waters, sediment supply and coastal changes of part of Lincolnshire: East. Mid. Geog., 3 Pt. 6 (22), p. 307-321.

Chapter 5

SAND BARS ALONG LOW ENERGY BEACHES PART 1 MULTIPLE PARALLEL SAND BARS OF SOUTHEASTERN CAPE COD BAY

Harold D. Nilsson

ABSTRACT

Multiple parallel sand bars are found in southeastern Cape Cod Bay and on other low-energy coastlines. The conditions under which they form are: sand-sized sediment supply, low slope gradients, repeatedly breaking and reconstructing waves, high tidal range, and low wave energy. The precise relationships between the conditions under which multiple parallel sand bars form are yet to be determined.

INTRODUCTION

The terms longshore bar, offshore bar, and sand bar have generally been applied to an entire group of intertidal and subtidal bottom features that form in varied ways and are quite different in appearance. Precise classification of bars according to their genesis and configuration has been attempted by King (1972), Shepard (1952), and Zenkovitch (1967). More detailed descriptions of bar forms and genesis are becoming available with recent interest in the subject (Bowen and Inman, 1971; Goldsmith and Colonell, 1970; Sonu, 1972; Niedoroda and Tanner, 1970; and Nilsson, 1972).

MULTIPLE PARALLEL SAND BARS

The intertidal and subtidal zones of southeastern Cape Cod Bay are characterized by several parallel sand bar systems which may extend as far as 2 km seaward. In some cases more than one parallel sand bar system may be developed in a particular area. At such locations the bar systems have different orientations, parallel and oblique to the shoreline, and form an intersecting pattern. Although they are similar to King's (1972) ridges and runnels, multiple parallel sand bars do not have the same character as the classical ridge and runnel system. Ridge and runnel topography is strongly asymmetrical, occurs in limited numbers (1 to 4) parallel to the coastline, is developed during periods of

Coastal Research Center, Department of Geology, University of Massachusetts, Amherst, Mass. 01002

relatively calm weather, and is destroyed by storms. Multiple parallel sand bars are commonly symmetrical, relatively stable during varied weather conditions, are as many as 30 in number, and are usually only found along low energy coastlines.

Zenkovitch (1967, p. 225-236) discussed the existence of multiple parallel bars in the Laptev and Chukchi Seas that closely resemble the bars of southeastern Cape Cod Bay. His conclusions concerning these bars are: (1) they develop only in sand-size sediment, (2) occur on slopes with gradients of 0.005 or less, (3) form as a result of repeatedly breaking and reconstructed waves, and (4) are relatively stable, forming a "profile of equilibrium" for the given conditions (Zenkovitch, 1967, p. 234).

The multiple parallel sand bars of southeastern Cape Cod Bay, in the vicinity of Brewster and Eastham (Fig. 1), have crest to crest distances which are highly variable, but are on the order of 60 m. Maximum bar height is about 1 m. The plan view of the bars may be linear, convex landward, or convex seaward. They may be symmetrical or asymmetrical landward in cross section. Algae trapped in the bar troughs contrast with the clean, light-colored crests and make the bars very distinct when viewed from the air.

The bars develop in the intertidal and subtidal zones, extending to depths ranging from 2-3 m (at mean low water). No bars form above mean high water. The processes that form multiple sand bars are not well understood. They appear to be related to breaking waves which may reform and break again. The resulting pattern is several rows of waves breaking over submerged bars and finally dissipating their energy completely by washing over exposed bars at higher elevations. Whether the bars are formed at the break point or in the swash zone is not clear. Perhaps the development of these bars is related to the processes at both the break point and the swash zone. The author has referred to multiple sand bars as swash bars (Nilsson, 1972), but with this paper, changes the name to eliminate the genetic connotation of the word "swash".

Items 1, 2, and 3 from above agree with the observations of Zenkovitch (1967). Items 4 and 5 are additional conclusions obtained in studying the multiple parallel bars of Cape Cod Bay.

The genesis of multiple parallel sand bars is intimately related to the above conditions. There are probably other factors, presently unknown which must be determined in order to give a complete explanation of their formation. The precise relationships between the conditions under which multiple parallel sand bars form are presently under investigation.

Along the eastern shore of Cape Cod Bay, near Wellfleet and Truro, irregular, intersecting, multiple sand bars occur as a result of two dominant wind directions (Fig. 2). Westerly wind-generated waves, dominant in winter, form a system of multiple sand bars parallel to the shore. Southwesterly wind-generated waves, dominant in the summer, form a system of bars whose crests are oblique to the shoreline, and intersect the bar system parallel to the coast.

Slope gradients along the tidal flats in southeastern Cape Cod Bay are on the order of 0.005, an order of magnitude less than Zenkovitch's maximum slope (0.005). Almost 100 percent of the material comprising the tidal flats is sand-sized sediment derived from the glacial debris of Cape Cod. Tidal range in the area is approximately 3 m.

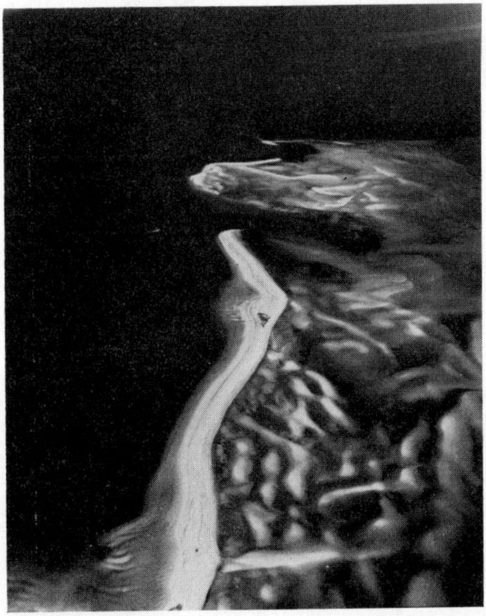

EGERTON
COLLEGE
LIBRARY

Figure 1. Multiple parallel sand bars in Cape Cod near Eastham. View is to the east. The bars at this location are exceptionally linear in plan view. Lake in background 0.9 km. across.

Figure 2. Intersecting bar systems near Jeremy Point in Wellfleet. Bars parallel to the sandspit are multiple parallel bars which result from westerly wind-generated waves. The intersecting oblique bars are also multiple parallel sand bars, and are formed by southwesterly wind-generated waves. Spit is 2.6 km. long. View is looking southward.

CONCLUSIONS

Multiple parallel sand bars form in areas which have a combination of the following conditions:

1. almost 100 percent sand-sized sediment
2. very low slope gradients
3. repeatedly breaking and reconstructing waves
4. relatively high tidal range (3m) over a low gradient slope (0.0005) (allows the bottom to be exposed to wave activity over a wide area (1-2km) and over a long period of time)
5. relatively low wave energy

SAND BARS ALONG LOW ENERGY BEACHES PART 2 TRANVERSE BARS

Alan William Niedoroda

ABSTRACT

The areas just seaward of low energy beaches often exhibit a series of subtle digital sand bars which point either perpendicular or at a high angle to the shoreline. These sand bars are called *tranverse bars.* In most areas the bars cause a distinct nearshore current to develop from refracted and oversteepened waves. It has been shown that this current system is competent to transport wave agitated sand so the transverse bars are in a state of dynamic equilibrium with the waves and currents. The pattern of the currents which develop about transverse bars is strongly controlled by their length. Relatively short transverse bars cause a relatively strong shoreward current to develop over the axes of the bars. Longer sand bars cause a more complex pattern of current. The long bars maintain an onshore current over their seaward portions and an offshore current over their landward portions.

INTRODUCTION

Low energy beaches are generally characterized by very gently sloping foreshore and offshore areas. This appears to agree with the ramp-slope concept developed by Price (1953). Such coastal environments often exhibit a series of subtle digital sand bars whose long axes are perpendicular or at a high angle to the beach. The name transverse bars has been given to these features by Shepard (1952). Transverse bars and similar related bars have been reported on by numerous authors including: Evans (1938), Bruun (1954), Tanner (1960a and b, 1967), Homma and Sonu (1963), Sonu (1968), Moody (1965), Goldsmith and Colonell (1970), Niedoroda and Tanner (1970), and Niedoroda (1972a and b). The names applied to these features often vary considerably and no attempt has been made to clarify the nomenclature. However, before clarification is attempted these bars must be fully described.

Coastal Research Center, Department of Geology
University of Massachusetts, Amherst, Mass. 01002

DESCRIPTION

The maximum relief of transverse bars seldom exceeds 1 m and is commonly less than 0.5 m. They occur on foreshores which slope seaward at 0.15% to 0.45%. The lengths of these bars vary from one location to another depending on the foreshore slope but are generally quite constant on any particular beach. Various authors have reported lengths ranging from 300-2000 m. The sand bars tend to have a quasi-regular spacing which is generally smaller than their length and are usually oriented with their long axes directed into the mean annual wave approach direction.

The correlation between the occurrence of these bars and gentle foreshore and offshore slopes precludes their development on high and moderate energy coasts. Therefore, they are most commonly located in lagoons behind barrier islands, in estuaries and bays, and along protected open coasts. Similar but not identical features such as rhythmic bars and sand waves do develop on higher energy open coasts. However, these features should not be classed with transverse bars.

Transverse bars are most common in areas of abundant sand size sediment. However, Zenkovitch (1967) shows an area on the Liaotung peninsula which has "shingle ridges" which may be transverse bars in shingle size sediment. The slope of the foreshore and offshore areas in areas of transverse bars are slight, but not as small as the slopes required to develop multiple parallel sand bars as reported in Nilsson (1972). In addition, there are indications that transverse bars are best developed in areas of small to moderate tidal range and may be modified or replaced by other features in areas of high tidal range.

The subtle relief of transverse bars often makes them difficult to observe from the ground except at low tide. However, they are often striking when viewed from the air. Figure 1 shows a typical group of transverse bars which occur on a protected stretch of open coast south of Tallahassee, Florida. Figure 2 shows an example of very long transverse bars on the southwest coast of Mansel Island, located in the northern reaches of Hudson Bay, Canada. This coastal ares is one of rapid uplift.

WAVE PATTERN

Transverse bars cause the nearshore waves to be refracted into a pattern of converging and crossing wave crests over the axes of the bars and a divergence of the wave crests between the bars. This results in producing a zone of spilling waves over a significant length of each bar and little or no spilling of the wave crests between the bars until the wave crests are very near the beach.

FIELD EXPERIMENTS

Field studies have mapped a discrete nearshore current component in the vicinity of transverse bars. This current system is subtle and almost impossible to

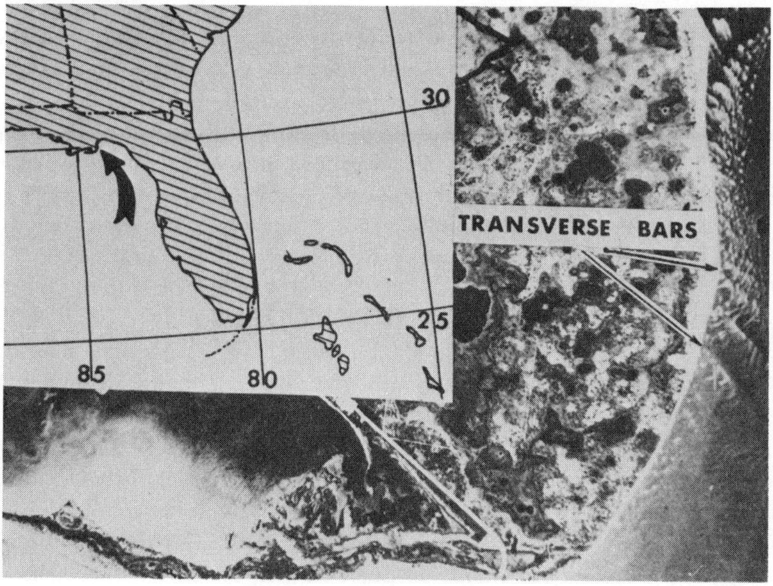

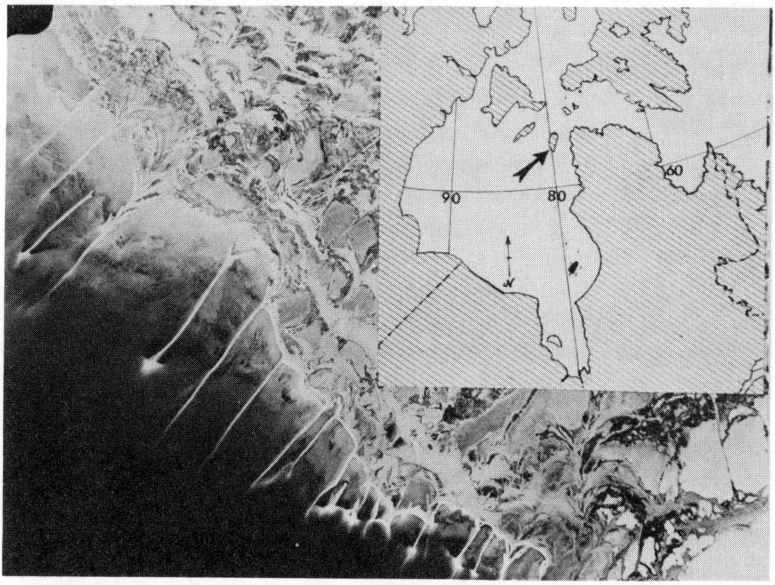

Figure 1. Vertical air photograph of short transverse bars along a beach in Franklin County, Florida (Repeated from Niedoroda and Tanner, 1970).

Figure 2. Vertical air photograph of long transverse bars on Mansel Island, Hudson Bay, Canada. Post-glacial isostatic uplift appears to have produced a series of raised beaches and truncated transverse bars. (Courtesy Canadian Government).

observe at or near high tide. However, as the tide falls this distinct nearshore system of currents becomes apparent. This current system consists of a relatively strong current directed toward the beach over the seaward part of the axis of the bar. The current returns seaward as a sluggish flow over the wider areas between the bars (Fig. 3). The existence of this nearshore current component is obviously an important factor in the origin and functioning of these bars. A variety of scale models have been employed to study the reason that this current component develops and to see how it varies under different nearshore conditions.

LABORATORY MODELS

The first series of experiments were carried out in a small scale model. These experiments showed qualitatively that when the transverse bars were short, a current system identical to that observed in the field could be produced solely from the refraction and spilling of the waves over the model bar. On the other hand, when the bars were relatively long a more complex system consisting of four meshed current gyrals developed as illustrated in Figure 5. It should be noted that this system of currents produces an offshore flow over the shoreward part of the bar axis and an onshore current at the seaward end of the bar.

A large (1:13.3) scale model of an idealized short variety transverse bar was constructed at the modeling facility of the Coastal and Oceanographic Engineering Department of the University of Florida. The currents which developed in this model were similar to those observed in the field and those which developed in the smaller scale models. The large model showed that the currents associated with a transverse bar originate from the spilling of waves over the axis of the bar. However, the geometry of the waves and currents in the 1:13.3 scale model was too complicated to be described mathematically. Another large scale model was constructed which simulated the dynamic conditions of the waves and currents about a transverse bar in a simplified manner.

The scale model consisted of a submerged horizontal platform, representing the transverse bar, resting in a 12.13 m long by 9.1 m wide wave tank (Fig. 4). The parts of the wave crests which encountered the submerged platform were caused to spill while the parts of the crests which missed the platform continued to the beach without spilling. This model eliminated the need to accommodate wave refraction and confluent currents over the bar in the mathematical model. The currents which developed in this model were qualitatively similar to those developed in the field and in previous model studies.

The complete results of the latter scale model study and the analysis of the mathematical study are too voluminous to be included in this paper. They are given in Niedoroda (1972a) and will soon be published elsewhere. In brief, the mathematical model was developed in an analogous manner to that used by Bowen (1969) and Longuet-Higgins (1970) in their explanation for the development of longshore currents on a plane beach. The analysis of these

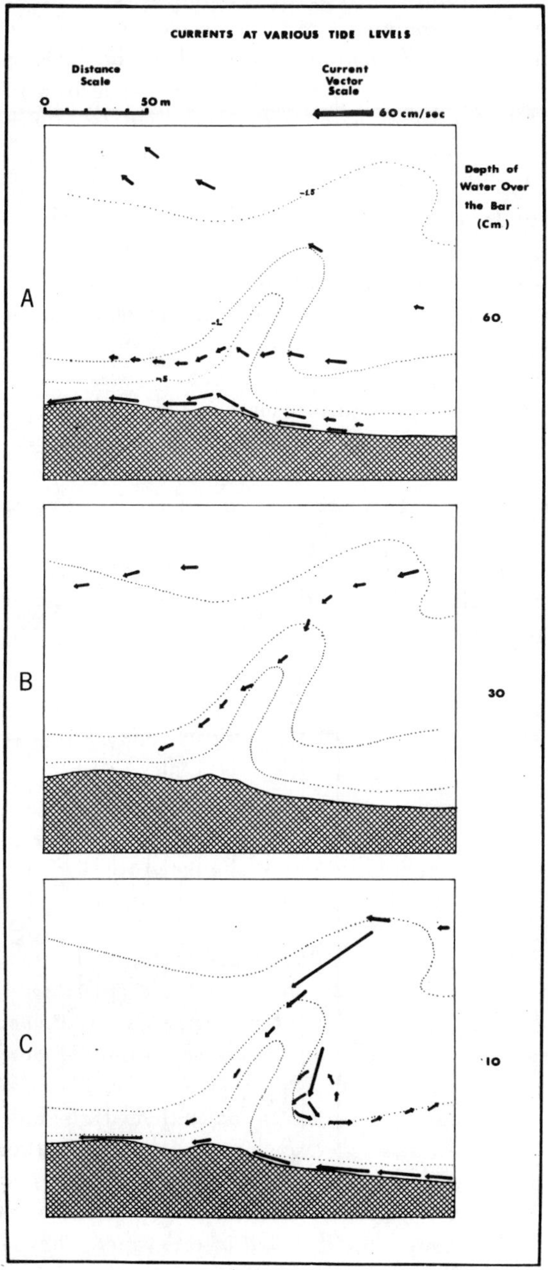

Figure 3. Streaklines of nearshore currents measured in the field at different levels of the tide. Notice that the wave driven current component is most apparent at low tide and that it is distinct from the longshore current.

authors shows that gradients of the components of the radiation stress tensor for the excess flux of momentum due to the waves, result in developing a set-up of the mean surface level and the longshore current in the zone where the waves are breaking and spilling. In the analysis of the currents which developed over the submerged platform of the 12.13 by 9.1 m model in the present study only the gradient of the normal component of the radiation stress tensor, the flux of momentum due to the currents, the slope of the mean surface level over the platform, and the effect of bottom friction appear as terms in the analytical expression. Comparative data for these variables (except bottom friction) were measured in the scale model. The estimated bottom friction coefficient necessary to produce agreement between the experimental and mathematical models was very close to most other published values for model studies. These results show there is a transfer of momentum from the spilling waves over the model transverse bar to a net rise of the mean surface level over the offshore part of the model bar which results in a down-towards-the-beach slope of the mean surface level over the shoreward part of the model transverse bar. It is this sloped mean surface level which drives the currents along the top of the bar, towards the beach and also eventually propels the return currents between the bars. The close agreement between the scale model measurements and the values obtained from the mathematical model clearly established that the entire current system developed from the momentum transferred from the spilling waves.

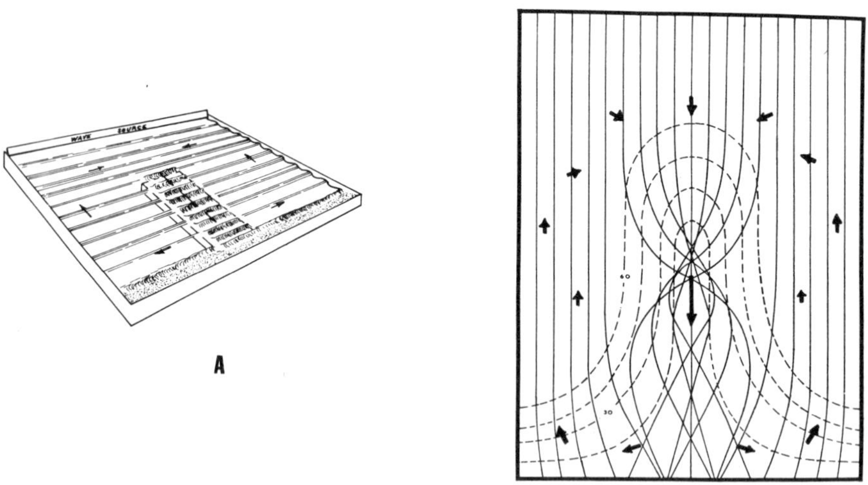

Figure 4. Large scale models employed in determining how the distinct current component develops from the waves about transverse bars. Sketch A showing the highly simplified model used in verifying mathematical model. Sketch B shows the 1:13.3 scale model with a computer drawn wave orthogonal diagram and arrows indicating the currents superimposed.

All of the larger scale model experiments were performed on models of the short variety of transverse bars. The more complicated current pattern associated with the longer transverse bars develops through a similar mechanism. In this case the same processes are responsible for the currents near the seaward end of the long bars as over the entire length of the short bars. However, shoreward of a given point along the bar axis the waves over the bar have been severely reduced in height due to spilling whereas adjacent parts of the same wave crests which are located between the bars just begin to spill. Apparently, wave refraction does not cause a significant flux of wave energy onto the bar over its shoreward portion. As a result, there is more wave spilling between the bars than on the bars in the shoreward regions. This produces a higher mean surface level set-up between the bars. The offshore slope of the mean surface level between the bars is exactly balanced by the gradients of the radiation stress due to the spilling waves and therefore cannot induce an offshore flow between the bars. However, this mean surface level set-up between the bars produces a higher mean water level at the beach between the bars than at the beach at the point where the bar axis intersects the beach. This results in an unbalanced offshore slope over the shoreward part of the bar crest which causes water to flow offshore along the transverse bar (see Figs. 5 and 6).

THE ROLE OF TRANSVERSE BARS IN THE NEARSHORE ZONE

Dyed tracer sand studies in the field have demonstrated that the nearshore current component which is derived from the spilling waves over a transverse bar is capable of transporting wave agitated sand. This indicates these bars are in a time averaged equilibrium state. The sand is moved to, and along, the bars by the currents in a pattern and at a rate which is correctly adjusted to maintain the bars. Therefore, the short variety transverse bars act as avenues of sediment transport away from the beach over most of their length.

THE ORIGINS OF TRANSVERSE BARS

As the concepts of this paper have been developed the reader may have been left with a distinct unhappy sense. There appears to be a question similar to the proverbial problem concerning the origin of the chicken or the egg. It appears that the shape of transverse bars causes the waves to be refracted and to break or spill. The change of the momentum flux due to the spilling of the waves causes the distinctive nearshore current component. But then this current component transports the sediment at just the right time averaged rate to maintain the shape of the transverse bar. The question of the origin of this system is really a question as to how any part of the system may be initiated. Once part of the system has developed, even to a very minor extent, then the rest of the system must develop and intensify through a positive feedback system until the whole system is in equilibrium.

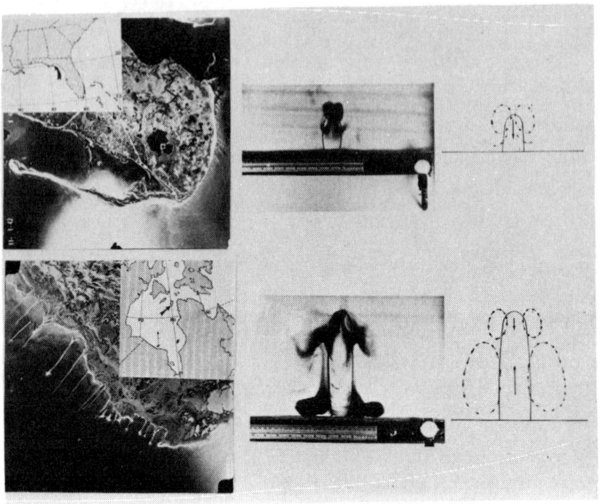

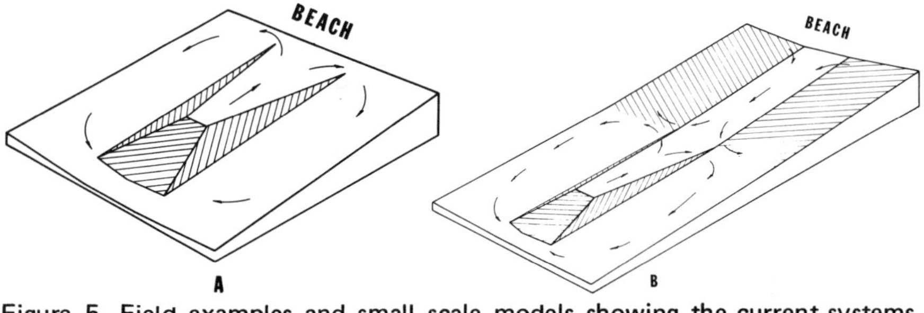

Figure 5. Field examples and small scale models showing the current systems which develop about short (top) and long (bottom) transverse bars. The middle photographs show the beach as a black horizontal bar and the outline of the model bar as a black line. The wave crest, moving from top to bottom, cannot be resolved. Black appearing dye has been used to show the current patterns.

Figure 6. Highly schematic diagrams of the mean surface level (the water level with the waves forms averaged out) over short (A) and long (B) transverse bars. The slopes of the surface propel the nearshore system associated with transverse bars. Shading shows areas where the gradient of the mean surface level is exactly offset by the gradients of the radiation stress tensor components which precludes flow down these sloped mean water level surfaces. The bar axis is located along the centerline of each figure.

There are several possible ways in which this system could develop initially. Any random irregularity in the shape of a gently sloping offshore area could produce a slight amount of refractive convergence of the nearshore waves. This in turn might start a weak current pattern around the initial irregularity which would transport sediment to begin the growth of the bar. Another possibility is that some offshore feature, such as ridge and swale topography on the continental shelf or an offshore shoal may cause waves to be refracted into a crossing pattern far offshore. As this pattern of crossing wave crests moves onto a gently sloping beach foreshore it would maintain an uneven distribution of wave energy density and hence uneven wave spilling in the nearshore region. The uneven spilling of the waves would then give rise to horizontal current gyres over the foreshore which could cause the development of transverse bars.

Finally, it should be pointed out that long sand bars with their long axes at a high angle to the beach are not all formed by the wave driven nearshore current component herein reported on. Sand bars of similar shape and orientation to transverse bars can be found around tidal inlets, particularly on ebb – and flood-tidal deltas. These are clearly the result of the tidal currents with little or no modification due to the wave driven nearshore current component discussed in the present paper. In addition, Nilsson (1972) discusses a variety of transverse bars which occur at relatively large distances from any tidal inlet but appear to form largely as a result of strong tidal currents draining very gently sloping sand-covered tidal flats in an area with a high tidal range.

CONCLUSIONS

Transverse bars cause, and in turn are caused by, a distinct nearshore current component which results from a transfer of momentum from the spilling waves over the bar to an elevated mean surface level over the bars. The currents associated with these sand bars result from gradients of the mean surface level which are not opposed by equal and opposite gradients of the radiation stress due to the waves. The equilibrium system of waves-currents-sediment transport-transverse bars can be initiated in areas with gently sloping nearshore areas by initiating any part of the system.

Relatively short transverse bars cause a current component and sediment transport which is directed toward the beach over the axis of the bar. Long transverse bars cause a more complex current pattern which is characterized by an offshore current and sediment transport along the portion of the bar axis near the beach.

Sand bars which appear very similar to wave produced transverse bars can be found near tidal inlets and on coasts which have very genly sloping nearshore areas and a great tidal range. It is herein recommended that the term *wave formed transverse bars* be used for those bars which are clearly formed as a result of the processes described in the present paper. The term *current formed transverse bars* should be used to describe transverse bars which result from the action of tidal currents.

ACKNOWLEDGEMENTS

I express appreciation to the following individuals for their help in various aspects of this research, the oral delivery of this paper, and the preparation of this manuscript: Dr. W. F. Tanner, Dr. R. C. Staley, Dr. R. Dean, Dr. R. T. Lai, Dr. B. J. Willmon, Dr. M. Hayes, Mr. J. Lee, Mr. D. Cassidy, Mr. S. Gustav, Mr. J. Lucey, and Mrs. I. Kubacki.

REFERENCES

Bowen,A.J. 1969. The generation of longshore currents on a plane beach: J. Marine Research, v. 27, n. 2, p. 206-215.

Bowen, A.J. and Inman, D.L. 1971. Edge waves and crescentic bars: Jour. Geophys. Res., v. 76, n. 36, p. 8662-8671.

Bruun, P.I. 1954. Migrating sand waves or sand humps, with special reference to investigations carried out on the Danish North Sea coast: Proc. of the 5th Conf. on Coastal Engineering, Council on Wave Research, p. 260-295.

Evans, O.F. 1938. The classification and origin of beach cusps: J. Geol., v. 46, p. 615–627.

Goldsmith, V. and Colonell, J.M. 1970. Effects of nonuniform wave energy in the littoral zone: Proc. Twelfth Conf. on Coastal Engineering, Washington D.C. p. 767-785

Homma, M. and Sonu, C. 1963. Rhythmic patterns of longshore bars related to sediment characteristics: Proc. 8th Conf. on Coastal Engineering, Council on Wave Research, p. 248-278.

King, C.A.M. 1972. *Beaches and Coasts:* St. Martin's Press, Inc., New York, 2nd. ed., 570 p.

Longuet-Higgins, M.S. 1970. Longshore currents generated by obliquely incident sea waves, 1 and 2: J. Geophys. Res., v. 75, p. 6778-6801.

Moody, D.W. 1965. Coastal morphology and processes in relation to the development of submarine sand ridges off Bethany Beach, Delaware: unpublished dissertation, John Hopkins Inst.

Niedoroda, A.W. 1972a. The generation of currents associated with transverse bars: Ph.D. dissertation, Florida State Univ., Tallahassee.

Niedoroda, A.W. 1972b. Waves, currents, sediments and sand bars associated with low energy coastal environments: Gulf Coast Assoc. of Geol. Soc., Trans., v. 22, p. 229-240.

Niedoroda, A.W. and Tanner, W.F. 1970. Preliminary study of transverse bars: Marine Geology. v. 9. p. 41-62.

Nilsson, H.D. 1972. Coastal and submarine morphology of eastern Cape Cod bay: M.S. thesis, Univ. of Massachusetts, 170 p.

Price, W.A. 1953. The low-energy coast and its new shoreline types on the Gulf of Mexico: Proc. Intern. Quaternary Congr., 4th, Rome, v. 1, p. 151-166.

Shepard, F.P. 1952. Revised nomenclature for depositional coastal features: Amer. Assoc. Petroleum Geologists Bull., v. 36, n.10, p. 1902-1912.

Sonu, C.J. 1968. Collective movement of sediment in the littoral environment: Proc. Eleventh Conf. on Coastal Engineering, p. 373-400.

Sonu, C. J. . . 1972. Field observations of nearshore circulation and meandering currents: J. Geophys. Res., v. 77, n. 18, p. 3232-3247.

Tanner, W.F. 1960a. Shallow water ripple mark varieties: J. Sed. Petrol., v. 30, p. 481-485,

Tanner, W.F. 1960b. Expanding shoals in areas of wave refraction: Science, v. 132, p. 1102-1103.

Zenkovitch, V.P. 1967. *Processes of Coastal Development:* Oliver and Boyd, London, 738 p.

Chapter 6

DEPOSITIONAL STRUCTURES AND PROCESSES IN THE MOUTHS OF SMALL COASTAL STREAMS, SOUTHWESTERN OREGON

H. Edward Clifton, R. Lawrence Phillips, and Ralph E. Hunter

ABSTRACT

The mouths of coastal streams form a unique depositional environment, in which sedimentation is controlled by a combination of waves, tides, and river discharge. Examination of processes and deposits in the mouths of small streams along the coast of southern Oregon provides considerable data on stream-mouth sedimentation on a coast characterized by high wave energy and moderate tidal range.

Depositional structures and processes in such a setting depend primarily on discharge and may be classified accordingly. The very smallest streams (discharge on the order of 0.001 m^3 sec^{-1} or less) modify the beach topography but slightly. Such streams flow in small intricately braided channels; the internal structure of the streambed material consists mostly of highly lenticular bedding, commonly defined by concentrations of heavy minerals. Larger streams (discharge on the order of 0.01 m^3 sec^{-1}) are more deeply incised into the beach. On the beds of such streams, sand ripples predominate, except in main channels that are occupied by standing waves and antidunes. Fluctuation between planar and antidune bedforms generates a pulsing flow in the lower reaches of such streams. During rising tide or periods of high waves, swash processes will either bury of destroy the stream deposit. If preserved, the deposit will consist of thin sheets of ripple or other lenticular bedding interlayered with parallel swash lamination.

Streams with larger discharge are deeply incised into the beach and have a bedform predominantly of megaripples. These structures may reverse their direction with incoming tide. Within the actual mouth of the stream, where wave action predominates, the bedform is generally planar.

The largest streams entering the coast of southern Oregon, such as the Rogue and Coquille Rivers, form complex tidal estuaries, the discussion of which is

Publication authorized by the Director, U.S. Geological Survey

U.S. Geological Survey, 345 Middlefield Road, Menlo Park, California 94025
U.S. Geological Survey, P.O. Box 6732, Corpus Christi, Texas 78411

beyond the scope of this paper.

Seasonal variation in discharge may cause different parts of some streams to evolve through several of the stages described above – an evolution that appears, over several years, to be relatively systematic. The resulting deposit will therefore show corresponding systematic transitions between depositional facies.

Ancient examples of stream-mouth deposits are rare, but can be recognized owing to their character and association.

INTRODUCTION

The mouths of coastal streams provide a unique environment for the accumulation of gravel, sand, and mud. River discharge brings sediment to the mouth, where it is modified and reworked by ocean waves. Tidal changes can reverse the currents in the mouth and effect a net upstream transport of sediment. They also raise and lower the zone of most intense wave activity.

Knowledge of the sedimentary processes that operate in the mouths of streams has manifold application. It is critical to resolving the problem of sedimentation in certain navigation channels. It bears on the problem of pollutant dilution and dispersal from streams that flow into the sea. It also pertains to the management of anadromous fish of commercial or recreational importance. Recognition of ancient stream-mouth deposits may be of considerable value in paleoclimatic or paleogeographic reconstruction.

Study of processes and deposits where small streams enter the ocean along the coast of southern Oregon (Fig. 1) provides much data on stream-mouth sedimentation on a coast characterized by high wave energy and moderate tidal range. During the summer, waves with a period of 8 to 12 sec form breakers that range from 1 to 3 m high along the open coast. During the winter storms the breaker height may be much greater. Tidal range approaches a maximum of 3 m. Precipitation and runoff depend greatly on the season of the year; average monthly precipitation ranges between 25 and 40 cm during the winter and drops to less that 3 cm during the summer.

Many of our observations of depositional structures in the mouths of these streams have been made underwater using scuba techniques. Box coring by the methods described in Clifton, Hunter and Phillips (1971) permits observation of internal structure of the sediment at a scale similar to that observed in outcrops.

Discharge was measured at segments of a stream where the channel maintained a relatively constant width and depth. Within this segment the average width and depth were established visually, and velocity was determined by the travel time of either neutrally buoyant drifters or concentrations of dye. Such discharge measurements are only approximate but are adequate for the purposes of this paper.

STREAM-MOUTH TYPES

The nature of stream-mouth sedimentation on the southern coast of Oregon differs widely, primarily as a function of stream size. Waves and tides have the potential for wide variation on the Oregon coast, but their effect at any one place tends, through time, to approach an average. Other factors, such as stream gradient, sediment load and grain size, and coastal configuration, are of unquestioned, but probably secondary, importance. The mouths of the observed streams can be roughly classified according to their discharge.

The very smallest streams that cross the beach have a discharge on the order of 0.001 m^3 sec^{-1} or less (typical average depth, 5 mm; typical velocity 10-30 cm sec^{-1}). They occur most commonly where uplands lie directly behind the beach; where the beach is backed by wide areas of sand dunes, streams of this size occur only where backshore runnels cut through a berm crest. Such streams tend to modify the beach topography but slightly. Flow occurs through an intricate system of minutely braided channels. (Fig. 2), some of which are occupied by tiny current ripples. Commonly the channel system is emphasized by concentrations of heavy minerals. The internal structure produced by such streams is a very lenticular form of bedding. Ripple lamination, oriented in a downstream direction, may or may not be apparent in X-ray radiographs through the sediment (Fig. 3).

Streams that are somewhat larger (discharge on the order of 0.01 m^3 sec^{-1}; typical average depth, 2-4 cm; typical velocity 30-50 cm sec^{-1}) cut as much as several tens of centimeters into the beach. Such streams show a relatively complex organization. In its upper reaches, where the stream crosses the backbeach, the gradient tends to be low (e.g. 1-2 mm/m), and the stream is confined to a single channel occupied by current ripples. But where the stream crosses the active foreshore, the gradient increases by a factor of 5 to 10. In these reaches the stream typically shows a braided structure (Fig. 4). The crests of the braid bars are planar if flow is very shallow, or microbraided or covered with small, flat ripples if flow is deeper. At the dowstream edge of the braided bar is a slip face, 1-4 cm high, which migrates downstream. The channels between the bars are occupied by ripples or, where flow is concentrated, by antidunes. A scour depression, up to several centimenters deep, usually forms where channels converge; it also migrates downstream. The internal structure produced by streams of this size is ripple bedding or small-scale crossbedding (Fig. 5).

Short-period (10-20 sec) fluctuations between plane bed and small antidunes are common to streams of this size, and produce flow pulsations having similar periods. Attempts to core the antidunes for internal structure have not been successful, probably because the features form on the surface of erosional channels. Placing a box into a swiftly flowing current generates eddies that destroy any thin surficial structure. Damming the stream to reduce the flow changes the flow characteristics, and the antidunes disappear before they can be cored.

During periods of high tide or particularly large waves, swash processes will either bury or destroy the small stream deposit. If it is buried, it will be

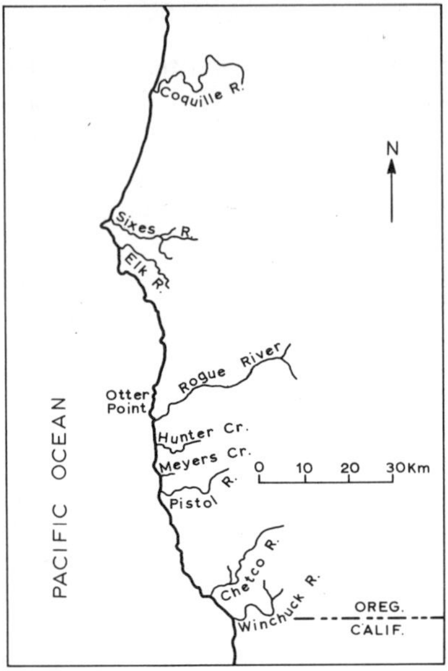

Figure 1. Index map showing location of the stream mouths studied on the coast of southern Oregon.

Figure 2. Minutely braided stream (discharge on the order of $0.001 m^3 sec^{-1}$). Note that stream cuts only slightly into beach. At constriction stream is about a meter wide.

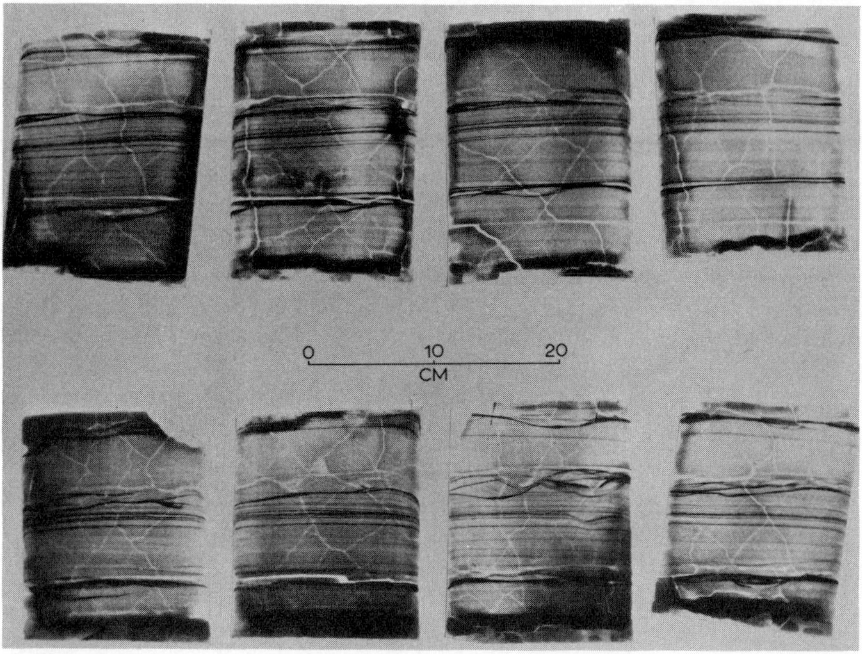

Figure 3. X-ray radiographs of internal structure produced by a tiny braided stream having discharge on the order of 0.001 m^3 sec^{-1} (three sets of lenticular bedding interbedded with parallel swash lamination). Upper set of radiographs is aligned parallel to flow direction (toward the left); lower set is aligned normal to flow direction. Intersecting light lines caused by fractures in the cores.

Figure 4. Meyers Creek (discharge on the order of 0.001$m^3$$sec^{-1}$) where it enters the ocean. Note braiding into separate channels characterized by ripple and antidune bedforms. Seagulls for scale.

preserved as a thin sheet of ripple or other lenticular bedding interlayered with parallel swash lamination. The radiographs in Figure 3 clearly show buried stream deposits of this type.

Because the smaller streams are not deeply incised into the beach, they flow directly down the foreshore slope. Larger streams (particularly those with a discharge exceeding 0.01 $m^3 sec^{-1}$) cut a channel 1 m or more into or through the beach, and the shape and direction of this channel can be strongly modified by waves that approach the shore obliquely. Shoaling waves affect the river mouth primarily by transporting sand landward and building a bar or sill at the point where the stream enters the ocean. This bar may grow in height, become attached to the shore at its updrift end, and extend itself in the direction of longshore drift, so that the stream is deflected and flows parallel to the strand line. Such a stream may flow parallel to the coast for a considerable distance; the Elk River, which has an annual average discharge of about 20 $m^3 sec^{-1}$, is at times deflected more than 2 km laterally (Fig. 6).

Channel deflection is greatest for streams with discharge on the order of 0.1 to 10 $m^3 sec^{-1}$. Bars are constructed at the mouths of streams with greater discharge, such as the Rogue, Coquille, and Chetco Rivers. But either runoff is sufficient to maintain the channel at a more or less constant position, or channel migration is restricted by areas of bedrock.

Streams in the area studied may be deflected laterally either north or south depending on the direction of wave approach during initial construction of the river-mouth bar. In winter, when waves approach the coast from the southwest, a stream tends to be deflected to the north. In summer, when waves approach the coast primarily from the northwest, a stream tends to shift to the south. The shoreline orientation and configuration, however, may be such that one of the seasonal wave regimes is much more effective than the other, so that a given stream may retain its deflection throughout the year. Also a stream may not be deflected in winter owing to its greater discharge at that time.

The channels of streams with discharge on the order of 0.1 to 1 $m^3 sec^{-1}$ are occupied by downstream-moving sand ripples and megaripples. If velocity is high relative to water depth (for example, water flowing at 1 $m sec^{-1}$ in a channel .25 m deep) a planar gravel bed may result. Pebbles in such a bed show an upstream-dipping imbication.

During a rising tide, directional structures in the channel may be reversed (Fig. 7) particularly in streams of relatively low gradient. Flow velocity during a flooding tide might be expected to be less than that during the tidal ebb, in which river discharge reinforces the seaward flow. The flood tide velocities, however, are increased by the effect of waves surging into the channel, and several factors discussed in the next section operate to decrease the velocity of tidal outflow. In some streams several lines of evidence suggest that velocities generated during an incoming tide exceed those generated during tidal outflow: sand in upstream-facing cross-strata is coarser and more pebbly than that in downstream-facing crossbeds, or pebble imbrication indicative of upstream

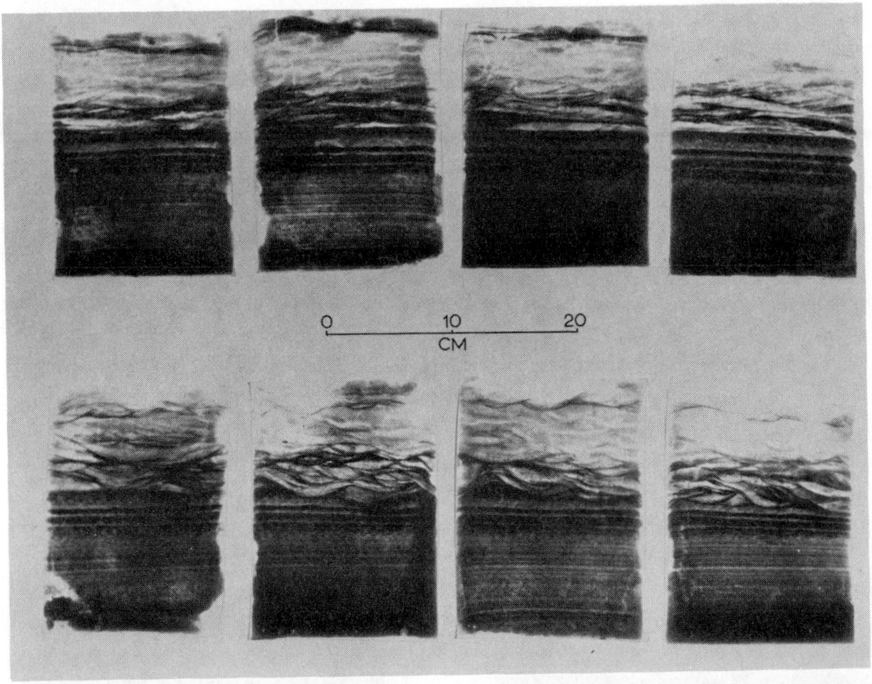

Figure 5. X-ray radiographs of internal structure from a stream with discharge on the order of $0.001 m^3 sec^{-1}$ (ripple cross-stratification in upper half of X-ray radiograph). Upper set aligned parallel to flow direction (to left); lower set aligned normal to flow direction.

Figure 6. Mouth of the Elk River in August, 1970. Note how stream has been deflected to the left (north) by beach-building processes.

Figure 7. Box cores taken from Pistol River channel downstream from first sill in early July, 1972. Downstream direction is to left. Upper set taken at low tide; cross-strata in upper part of cores indicates downstream sediment transport during falling tide (but note upstream-facing foresets in lower part of core left from previous tidal cycle). Lower set in same part of channel at the subsequent high tide; crossbedding faces upstream, indicating upstream sediment transport during rising tide. Note coarser grain size of flood-tide deposit.

current movement lies in the troughs of downstream-migrating sand waves.

In streams of relatively steep gradient or where the rise in tide is low relative to discharge, flood tides may not cause a complete reversal of directional structures throughout the stream. Individual waves surge upstream, but the water itself does not move upstream except at the channel margins, where it is less opposed by downstream flow. At low tide, ripples, megaripples, or pebble imbrication at the margins of such streams commonly reflect upstream transport in contrast to structures oriented downstream within the channel itself.

The largest streams that enter the ocean along southern Oregon – the Coquille, Rogue and Chetco Rivers – form complex tidal estuaries that are beyond the scope of the present paper. They have, however, certain attributes that are consistent with the foregoing discussion. Each, except during flood, is silled by a river-mouth bar (which in these streams constitutes a hazard to navigation). Directional structures (ripples, sand waves, scour crescents and pebble imbrication) exposed at low tide on the margins of the Rogue and Coquille Rivers (and atop a high central bar in the Rogue River) face in an upstream direction.

SEASONAL CHANGES

Precipitation and runoff in southwest Oregon change greatly with the seasons. Since discharge strongly controlls the nature of sedimentation in the small stream mouths, systematic seasonal variations in discharge produce attendant changes in depositional structures and processes there. Observations at the mouth of the Pistol River (Fig. 1) over a period of 3 yr show that predictable, systematic change does occur.

During the course of a year, the Pistol River mouth passes through all of the types described in the foregoing section. Discharge measurements throughout the year are not available for the Pistol River but do exist for the Sixes River, which undergoes a very similar pattern of sedimentation. The maximum, minimum and mean daily discharges of the Sixes River are summarized by month in Table 1. Discharge for the Pistol River is probably comparable.

During flood stage, the Pistol River has a discharge on the order of a few hundred m^3 sec^{-1}. At such times it behaves as a large stream and flows without deflection into the ocean (Fig. 8). At such times the bed consists of seaward-moving sand and gravel. Paradoxically, a layer of fine mud may accumulate in the stream channel at the mouth during flood stage. Such layers are formed by deposition at high tide, when downstream currents diminish and and bed-load movement ceases. Although the water remains relatively turbulent, silt and flocculated clay may settle from suspension and form a film or thin layer on the bottom. This layer may inhibit further sand transport, and if the bed is not eroded, additional mud can be deposited at successive high tides. In this way, a layer more than a centimeter thick may form in the stream channel, particularly during the waning stages of a flood. Strong currents locally scour through the mud layer and may undercut it by several centimeters (even though the consistency of the mud resembles that of toothpaste). Such a layer may be buried and preserved by further deposition of coarser sediment.

As the flood stage passes, the discharge drops to an order of 10 m^3 sec^{-1}. At this time, a river mouth bar becomes obvious near the beach, and the stream is deflected to either the north or the south. This deflection may result from the development of a new channel off to the side rather than by a lateral shifting of the main channel. Such a diversion was observed to occur in 1971 and is suggested by the configuration of the estuary in 1972 (Fig. 9).

With further waning of stream discharge, the combination of waves and tide creates a topographic sill to the channel above low tide level in the lower reaches of the stream (Figs. 9 and 10). In 1972, the first sill may have formed attendant with the development of the lateral channel; in 1971, it formed later within the channel.

The development of the sill produces the particular estuary configuration that is common to the streams with summer discharges on the order of 0.1 to 1 m^3 sec^{-1}. A broad estuarine basin lies immediately inshore from a narrow outlet to the ocean. This basin may be 100 to 200 m across and extends upstream for several kilometers in the larger streams; it appears to have been cut by flood channels that occupied different positions through the years. Maximum water

Figure 8. Mouth of Pistol River at flood stage, February, 1972. Compare position of the channel here with that in 1970 (Fig. 18). Subsequent photos of this river mouth in 1972 (Figs. 10, 11, 12, 13, 15, 16, and 20) taken atop highest sand dune at river margin in center of the photograph (taken by R.J. Janda).

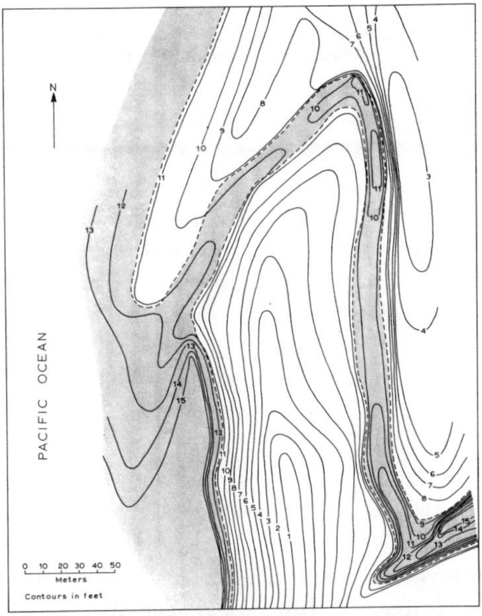

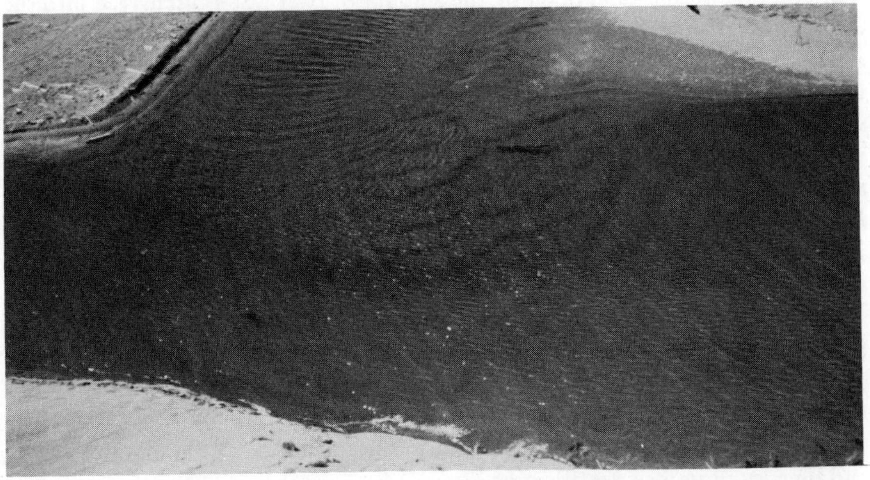

Figure 9. Topographic map of the Pistol River mouth, July 6, 1972. Main estuarine basin lies to the east. Note position of previous channel directed toward the southwest and now partly marked by a westerly appendage of the estuary (A). Also note first sill to the estuary where the outflow turns to the north (B). Water level in the estuary at low tide.

Figure 10. First sill developed in the mouth of the Pistol River; photographed at low tide, July 6, 1972. Note upstream-facing sand waves across the upstream face of the sill; these sand waves have been preserved during the falling tide. Also note the appendage of the estuary to the left that marks the February channel of the stream (Fig. 8). Structures seaward from this sill face downstream at low tide.

depth in the basins ranges from 2 to 3 m, in contrast to the outlet depth, which generally is less than a meter.

The tidal range and timing within the basin (Figs. 11 and 12) may differ markedly from that in the ocean outside. The presence of a basin having a sufficiently constricted mouth in effect establishes a hydraulic level thal lies alternately above and below the tidally varying level of the adjacent ocean. In such a basin, the two water levels correspond only for a short time twice each tidal cycle. One of these occurs at some time during the rising ocean tide; it is marked by reversal of the outflowing currents from the estuary. The other occurs at some time during the falling tide. At high tide, the estuarine level lies below the adjacent sea level. Water therefore continues to pour in from the ocean during the initial fall of the tide. This inflow may continue for more than an hour after high tide occurs in the ocean.

As long as the channel bottom slopes more or less uniformly seaward, directional structures in the tidally affected part of the channel reverse with the rise and fall of the tide. The presence of a sill at the mouth of the estuary, however, profoundly affects the orientation of the directional structures in the channel. The discharge from the estuary during ebb flow should generally exceed that entering during tidal flood by the amount of river water discharged into the estuary. Intuitively, one would therefore expect the structures in the channel of a stream mouth at low tide to be oriented in a downstream direction. But once a topographic sill develops, upstream-facing megaripples within the channel upstream from the sill become persistent features. These structures actively migrate only during the rising tide; they are inactive but not destroyed during the falling tide. A number of factors may contribute to the preservation of landward-facing megaripples upstream from the sill. Stratification of fresh and saline water is of prime importance. Other factors relate to sill-induced changes that reduce the velocity of the ebb tidal flow relative to that of the flood tide.

Stratification of fresh and saline water occurs in the lower reaches of an estuary in which water masses are incompletely mixed. In a stratified estuary or stream mouth, water of oceanic salinity enters from the sea. This saline water flows upstream beneath less dense fresh water discharged by the stream. The development of a topographic sill tends to reduce mixing of these water masses and thereby promotes stratification.

Water stratification within a stream such as the Pistol River is quite evident once a sill develops. It may be seen, particularly during initial stages, as a color difference – algal blooms in the saline water make it appear as a green mass beneath the clear fresh water. The interface is marked by visual blurring owing to a difference in refractive index between the two water masses. A concentration of organic debris, particularly algae, hangs suspended along the interface, denser than the overlying fresh water, lighter than the underlying saline water. The stratification may also be felt. Water temperature at the surface (at 62°F) contrasts markedly with that in the underlying saline water (at 51°F), which is nearly as cold as the parent oceanic water (at 50°F).

Figure 11. Lower estuary and river mouth at low tide, July 7, 1972. Large bar in upper right quarter of photo was formed by wave activity in February (Fig. 8).

Figure 12. Lower estuary and river mouth at high tide, July 4, 1972. Compare with Figure 11. Note topmost foam line in center of photograph; line marks the division between fresh and saline water at the surface.

During a rising tide the underflow of saline water is clearly visible near the sill. The point at which it sinks is marked on the surface by a concentration of foam and oceanic flotsam (Fig. 13). The underflow is a site of intense turbulence. Organic debris (jellyfish, seaweed, etc.) can be seen moving swiftly upstream beneath a layer of fresh water in which suspended material moves slowly downstream. Dye introduced into the saline water traces its passage into the estuary beneath the fresh surface layer.

As the tide begins to fall, the clearer fresh surface water can be seen pushing its way downstream against the waning pulses of sea water pushed into the estuary by wave surge. Ultimately this surface water flows directly out to sea. Downstream from the sill the fresh-water outflow extends to the bottom and causes a reversal of the orientation of directional structures. But upstream from the sill, a motionless layer of saline water persists and thereby protects the upstream-facing megaripples formed during the tidal rise.

During 1972, the Pistol River developed three separate sills. The first of these (Figs. 9 and 10) exceed a meter in height and apparently formed when the main channel was blocked and a shallower lateral channel developed. Stratification upstream from this sill was quite noticeable, and landward-facing megaripples persisted upstream from it throughout the tidal cycles. A second sill, about a half meter high, developed in mid-July approximately halfway between the first sill and the ocean (Fig. 14). Prior to the development of this sill, megaripples in the channel downstream from the first sill reversed orientation as the tide changed (Fig. 7); once the second sill formed, upstream-facing megaripples 30-40 cm high became persistent bedforms upstream from it. Stratification was evident for the first time in this part of the stream; at low tide saline water could be detected in the troughs of these megaripples.

A sill may also promote the preservation of upstream-facing structures if it grows during the rising tide and causes a lasting rise in water level in the estuary. In such a case, less water flows out on the falling tide than entered during the previous tidal rise. Structures produced by inflow during this rise are thereby preserved through the following low tide and are likely to remain inactive except during successively higher tides (or larger waves). The effects caused by this situation are particularly noticeable in the lowermost reaches of the stream mouth.

The change in estuary level due to a growing sill may be dramatic. Overnight on July 19, 1972, a third sill developed near the mouth of the Pistol River, and the low tide within the estuary rose half a meter over its position the previous day (Figs. 15, 16, and 17). Discharge from the estuary dropped from 1.5 m^3 sec^{-1} at low tide on July 19 to .025 m^3 sec^{-1} at low tide on July 20. With the development of this third sill, ripples and megaripples in all parts of the stream but the final outlet faced upstream. In the outlet, downstream facing ripples constituted the dominant bedform, as they generally do in streams of this size.

Some of the water that is impounded in the estuary by a developing sill is gradually returned to the ocean by a process other than stream outflow. When

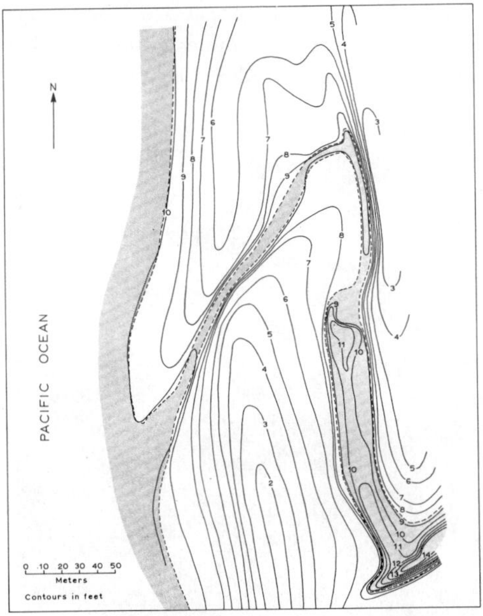

Figure 13. Underflow of saline water into the Pistol River during a rising tide (July 5, 1972). Concentration of foam marks the site of the underflow.

Figure 14. Topographic map of the Pistol River mouth, July 19, 1972. Note second sill which has developed about half way between the first sill and the westward bend in the stream. Note also that the bend in the stream has migrated to the south since July 6 (Fig. 9). Water level in estuary at low tide.

water level in the estuary is higher than that in the ocean, a certain volume of water seeps from the estuary through the sand at its margins. As noted previously, the river-mouth sill or bar tends to deflect the stream laterally, inducing it to flow parallel to the beach. As this lateral course develops, an increasingly broad surface of sand confronts the stream, and an increasingly large volume of water is diverted from the discharge. In this way the volume of water flowing through the channel into the estuary at tidal flood is increased relative to that flowing out during the ebb.

The loss of water by seepage through the sand becomes greater as the sill height increases and the water level in the estuary rises. Ultimately, streams such as the Pistol River becomes totally dammed at the beach during the late summer (Figs. 18, 19, and 20). At this point all of the river runoff is emitted to the ocean by seepage. The volume may be considerable. Measurements of discharge in the Pistol River above the range of tidal influence compared to that in the mouth of the estuary just before damming indicate that as much as 1 m^3 sec^{-1} or 10^6 gal/hr may seep from the estuary.

Once a river mouth becomes dammed, the water stratification in the estuary breaks down and the water column becomes relatively homogenous. The change in salinity drastically affects the marine benthos that became established earlier when fresh ocean water was added daily to the estuary. Shortly after damming,

Figure 15. Pistol River mouth at low tide, July 19, 1972. Second sill has developed downstream from the first, at about the point where the stream channel narrows just before the westward directed bend. Large upstream-facing sand waves can be faintly seen in the channel between the first and second sill.

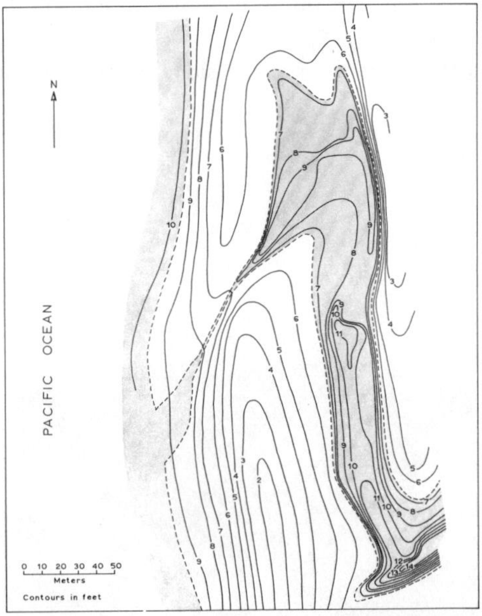

Figure 16. Pistol River mouth at low tide, July 20, 1972. Third sill has developed overnight past the bend. Note rise of estuary level over that of previous day (Fig. 15). Upstream-facing sand waves now extend down to the third sill (located at the constriction of the stream past the bend).

Figure 17. Topographic map of the Pistol River mouth on July 20, 1972. Note development of third sill seaward from the bend in the river. Note also that low tide in the estuary has risen considerably from previous day (Fig. 14).

the estuary floor is littered with the remains of immature barnacles, crabs, and the razor clam *Siliqua patula*.

We have considered thus far only those depositional structures and processes that result from stream discharge or from tidal flow in an out of the estuary. Sediment can also be introduced into the estuary from its margins by berm overwash and by wind. These mechanisms may be important throughout the development of the estuary, but are the dominant depositional mechanisms once the estuary is dammed.

Berm overwash results from waves that top the berm or ridge that develops between a laterally deflected stream mouth and the ocean. Overwash occurs predominantly during high spring tides or during periods of particularly large waves. Sediment is transported from the beach and is deposited on an avalanche slip face at the estuary edge that extends from water level to the stream floor (Fig. 21). Additional overwash causes the slip face to migrate, producing crossbedding that dips in the direction of the overwash, primarily landward. The scale of this crossbedding may be large, for the unit produced is as thick as the water depth at the time of formation. If the stream is actively flowing, these deposits may soon be eroded, but if the stream is dammed, they will be preserved at least until the next large flood.

Wind-blown sand similarly enters the estuary from the side. On the Oregon

Figure 18. Pistol River mouth, August, 1970. Stream is completely dammed. Note partial closing of stream channel by berm overwash. Also note appendage of estuary extending toward the camera, probably an earlier channel of the river formed at flood stage.

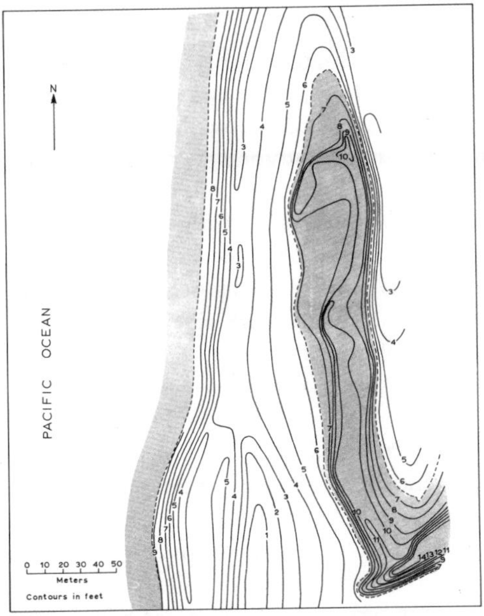

Figure 19. Topographic map of the Pistol River mouth on August 25, 1972. Note the amount of lateral filling by berm overwash since July 20 (Fig. 17).

Figure 20. Pistol River mouth on August 25, 1972. Note figures at far end of stream for scale. Note also berm overwash pouring across the beach into the river.

coast, eolian transport of sand is predominantly by northerly and northwesterly summer winds. Continuous addition of this sand produces avalanche slip faces at the estuary margin similar to those produced by berm overwash. The two types of deposits differ in their orientation and position and their grain size. Material brought in by berm overwash consists of sand and fine gravel; it lies along the seaward margin of the estuary in large-scale foresets that dip predominantly landward. Sand introduced by wind is better sorted, and the largest deposits lie on the northern and northwestern side of the estuary in large-scale foresets that dip predominantly to the south and southeast. Where the channel or estuary shorline lies oblique to wind direction, the foresets will strike parallel to this shoreline, not normal to wind direction; the rate of outbuilding will of course be less than where the wind is normal to shore.

The degree of lateral deposition differs from year to year, but would rarely fill the entire estuary. The amount of lateral filling during a short period in late summer 1972 can be seen by comparing Figures 17 and 19.

Once a river mouth is dammed, it may, on occasion, overflow during the summer and cut a new channel through the beach. Flow in such a channel is intense, but generally short lived. The next series of high tides or large waves tend to rebuild the berm and re-dam the stream. Ultimately, increased runoff due to fall or winter rains will raise the stream level to the point where it breaks

Figure 21. Avalanche slip face formed by berm overwash, extending from near the surface to the floor of the channel. Ocean is in background behind figure standing at stream bank. Note coarse grain size of the sediment on the slip face.

through. Winter and spring floods erode much of the previously deposited sediment and start the cycle anew.

DEPOSITIONAL FACIES IN A SMALL STREAM OF VARIABLE DISCHARGE

During the summers of 1970, 1971, and 1972, the shape of the Pistol River mouth differed somewhat, but it evolved each year through the stages described in the the previous section. Streams of equivalent size, such as the Sixes and Winchuck Rivers, as well as many of the smaller streams, evolve through the summer in a similar fashion. Their similarity suggests that a general model of depositional facies may be developed for such streams.

For the purpose of the model, the lowermost reaches of a stream can be considered in three separate segments: "upper estuary," "lower estuary," and "mouth." The upper estuary is that part of the estuarine basin that lies upstream from significant sedimentation by tidal, eolian, or overwash processes. The lower estuary is the seaward-most hundred meters or so of estuarine basin; sedimentation there is strongly influenced by tides, winds and overwash. The mouth is where the stream flows directly into the ocean. It generally lies seaward from the estuarine basin and includes the area in which the stream is deflected laterally.

The upper estuary mostly contains sediment brought in by the stream. The basal part of this material consists of flood deposits – primarily gravel in which imbrication dips upstream. The upper part of the flood deposits should contain sand, crossbedded with downstream-facing forests, as well as thin, sharply defined layers of mud. The flood deposits are overlain by a film of silt and clay, deposited in quiet water during the summer time. Flood gravel forms the basal layer in the lower estuary as well, but there, it is overlain by sand deposited by incoming tidal currents. This sand forms units, 10 to 20 cm thick, of upstream-facing crossbeds. These deposits in turn are overlain by a thin layer of silt and clay, or, if near the estuary margin, possibly by sediment deposited by wind or overwash. Along the seaward margin, sand and gravel may constitute the topmost facies, deposited by berm overwash in thick (2-3 m) crossbeds that dip predominantly landward. Along the northern and northwestern sides of the estuary, medium-grained sand may be found, brought in by wind and deposited in equally thick crossbeds that dip predominantly to the south.

The mouth area is complex and includes interlayered beach and stream deposits. In the area of the main winter channel, flood gravels again form the basal facies. River deposits above the basal facies tend to consist of channels filled with sand and gravel. These channels may trend parallel to the beach, and the directional structures within them show a bimodal orientation, with both modes parallel to the channel axis. These deposits may be covered by berm overwash deposits of sand and gravel in foresets that show a landward orientation.

ANCIENT EXAMPLES OF STREAM-MOUTH DEPOSITS

Examples of stream-mouth deposits in the stratigraphic record are rare. Shoreline deposits themselves are relatively uncommon, and stream mouths comprise but a limited part of the total shoreline facies assemblage. Nevertheless, examples do exist; they can be readily recognized owing to their unique character and associated facies. Identificaiton of such deposits provides further data on paleoclimate and paleogeography at the time of their deposition.

An excellent example of a deposit from an ancient stream with discharge on the order of 0.01 m^3 sec^{-1} occurs in Pleistocene terrace deposits at Otter Point, 4.5 km north of the present mouth of the Rogue River (Fig. 1). This deposit overlies a nearly horizontal platform, about 10 m above present sea level, that is cut into the Otter Point Formation of Late Jurassic age (Koch, 1966). The Otter Point Formation here consists of a very steeply dipping sequence of alternating sandstone and shale. Beds of resistant sandstone stand as much as a meter in relief above the platform; clam borings in such beds imply that the beds lie on an ancient surf-cut platform.

The lowermost 3-4 m of the terrace deposit consists of loosely consolidated, well-sorted sand that contains scattered pebbles and pebble layers. Sedimentary structures in this material indicate deposition in a beach environment. Evenly parallel lamination that dips gently seaward is the predominant structure. Many of these laminations show the inversely graded bedding characteristic of swash lamination (Clifton, 1969). As with modern beach deposits, the gently dipping foresets grade finer and thinner in an updip direction, and heavy minerals are concentrated near their upper ends. Pebble imbrication in the coarser beds dips in a seaward direction, as do pebbles on modern foreshores. In the lowermost part of the deposit, near its seaward-most exposure, medium-scale crossbedding occurs, dipping predominantly seaward. This crossbedded sand, which locally is intensely burrowed, resembles that deposited at the bases of modern beaches. (Clifton and others, 1971).

About 1.5 m up from the base, the deposit locally contains a zone about a meter thick that is characterized by channels and abundant ripple bedding. The floors of many of the channels are strewn with well-sorted discoidal pebbles mostly 5-10 cm in maximum dimension (Fig. 22). Crossbedded sand units 10-20 cm thick indicate seaward transport, as do imbricated pebbles. Between the channels, the sand shows aboundant ripple bedding (Fig. 23); concentrations of heavy minerals define most of these ripples, which also indicate sediment transport in a seaward direction.

The channeled and ripple-bedded zone shows much similarity to deposits produced by small streams that cross the present-day beaches in this area. The directional aspects of structures in the zone support this interpretation. Its position in the middle of what clearly appears to be an ancient beach deposit adds further corroboration.

Other probably examples of stream-mouth sediment occur in other shoreline

Figure 22. Probable example of an ancient deposit produced in the mouth of a small stream, Pleistocene terrace deposit, Otter Point, Oregon. Lower and uppermost parts of exposure, characterized by parallel lamination, are inferred to be an ancient swash zone deposit. The cobble-lined channels above the hammer mark the probable position of a small stream crossing the beach.

Figure 23. Detail of internal structure from the probable stream-mouth deposit shown in Figure 22. Note marked similarity of this structure to that produced by small streams that cross presentday beaches (Fig. 5).

deposits – middle Miocene strata in the southeastern Caliente Range of California and Pleistocene terrace deposits at Montara Beach, about 20 km south of San Francisco, for example. Stream discharge can be estimated within an order of magnitude at both places. Such data aid in the interpretation of paleoclimates; they also facilitate reconstruction of the ancient environmental setting. It is likely that many ancient shoreline deposits, if examined closely, would yield similar examples of stream-mouth deposition.

CONCLUSIONS

1. The small streams along the coast of southern Oregon produce distinctive deposits where they enter the ocean. The nature of these deposits depends primarily on the size of the stream.

2. The smallest streams (discharge on the order of 0.01 m^3 sec^{-1} and less) tend to cut but slightly into the beach. The predominant internal structure of their deposits is ripple bedding that reflects sediment transport in a seaward direction. Medium-scale crossbedding may occur in the deposits of the larger of these small streams. In the very smallest (discharge less than 0.001 m^3 sec^{-1}), bedding may be quite lenticular, and the ripple foresets difficult to detect, even in X-ray radiographs. The deposits of such streams are likely to be preserved in thin sheets between upper or middle swash zone deposits, characterized by fine evenly parallel lamination.

3. Streams on the order of 0.01 m^3 sec^{-1} commonly exhibit a pulsing flow. The pulses are caused by the transformation between antidune and planar bedforms within the main channels. The transformation seems to occur because the bedform modifies the water flow so that it continually fluctuates between the antidue and plane bed fields. As antidunes break, water is momentarily dammed then released as the bedform shifts to planar form. The release of water forms a pulse that is reinforced as it travels downstream, overwhelming the breaking antidunes in its path.

4. Larger streams are more deeply incised into the beach. Wave activity tends to build a topographic bar at the point where the stream channel contacts the ocean. The smaller of such streams (discharge on the order of 0.1 to 1.0 m^3 sec^{-1}) are thereby likely to be deflected laterally along the coast. The beds of streams with discharges greater than 0.01 m^3 sec^{-1} are shaped into ripples and megaripples. These structures may reverse their directions with the rise and fall of the tide.

5. Seasonal variation in rainfall causes systematic changes in discharge of the small streams along the southern Oregon coast. The discharge of a single stream may differ by two orders of magnitude during the year. The seasonal changes produce systematic changes at the mouth that are roughly repeated year after year.

6. During flood stage, the streams transport a large volume of silt and clay in suspension. This material can settle out onto the bottom during the periods of

high tide when movement of the bed load momentarily ceases even though the water remains relatively turbulent. The layer of mud thereby deposited tends to armor the bottom against further bed-load transport.

7. As discharge waned following the winter floods, a bar or sill is constructed at the point where the stream enters the ocean. This bar tends to divert the stream laterally. If the bar also decreased water depth at the stream mouth, it will promote a stratification of fresh and saline water. Water stratification is a major factor in the formation and preservation throughout the tidal cycle of upstream-facing megaripples upstream from the sill.

8. With waning discharge, a stream mouth may be filled or partly filled by berm overwash or wind-blown sand. Each of these processes fills the stream channel laterally, producing large, steeply dipping foreset units as thick as the water is deep. The cross-strata produced by berm overwash tend to dip landward. Those produced by wind-blown sand tend, on the Oregon coast, to dip southward or southeastward.

9. Each year, a new flood channel may be carved at the stream mouth. Through time these channels merge to produce a relatively broad, deep estuarine basin upstream from the area of filling by wave activity and eolian processes.

10, If discharge diminished sufficiently, the stream may be completely dammed off. Runoff is then accommodated by seepage through the sand into the ocean. The amount of seepage is controlled by several factors, including grain size of the beach sand, the difference between river and sea level, and the size of the area along which the stream contacts the dam; it may approach 1 m^3 sec^{-1} for southern Oregon streams.

11. The sequential changes produced by seasonal variation of the southern Oregon streams result in a systematic distribution of depositional facies. The upper estuary facies is characterized by sand and gravel that show evidence of only downstream transport and contain interlayers of fine mud. The lower estuary facies has similar deposits at its base, but these are overlain by crossbedded sand in which the foresets dip upstream. This sediment may be overlain by coarse pebbly sand brought in by berm overwash that forms large-scale landward-dipping crossbedding or by well-sorted medium to fine sand brought in by eolian processes, that forms large-scale south-dipping cross-bedding. The mouth of the estuary – the part downstream from the landwardmost sill – is represented by a complex facies. Many of the directional features in its lower part may be bimodal parallel to the coast. The upper part may consist of large-scale foresets, particularly those produced by berm overwash.

12. Ancient examples of small stream-mouth are rare but are readily recognizable. Their character will indicate the size and orientation of the stream that formed them. Such data are valuable for paleoclimatic and paleogeographic interpretation.

ACKNOWLEDGEMENTS

A number of different persons ably assisted us in this research: R. J. Anima, J.A. Goodale, Charles Jones, A. G. McHendrie, and J. C. Scheihing. Paula Quinterno and H. C. Wagner reviewed the manuscript and provided many helpful comments.

REFERENCES

Clifton, H.E. 1969. Beach lamination–nature and origin: Marine Geol., v. 7, p. 553-559.

Clifton, H.E., Hunter, R.E. and Phillips, R.L. 1971. Depositional structures and processes in the non-barred high-energy nearshore: Jour. Sed. Petrology, v. 41, p. 651-670.

Koch, J.C. 1966. Late Mesozoic stratigraphy and tectonic history Port Orford–Gold Beach area, southwestern Oregon coast: Am. Assoc. Petroleum Geol. Bull., v. 50, p. 25-71.

U.S. Geological Survey. 1969. Water resources data for Oregon, 1968, Part I. Surface Water Records: 372 p.

Chapter 7

BEACH PROCESSES IN AN ARCTIC ENVIRONMENT

S. Brian McCann

ABSTRACT

Data on ice conditions, ablation sequences, beach profiles, beach sediments, wave conditions, and freeze-up sequences have been collected during four field seasons, 1968-71, in the Radstock Bay area of S.W. Devon Island (74°N, 91°W) in the Canadian Arctic Archipelago. From these observations it is possible to build up a reasonable picture of the beach conditions and annual regime in this particular area of the Arctic and, perhaps more important, to gain an insight into the year to year variation in the intensity of operation of beach processes. In this latter regard the field data have been supplemented by wind data from the Resolute meteorological station and ice reconnaissance data for the Barrow Strait-Lancaster Sound sea area, for a ten year period 1959-1968.

The paper considers the special characteristics of Arctic beaches and discusses the annual beach regime in the study area. Considerable importance is attached to freeze up conditions in the fall as one determinant of beach conditions and the operation of beach processes during the following summer. The magnitude and frequency of periods of significant wave action are considered in relation to the probability of simultaneous occurrence of ice free ocean, suitable winds and ice free beaches, and the effects of three major storms during the study period are evaluated. The study has applications in the selection of landing beaches and in the construction of coastal installations.

INTRODUCTION

A universal characteristic of Arctic beaches is the relatively short period of the year during which normal wave processes can operate. The presence of ice at sea, either as a solid cover or as masses of pack ice, prevents the generation of waves for much of the year, and the presence of fast ice on the beaches may further restrict the action of waves even during the open water season. This inhibitive role of ice is more important in considering Arctic beaches than the physically more obvious positive role, which produces ice push and ice rafting of beach

Department of Geography, McMaster University,
Hamilton, Ontario, Canada

sediments. The presence of frozen ground below the beach surface is also important in some areas.

The effect of many months of near total ice cover each year means that Arctic beaches are low energy beaches, with relatively low rates of longshore sediment transport and a low degree of reworking of beach material. Because in most areas there are considerable variations in ice cover conditions from year to year, there are similar variations in the amount of wave action from year to year. During certain years many Arctic beaches undergo scarcely any wave action due to the inhibiting effects of ice, and this makes the occasional large storm, when strong onshore winds blow over open water, an important component of the beach regime. One such catastrophic storm recorded by Hume and Schalk(1967) in the Point Barrow area of Alaska moved the equivalent of 20 yr normal transport of beach sediment. For an area, namely the Arctic, about which little is known of the rate of operation of geomorphic processes, the point is well made – long term observations are necessary to understand the magnitude and frequency of such processes.

Though Arctic beaches have as a common factor the central role of ice, there are considerable differences in beach regimes and beach types across the North American Arctic which are related to length of open water season, local important wind directions and lengths of fetch, and differences in tidal range. There are many different types of coast also, from high rock coasts to low prograding depositional coasts, and in certain areas thermal erosion of permafrost ice, rather than wave action, is the dominant process. One's view of Arctic coastal processes tends to be formed, and perhaps limited, by direct acquaintance with a particular area of study, in this case Radstock Bay, Devon Island, though I have now broadened the area of reference to include contrasting sites within the Queen Elizabeth Islands.

INVESTIGATIONS IN THE RADSTOCK BAY AREA OF S.W. DEVON ISLAND (74°N, 91°W)

The southern coastline of Devon Island is essentially one of high cliffs and talus slopes, developed in lower Paleozoic limestones, which form a high plateau inland to the north. At the entrance to Radstock Bay a detached part of this plateau, defined by steep cliffs and talus slopes, is linked to Devon Island proper by a series of great raised beaches of gravel developed with isostatic uplift since deglaciation (Fig. 1B). The 5 mi of modern beach at the foot of the raised beach sequence have been the site of investigations since 1968, during which time data on ice conditions, ablation sequences, beach profiles, beach sediments, wave conditions, and freeze-up sequence have been collected (McCann and Owens, 1969, 1970; McCann and Hannell, 1971; McCann, 1972). The site was selected because it has a longer open water season, during which wave processes may operate, than the logistically more suitable site at Resolute and because it is open to waves generated across Lancaster Sound to the south and east. It is clear

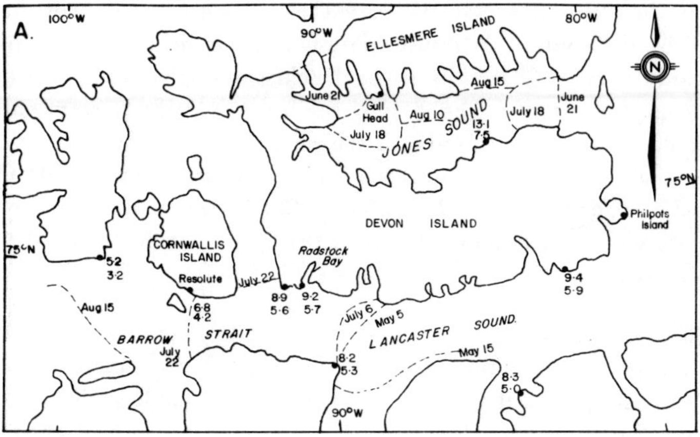

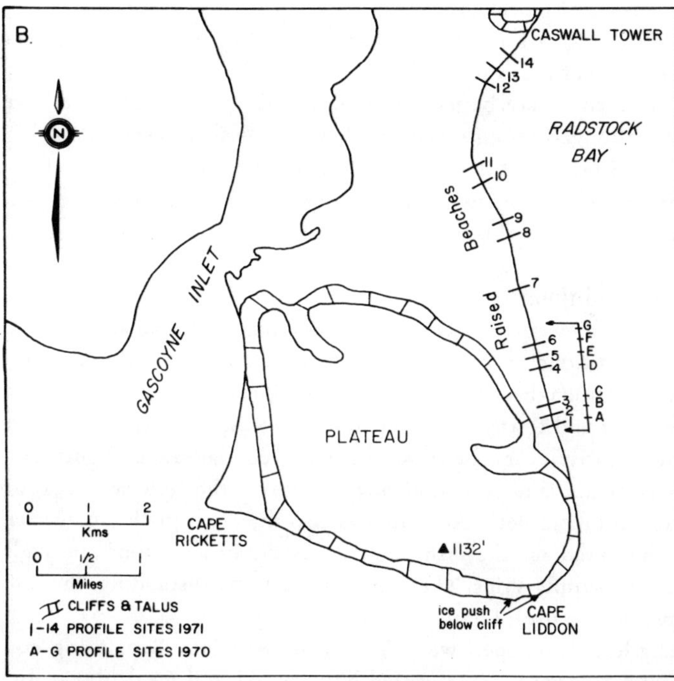

Figure 1A. Southern Queen Elizabeth Islands: dates of ice break-up in Barrow Strait-Lancaster Sound and Jones Sound after Lindsay (1968), for 1966—"an average year": tidal ranges for mean tide and large tide in feet taken from Pilot of Arctic Canada and Canadian Tide and Current Tables.

Figure 1B. Location map of study beach on Radstock Bay, Devon Island, showing positions of profiles illustrated on Figures 4, 5, and 6.

also from the disposition of the raised beach ridges that they have been developed by the continuous transportation and deposition of beach material in a northward direction from a source in the cliffs of Cape Liddon. The dominant northward movement of material along the beaches at the western entrance to Radstock Bay which occurs today has been continuous for the past 9000 yr. (Shell material taken below the marine limit, at a height in excess of 350 above present sea level, dates at 9230 ± 150 yr, GSC – 1502).

The basic environmental conditions of the beaches in Radstock Bay may be listed as follows:

1. They are situated in an area that has undergone considerable isostatic uplift since deglaciation, which probably continues today.
2. There is a plentiful supply of new material available at the southern end from the erosion of cliffs and talus slopes, and to some extent of older, now raised, beach deposits; the beaches are mainly built of limestone fragments of gravel and small cobble size.
3. The tides are of a mixed semi-diurnal type, with a range at large tides of 9.2 ft and at mean tides of 5.7 ft.
4. The beaches may be ice free and open to wave action for 8-9 weeks each year, though there is considerable variation from year to year.
5. There is maximum fetch to the southeast of approximately 100 mi.

This paper focuses on two related topics – the length and variation of the open water season, and the occurrence of effective storms during this period.

Open water conditions

In an earlier paper paper (Owens and McCann, 1970) an attempt was made to assess the conditions at Resolute, Cornwallis Island, by reference to published data on conventional break-up and freeze-up dates, to aerial reconnaissance maps of sea-ice cover and to the local wind data for Resolute airport. The procedure is somewhat unsatisfactory because the ice reconnaissance flights are made at irregular intervals and do not always cover the relevant sea areas. The information is incomplete. Attention is also directed on the main channels and straits, rather than on the inshore areas, as the observations are made to assist the passage of supply ships. These limitations notwithstanding, the results of the analysis provided a useful basis for further study.

The length of the open water season in Resolute Bay, using conventional break-up and freeze-up dates, varied between 41 and 64 d (days) for the years 1948 through 1963. These figures set an outside limit to the period of potential wave action, which is greatly curtailed. if available fetches of ice free water in Barrow Strait to the south are considered in conjunction with information about wind speeds and directions. For the years 1961, 1962 and 1963 there were only 3, 5 and 2 short periods, respectively, when ice and wind conditions were such that effective waves could have been generated which would affect the Resolute beaches.

Radstock Bay has a longer open water season than Resolute, though there are difficulties in the long term assessment of the number of days of potential wave action each year, because a complete cover of sea ice usually remains in the bay long after the ice in Lancaster Sound to the south has broken. It is not unusual to find the bay completely locked in ice and much of the Sound with only a small amount of pack ice. Again the ice reconnaissance maps are incomplete, in that attention is focused on the main sea channels and detailed recording of ice conditions in the bays and inlets is lacking, but with careful analysis some general inferences can be drawn and an attempt made to compare the study site with other localities in the Queen Elizabeth Islands. Table 1 presents the results derived from a study of the relevant ice reconnaissance maps for 12 yr, 1960-71, and it should be noted at the outset that certain assumptions have been made in its preparation. The most important is that an event, say the beginning of open water conditions, which occurred between the dates of two successive reconnaissance flights, is assumed to have occurred midway between them. The dates listed are break-up, open water, reclosure, reopening, final reclosure and dates of other significant events. In addition the number of days of open water are calculated. This figure refers to the number of days between first open water, not break-up, and reclosure, minus and mid-season closed periods. The plus sign after some length of season values indicates that the season is open ended; that is, no final date of closure is available, and thus the calculated value for length of season is a minimum value.

There is considerable variation in the date of first open water, from as early as July 5 to as late as August 26, and also in the mean length of the open water season. The mean length of the season is 56 d, though out of the 10 yr for which there is information 5 have open-ended seasons (that is, reconnaissance flights do not cover the entire season) so this figure is probably too low. However, as comparison of the first dates of open water derived from reconnaissance maps (Table 1) with the actual dates recorded in the field (Table 2) will show, the former have a tendency to be too early.

Study of the ice reconnaissance maps and calculation of mean length of open water season for three other sites in the southern Queen Elizabeth Islands – Philpots Island on the east coast of Devon, Gull Head on the south coast of Ellesmere Island, and Resolute – give values of 62, 42 and 44 d, respectively. The beaches of the Radstock Bay site may be considered fairly typical, then, in terms of length of open water season, of many of the beaches of the area shown in Figure 1A.

Storm wave action

Only one major storm, producing waves capable of really significant action on the beach, has occurred during the various periods of field observation, between 1968 and 1971, at the Radstock Bay site. It can most probably be considered to be the only event of its kind to have occurred during the four open water seasons, 1968-71, though several smaller storms and their effects have been

TABLE 1. OPEN WATER CONDITIONS, RADSTOCK BAY, DEVON ISLAND

YEAR	BREAK-UP	OPEN WATER	RECLOSURE	REOPENING	RECLOSURE	OTHER	SEASON LENTH IN DAYS
1960	May 1	No Data	No Data	No Data	No Data		No Data
1961	June 13	July 6	No Data	No Data	No Data	Sept 13 (open)	98+
1962	June 14	July 5	Oct 11 (9/10)	No Data	No Data		98
1963	June 7	Aug 26	Sept 16 (10/10)	Sept 28	Sept 24		20+
1964	May 7	Aug 9	Oct 14 (10/10)	Aug 27	No Data	Sept 30 (8/12)	52
1965	July 22	July 27	Aug 8 (10/10)	Aug 15	Oct 16	Sept 29 (9/10)	64
1966	Aug 1	Aug 7	Aug 26 (7/10)	No Data	No Data		No Data
1967	July 11	Aug 17	Aug 22 (9/10)	Aug 26	No Data	Aug 26 (open)	5+
1968	July 11	July 14	No Data	No Data	No Data	Sept 20 (open)	68+
1969	July 20	Aug 2	Sept 28	No Data	No Data		57
1970	July 31	Aug 9	Aug 15 (4/10)	Aug 12	Sept 23		43
1971	July 18	July 22	No Data	No Data	No Data	Sept 21 (open)	61+

Source:- Ice reconnaissance maps, Meteorological Branch, Canadian Department of Transport.
Compiled by R. J. Carlisle (1972)

observed. It becomes important then to assess the effects of the major storm and to set it in context by determining the frequency of such events.

The major storm effects, which occurred on August 11-12, 1969, form part of a longer sequence of events extending over a period of a month, during which winds over Lancaster Sound and Radstock Bay blew mostly from directions between east and south, inclusive. This preponderance of winds from the southeast quadrant is well shown in Figure 2, a wind rose for the period based on data from Resolute, Cornwallis Island. The sequence of events at Radstock Bay during this period is indicated in Figure 3, superimposed on the graph of hourly wind speeds at Resolute for winds blowing from the southeast quadrant. Again the preponderance and continuity of these winds is very apparent.

Lancaster Sound was relatively ice free in 1969 by July 21, but until August 7 Radstock Bay, outside the nearshore zone, had a complete cover of unbroken sea ice. Two events had occurred, however, before that date which are of importance. Firstly, on July 31, a reversal of winds to a northerly direction opened a wide shore lead along the beach, clearing the ice which was ablating *in situ* in the zone of tidal cracks. No fast ice remained on the beach and the situation was right for the bay ice proper to move out with the next series of northerly winds on August 7. Secondly, strong southeasterly winds, blowing continuously for a period of 24 hours on August 3-4, drove masses of pack ice aground below the cliffs around Cape Liddon, which was then piled and pushed to heights of 60 ft up the talus, causing considerable erosion. The strong southeast winds of August 6 added to this ice push effect which was the most impressive we have seen in the area.

The day of northerly winds which moved the ice *en masse* out of Radstock Bay was followed almost immediately by brisk southeast winds on August 8-9, which produced the first significant wave action on the beach, rapidly replacing

TABLE 2

FIELD OBSERVATIONS OF DATES WHEN ICE MOVED OUT OF RADSTOCK BAY AND BEACHES BECAME OPEN TO WAVE ACTION

1968	August 10
1969	August 7
1970	August 9: ice foot present until August 15
1971	Last week in July, field observations commenced August 20
1972	Solid ice cover still present August 22

the somewhat irregular pitted profile due to ice effects with a smoother wavebuilt profile. After one high tide a 1-2 ft high beach ridge had been constructed at high water mark by short, steep waves which were about 3 ft high at the break point and of 3-4 sec period. Somewhat larger waves of 5 sec period occurred on the night of August 9-10 (though this is not apparent from the Resolute wind data) and incorporated small blocks of ice in the beach ridge which was pushed further up the beach. The earliest beach profiles for August 10 (Fig. 4) show this ridge, which represents the results of more vigorous wave action than had been observed in 1968 or were to be observed in 1970. The largest waves and greatest effects occurred, however, two days later on August 11-12 when wind speeds increased steadily throughout the day to speeds of 40 mph, with gusts up to 50 mph, and winds came constantly from the southeast quadrant, blowing over 80-100 mi of open ocean with very little pack ice. Wave period was 6-7 sec and wave height was nearly 5 ft at the breaking point. The beach ridge built up on the August 8-9 was rapidly removed and the whole beach face combed down, the profile being lowered by 1 to 1.6 ft in the mid-to-high tide zone. Erosion of the low cliff in raised marine deposits which backs the modern beach was considerable, due to the powerful swash which reached over 3 ft above normal higher-high water level, and large amounts of material were transported alongshore to the north and several marked pebbles travelled over 300 m over one high tide period. All the profiles measured on August 12 (Fig. 4)

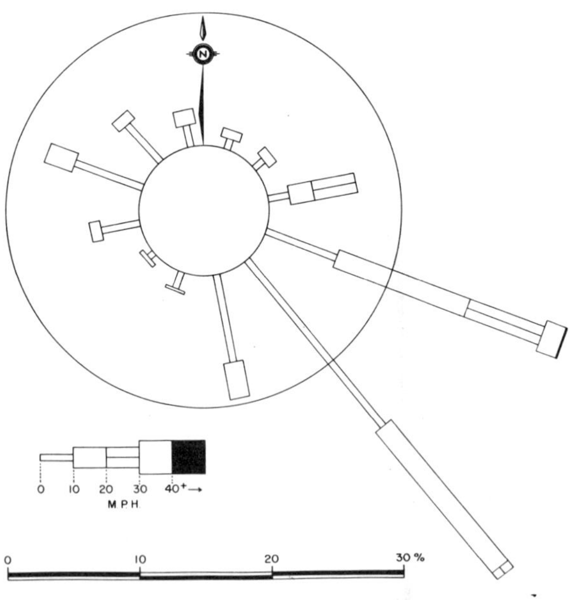

Figure 2. Wind rose for Resolute, N.W.T., for period July 21—August 23, 1969.

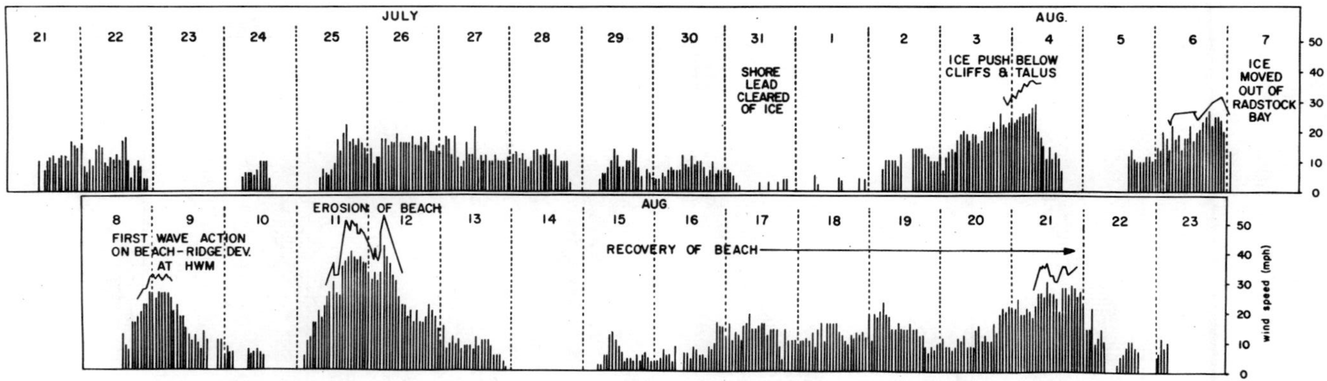

Figure 3. Hourly wind speeds for winds from directions between E and S inclusive, July 21-Aug. 23, 1969, at Resolute, N.W.T. (Source—Arctic Summary, Canadian Meteorological Office).

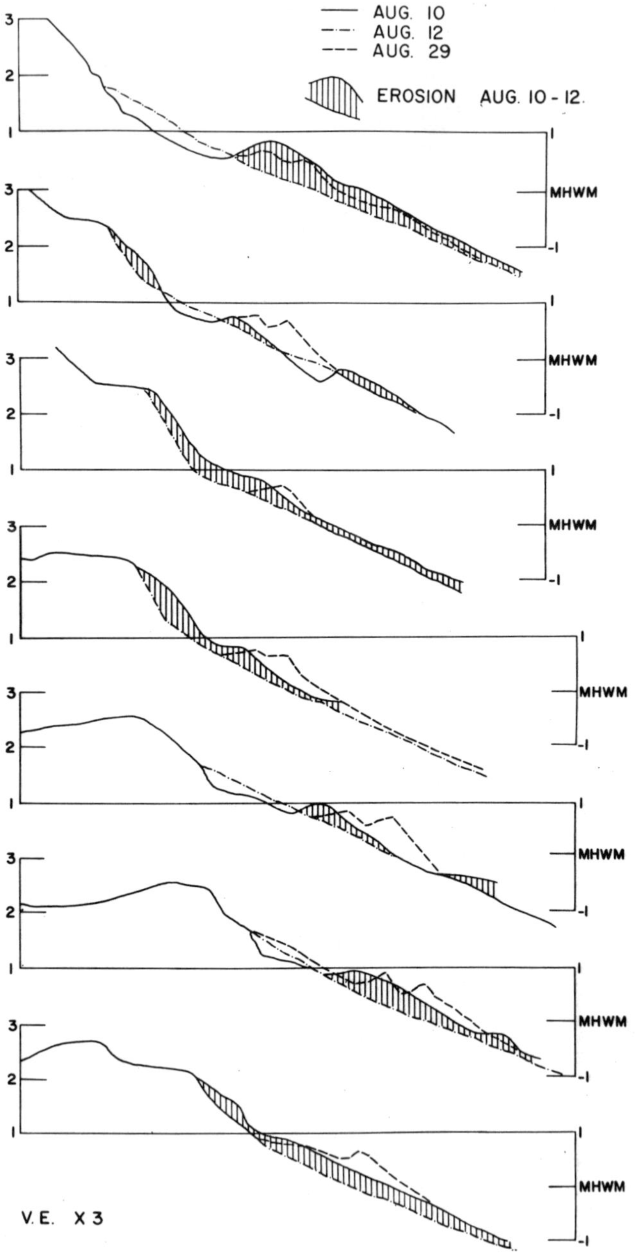

Figure 4. Beach profiles, Radstock Bay, showing storm effects between Aug. 10-12, and beach recovery between Aug. 12-29, 1969.

show the extent of the erosion of the beach face slope, those at sites B, C, D and G show also the more permanent erosion of the raised beach cliff at the rear of the beach.

As wave height decreased on August 13 there was some return of beach material from immediately offshore and this was followed by a quiet period of 2 d with very little wave action. However, the wind returned to the southeast quadrant again on the August 15 for a period of a week and continued to generate effective waves, which were up to 3 ft high with a period of 4 sec on August 21. The effects on the beach during this time were measured by a series of profiles surveyed on August 13, 19 and 22, which show a continuous recovery of the beach. A low ridge built up at the level of lower high tide was each day pushed up the beach on the succeeding higher high tide to form a more substantial ridge similar to that of August 8-9. There was some further longshore transport to the north during this period, though at a much slower rate. The maximum distance travelled by a series of marked pebbles, in an experiment with good recovery rates, was only 550 ft in a week. The recovery of the beach was completed by August 22, for after that date there were only small waves from the north which produced small incipient cusps on the beach face slope. The final profile in Figure 4, dated August 29, is basically the same as the profile of August 22.

An attempt to rank the storm of August 11-12, 1969, in relation to similar events which had occurred in the previous ten years, using the wind data for Resolute is described in a paper presented at the International Geographical Conference in Montreal (McCann, 1972). In order to define a "storm" in the context of wave action on the study beach it was assumed that to generate significant waves the winds must blow from the southeast quadrant, the effective fetch direction, at a speed of at least 20 mph for at least 6 h, during July, August and September. The largest 'storm' in each of the ten years was then isolated, and computing and plotting of storm frequency curves in a manner similar to the calculation of flood frequency curves for flood discharge was carried out. If only wind strength and duration are considered then a storm with waves similar to those documented in 1969 might be expected to occur every year. However, in at least five of the years 1959-68, the maximum storm would have had little effect on the beaches of Radstock Bay, due to the presence of ice either on the beach or at sea. The ice reconnaissance data is unfortunately not adequate enough to carry out an analysis of the dual probability of occurrence of effective winds over open ocean. At best it may be concluded that the storm effects recorded on the beach in 1969 may be a once in 2 or 3 yr occurrence, which in our field experience has occurred only once in 4 yr, 1968-71. The 1969 maximum 'storm' is, however, somewhat unusual in that the event, as defined, was part of much longer period of effective wave action during the period August 8-21.

The results of two smaller storms were observed in 1971, on August 23-24 and September 10-11, but neither were so effective as the events of August,

1969, described above. The daily mean velocity of winds from the southeast quadrant on August 23 and 24, 1971, was 35.5 mph and 36.8 mph, respectively, and waves of 6 sec period in excess of 3 ft in height were generated across the long fetch of virtually ice free water. However, the winds carried pack ice into the bay which was concentrated in a 60-320 ft wide band alongshore, so that in spite of large inputs of wave energy virtually no waves reached the shore during August 23. It took almost 24 h for the soft, ablating pack ice to be severely ground and broken, and to move alongshore farther up Radstock Bay. Effective wave action then occurred over a period of 18 h on August 24, the results of which are shown by the beach profiles in Figure 5. In 11 out of the 14 profiles there had been deposition in the upper part of the beach, usually well above mean high water mark, the material consisting of a mixture of gravel and brash ice. In 12 out of 14 profiles the lower beach had been combed down, though there are differences along the shore, with considerable build up of the whole beach at profile 13. At the exposed site of Cape Liddon, to the south of the porfiles, the talus was markedly undercut and everywhere it was obvious that sediment was rapidly moving alongshore to the north.

After a period of light northerly winds from August 25 – September 9, with only small waves and little change in the beach, larger waves from the south occurred again over a period of two high tides on September 10-11. Wind speeds were lower than on August 23-24 and the waves were of 4-5 sec period and 2.5-3 ft high. The beach changes (Fig. 6) again involved a lowering of the lower beach and deposition at many places above normal high tide level. The maximum distance of travel northwards along the beach of injected marked pebbles over the period of the two high tides was 1460 ft; none were recovered south of the injection point.

CONCLUSION

The most important waves to operate on the study beach at Radstock Bay are generated by strong winds from the southeast quadrant. Open water conditions in Lancaster Sound provide a maximum fetch of 100 mi but for waves to be effective the bay must not only be clear of solid ice cover but also free of pack ice. The combination of adverse sea, inshore and beach ice conditions means that many periods of storm force winds in the potential open water season, which has an average length of 56 d, either do not generate large waves or produce waves the energy of which is expended on ice. The dominant longshore movement of material northwards along the beach would appear to take place very largely during occasional storms such as that which occurred August 11-12, 1969.

ACKNOWLEDGEMENTS

The work reported above has been supported by grants-in-aid of research from the National Research Council of Canada, the Defence Research Board, Canada, and McMaster University. It could not have been carried out without the considerable assistance of E. H. Owens, R. J. Carlisle and R. B. Taylor.

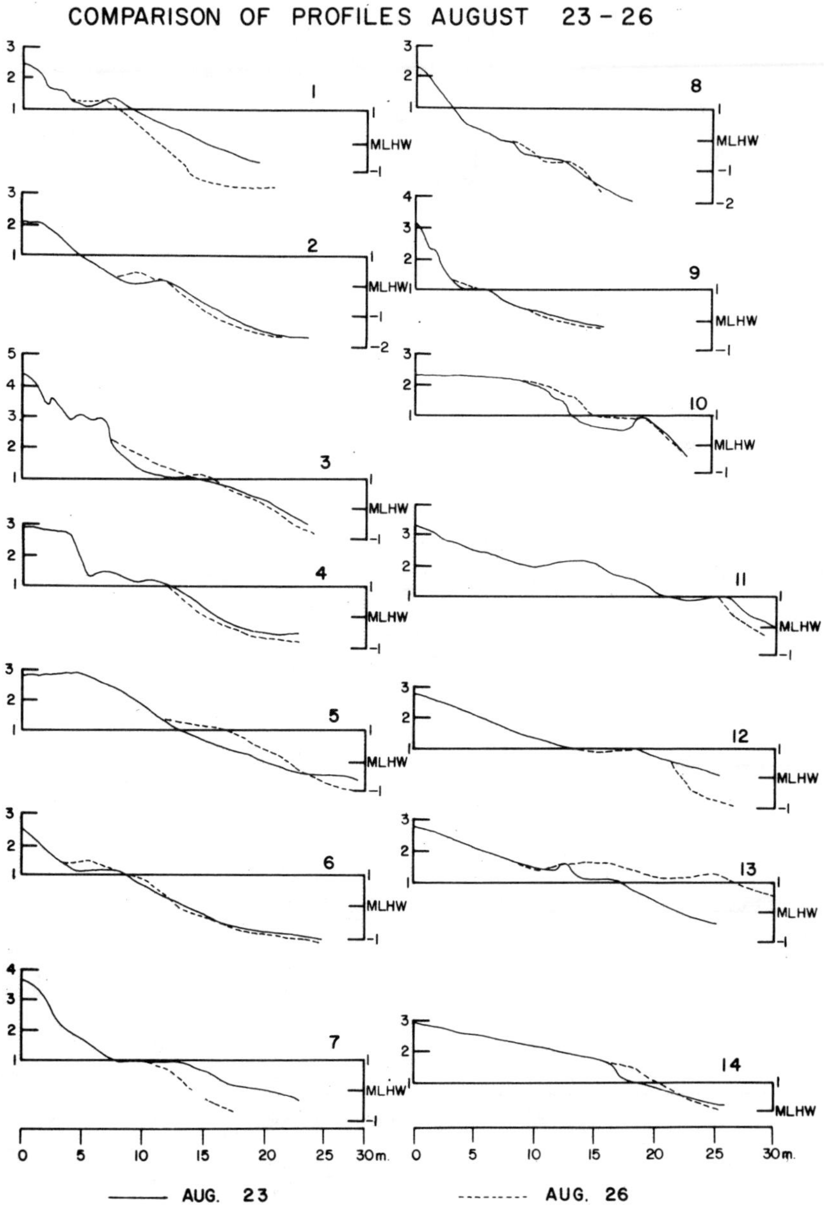

Figure 5. Beach profiles, Radstock Bay, Aug. 23-26, 1971.

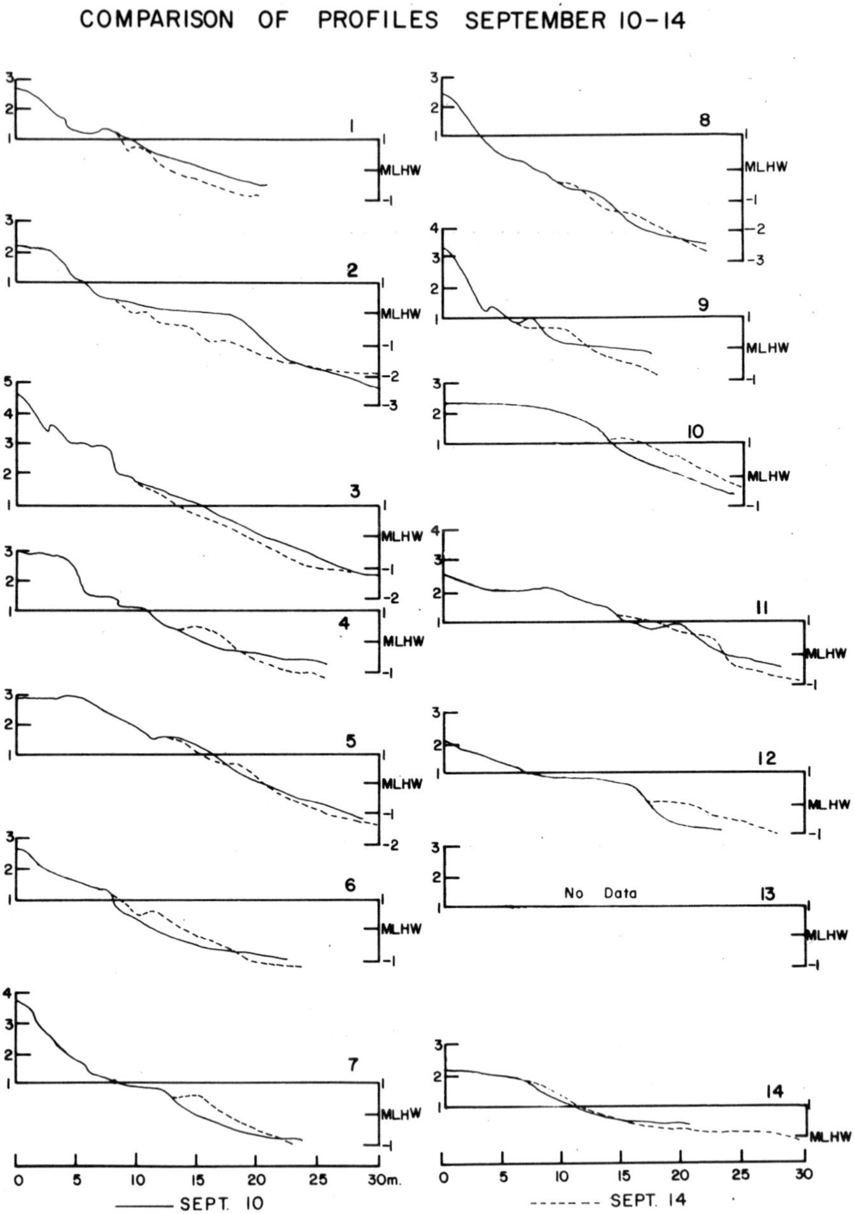

Figure 6. Beach profiles, Radstock Bay, September 10-14, 1971.

REFERENCES

Hume, J.D. and H. Schalk. 1967. Shoreline processes near Barrow, Alaska: a comparison of the normal and the catastrophic : Arctic, v. 20, p. 86-103.

McCann, S.B. 1972. Magnitude and frequency of processes operating on Arctic Beaches, Queen Elizabeth Islands, N.W.T., Canada: in International Geog., W.P. Adams and F. Helleiner, eds., p.41-43.

McCann, S.B. and R.J. Carlisle. 1972. The nature of the ice foot on the beaches of Radstock Bay, S.W. Devon Island, N.W.T., Canada in the spring and summer of 1970: Trans. Inst. Brit. Geogr., Sp. Public. 4, p. 175-186.

McCann, S.B. and F.G. Hannell. 1971. Depth of the "frost table" on Arctic beaches, Cornwallis and Devon Islands, N.W.T., Canada: Jour. of Glaciology, v. 10, n. 58, p. 155-157.

McCann, S.B. and E.H. Owens. 1969. The size and Shape of sediments in three Arctic beaches, Southwest Devon Island, N.W.T., Canada: Arctic and Alpine Research: v. 1, n. 4, p. 267-278.

McCann, S.B. and E.H. Owens. 1970. Plan and Profile Characteristics of Beaches in the Canadian Arctic Archipelago: Shore and Beach, v. 38, n. 2, p. 26-30.

Owens, E.H. and S.B. McCann. 1970. The role of ice in the Arctic beach environment with special reference to Cape Ricketts, Southwest Devon Island, Northwest Territories, Canada: Am. J. Sci., v. 268, p. 397-414.

PART 2

BARRIER ISLANDS

The reader may wonder why one type of coastal landform has been singled out for special treatment, since an entire section of the book is devoted to the topic of barrier islands. The reason is threefold: (1) they are of world-wide extent and constitute one of the major landform on today's coastlines, (2) their origin and developmental history has been argued more than 125 yr with increasing amount of research during the past 12 yr, (3) man's utilization of these areas, is greatly expanding as they are favored sites for buildings and recreation.

The term "barrier island" has become the preferred nomenclature for this narrow strip of land that occurs at some distance from the mainland. G.K. Gilbert first used the term barrier in 1890, but the landform had been defined and analyzed as early as 1845 byE. de Beaumont. D.W. Johnson (1919) called them *offshore bars* and his terminology was adopted by most authors of physical geology texts for the next 40 yr. Other terms used to describe this feature include *barrier beaches, barrier bars, sand reefs, sand bars, sand banks,* etc. A distinction must be made between barrier islands, and barrier reefs which occur in tropical seas, are largely solid coral rock, and formed through biochemical processes. It is of interest to note that Louis Agassiz believed the barrier islands of the eastern coast of United States were sand bars on top of coral reefs. In Part 2, barrier islands are restricted to those elongated ribbons, mainly parallel to the coast, of unconsolidated sediments, largely sand, which are separated from the mainland by marine lagoons and marshes. The islands may be 10's of mi long, rarely are a mile wide, contain dunes that normally reach 20-30 ft and occasionally more than 100 ft. Distance from the mainland shows a wide range from less than 1 mi to more than 35 mi such as at Cape Hatteras.

Although barrier islands are common and occur throughout the world, several generalizations can be made concerning their location. The largest string of barrier islands occurs on the east and southeast coast of the United States where many large cities such as Atlantic City, Daytona Beach, Miami Beach, and Galveston are built on the islands. Other large systems occur in South America, India, Africa, southeast Australia and the Baltic region. Conditions prerequisite for their development include plentiful supply of sand-size or large particles, coastal setting of low sloping plains, and littoral regime of moderate wave energy with tidal range less than 10 ft. Barrier islands are scarce to non-existant in areas of rocky coasts, tropical climates where dunes are rare, and high tidal ranges.

The origin of barrier islands is exceptionally controversial and has been debated as thoroughly as nearly any other single landform. The major issues concern: (1) the source of sand, whether offshore or from cliff erosion or from inland via the rivers (2) the type of littoral process responsible for the sand accretion, longshore currents or whether the sand is brought along shore or from offshore by the waves (3) the importance of sea level position, whether the relative position of the coastline is emerging, submerging, or changes occur at an accelerating or decelerating rate. Fisher supplies some perspective to the problem and concentrates on the views of D.W. Johnson (1919) since the Johnson theory of emergence became so widely accepted and was the ruling explanation for several decades. For example, it received no serious challenge until 1942 (O.F. Evans), and remained in vogue for another 10-15 yr. It is now appropriate to indicate with a partial listing the large number of scientists with publication dates who have written on the origin of barrier islands since 1960: S. Jelgersma, 1961, and et al 1970; E.D. McKee and T.S. Sterrett, 1961; W.A. Price, 1961, 1962; F.P.Shepard, 1963; J.R. Curray, 1964; O.K. Leontyev and L.G. Nikiforov, 1966; J.H. Hoyt, 1967, 1968; J.H. Hoyt and J.R. Hails, 1967; J.H. Hoyt and V.J. Henry, 1967; D.J.R. Swift, 1968; C.W. Cooke, 1968; J.J. Fisher, 1967, 1968; T.W. Todd, 1968; J.W. Pierce, 1969; J.D. Smith, 1969; W.P. Dillon, 1970; E.G. Otvos, 1970. Hoyt for example, provides four fundamental conditions that a valid theory for explanation of barrier islands must incorporate; (1) absence of marine or shallow-water neritic sediments on landward side of lagoon (2) ability of the system to reform after growth has been terminated by an emergence (3) absence of world-wide higher than present sea-levels during the Holocene (4) development and maintenance of system during a slow submergence.

Other authors do not agree with Hoyt's views, but it is probably true that barriers can only develop on gently sloping coasts where there is an abundant supply of sediment that can reach the barrier either from alongshore or offshore. Work on the barriers of the Netherlands suggests that about half the sand came from each source. If sea level maintains its relative stability, barriers are liable to erosion as the sources of sand become exhausted, either by steepening of the offshore gradient or be reduction of sand moving alongshore.

The four chapters that compose Part 2 are included here because their main focus is on the barrier island as an entire system. Other authors in this volume, such as Everts and McCormick analyze beaches that are part of barrier islands, and Dolan's chapter presents an aspect of barrier islands that more appropriately fits with the section on geomorphic applications.

Chapters 8 and 9 are concerned with the form of barrier islands and processes that shape them. Fisher provides an historical base for development of ideas about the origin of barrier islands. Since the work of D.W. Johnson has been so influential, it is appropriate that the evidence upon which he based his theory of emergence for barrier island genesis should be thoroughly scrutinized. The main argument that Fisher seeks to settle is whether the barrier sediment was derived from offshore as suggested by de Beaumont or alongshore as suggested by Gilbert. The evidence Johnson sites is the position relative to the lagoon margin that is reached by an extension of the offshore profile shorewards. By using a quantitative assessment of this classic method, Fisher shows that offshore profiles of barrier and non-barrier areas are not significantly different. If such techniques had been used by Johnson and other workers, perhaps some of the confusion about origin could have been reduced. Such evidence, as profile similarity in diverse coastlines, provides an important reason for reexamining the source of sediment supply.

In Chapter 9, the detailed work of Orme provides the first comprehensive investigation of the Zululand coast of east Africa. This unusual area with 180 m high dunes, with swamps, lagoons, and old dune ridges, is shown to have developed as response to high energy and material transfers since the Holocene transgression. The present barrier is complex, it rests on remains of earlier Pleistocene barriers, and rapid changes occur in beach configuration, lagoon orientation and inlet position. Man-induced erosion of inland watersheds has caused a three-fold increase in sedimentation rates which has greatly accelerated the growth of the barrier, its dunes, and has threatened the ecosystem of the area.

Chapters 10 and 11 deal with erosional aspects of barrier islands. Ruzyla summarizes the processes that are involved in beach and dune changes on Fire Island, N.Y. A 1-yr quantitative study of beach and shore features provided information on the magnitude and nature of profile changes along a 1660 ft stretch of coast. During 1970-1971, 1 ft of beach lost about 11 yd^3 and in the accompanying dune field 35 yd^3 of sand were eroded. Sand wave position and offshore topography seem to be the major controls that influence the total amount of material lost in this coastal reach. Ravinement of the island is occurring and the beach position is moving toward the mainland at rates of 3-5 ft per yr.

Two main points are made in Chapter 11. One is that barrier islands provide an excellent model to show the relationship and harmony that exists in natural bio-geomorphic systems. The Godfreys also evaluate man as a geomorphic process and then document the biologic and physical changes he creates. Additional information on this topic is contained in the related study of Dolan (Chapter 12). These chapters compare two barrier islands of the North Carolina coast; one has been extensively altered by man as he attempted to stabilize the terrain, whereas the other is largely free of man-induced changes. Chapter 11 shows the character of the ecological succession of flora necessary to provide natural stabilization of the barrier island system. This is contrasted with the upset of the vegetative balance that is initiated by the artificial construction of dunes. Emphasis is placed on the overwash cycle as a major part of the natural maintenance of equilibrium, both in terms of sediment and the vegetation canopy. When overwash is prevented by artificial structures, the entire biological and physical ecosystem is either disturbed or destroyed. Similar conclusions of man's interference of barrier islands were reached by Dolan using a different approach through the physical sciences. Thus, those that would manage terrain of barrier islands, such as the National Park Service, are faced with an important decision, whether to allow the islands to maintain their own natural balance, or whether to permit man to alter nature's scheme and in so doing become an added littoral process of change.

Donald R. Coates

BATHYMETRIC PROJECTED PROFILES AND THE ORIGIN OF BARRIER ISLANDS – JOHNSON'S SHORELINE OF EMERGENCE, REVISITED

John J. Fisher

ABSTRACT

Many geologists believe that Johnson developed his classical theory relating the origin of barrier islands to shorelines of emergence by only deductive reasoning. However, he originally considered two hypotheses for the origin of barrier islands: one, in response to wave energy establishing a new profile of equilibrium on an emerging shoreline; and two, in response to longshore movement along a stable shoreline. Analysis by Johnson of 18 offshore "projected" profiles along United States and European barrier islands indicated to him that his first hypothesis was the correct one. Coastal geomorphic field studies along the Virginia-North Carolina "Outer Banks" coast by the author have suggested that parts of this barrier island chain developed in response to longshore movement. Restudy of Johnson's bathymetric evidence seemed called for and, whereas Johnson had constructed only 3 projected offshore profiles across the barrier island portions of this coast, 16 projected bathymetric profiles were constructed. In addition, profiles were also constructed across non-barrier shorelines along this coast, including cuspate spits and headland mainland shorelines.

Analysis of the mean gradient of the offshore slope of these profiles shows that, in general, all types of shorelines along a particular segment of coast exhibit the same offshore characteristics, whether it be a barrier island, a spit, or a mainland shoreline. In addition, when quantitative values are applied to his profile geometry for the North Carolina coast, his results are both incorrect and ambigious. If Johnson had originally constructed projected profiles on both barrier and non-barrier shorelines in the testing of his hypotheses, he might have noticed this discrepancy in his theory of barrier islands as features of a shoreline of emergence.

"He who does not study the past, is doomed to repeat it."

Department of Geology, University of Rhode Island, Kingston, Rhode Island 02881

INTRODUCTION

Recent field evidence from several coastal areas has indicated that the classical Johnsonian theory relating the origin of barrier islands to "shorelines of emergence" is in many cases no longer valid (Otvos, 1970; Fisher, 1967 and 1968; Hoyt, 1967) and that they actually often develop on a "shoreline of submergence." Consideration of these and other studies has resulted in the concept of a "multiple causality of barrier islands" (Schwartz, 1970) with barrier islands perhaps developing during both rising and falling sea levels (Zenkovitch, 1967).

It is known by most coastal geomorphologists that D.W. Johnson considered barrier islands ("offshore bar" of Johnson) as indicators of a shoreline of emergence based on Elie de Beaumont's early discussion (1845) of wave action on a shallow bottom which builds a ridge (bar) in readjusting the submarine slope to a "profile of equilibrium" (see Johnson, 1919, p. 351). Before Johnson's classical work W.M. Davis (1896) discussed the origin of barrier islands by application of de Beaumont's work in his study of the Cape Cod shoreline. Many geomorphologists further believe that Johnson developed his theory relating this origin of barrier islands to shorelines of emergence by only deductive reasoning, in the often criticized mode of W.M. Davis, rather than by experimental studies and inductive reasoning. Actually Johnson did apply the then well-known geologic "systems" analysis technique of "multiple working hypotheses" to his theory for the origin of barrier islands. He originally considered two hypotheses: (1) barrier islands developed in response to wave energy establishing a new profile of equilibrium on an emerging shoreline, and (2) barrier islands developed in response to longshore action on a stable shoreline. To test these two possible hypotheses, Johnson constructed 18 offshore "projected" profiles 5 mi wide along United States and European barrier island coastlines. Projected profiles were then, and still to some extent are, an important geomorphic analytical technique in geomorphic terrain studies ("peneplain surfaces"). Analysis of these profiles, extending from the mainland across lagoon and barrier island to oceanic depths of over 100 ft, was conducted by Johnson for comparison with his deductive study of several idealized "cases" (alternate hypotheses). This analysis indicated to Johnson that his latter, now well-known, hypothesis was correct – that barrier islands develop in response to wave action due to adjustment to a new profile of equilibrium on an emerging shoreline.

For several decades this concept was unquestionably accepted, perhaps more for its pedagogic value in the teaching of the cycle of shoreline development, similar to the Davisian cycles of landscape development, as well as for its "yin-yang" qualities of a shoreline of emergence counterpoised by a shoreline of submergence ("yin-yang"-the Buddhist philosophy of polar extremes). The only questioning comment on this experimental proof of the deductive reasoning was by Zenkovitch (1957, 1967) who felt that although the results (of the profile analysis) "were not entirely convincing," they demonstrate that Johnson's suggested origin for barrier islands was confirmed in the "majority" of cases. Shepard (1963) was the only one in this country who said that, while the Johnson profile analysis was found favorable to the de Beaumont concept, Johnson did not have information from recent borings showing that some barrier islands (primarily the Gulf Coast in 1963) have grown upward as sea level rose. Shepard felt that this stratigraphic evidence did not support Johnson's profile analysis and that, in addition, other evidence supported both hypotheses for barrier development. He therefore did not undertake further analysis of Johnson's profile technique.

NORTH CAROLINA - VIRGINIA BARRIER STUDY

Coastal geomorphic field studies along the Virginia - North Carolina "Outer Banks" coast (Fig. 1) by the author suggested that parts of this barrier island chain developed in response to longshore movement (Fisher, 1967), a theory tested and discarded by Johnson several years before. The classical theory, that barrier islands develop on a shoreline of emergence in the manner described by Johnson, has been applied to the origin of the Outer Banks of North Carolina by numerous writers (e.g., Abbe, 1895; Shaler, 1895; Johnson, 1919; Fenneman, 1938; and Feltner, 1948). However, recent subsurface studies along the Hatteras Island portion of the Outer Banks north of Cape Hatteras (Pierce and Colquhoun, 1970) lend support to the development of this part of the barrier island by longshore movement (Swift, et. al., 1971).

"Outer Banks" is the local geographic name for these offshore islands which extend some 225 mi, from just south of Cape Henry, Virginia, to Bogue Inlet, North Carolina (Fig. 2). South of Bogue Inlet, the distinctive barrier island character disappears as the lagoons narrow, becoming for the most part only tidal marshes with many of the "islands" actually being part of the mainland at low tide. In addition to Cape Henry, two other prominent cuspate shoreforms exist along the Outer Banks, Cape Hatteras and Cape Lookout.

BATHYMETRIC PROJECTED PROFILES, NORTH CAROLINA-VIRGINIA COAST.

Thinking perhaps that Johnson's original profile study may have suffered from early inaccurate maps or the lack of offshore surroundings, I decided to restudy Johnson's work in this region and construct offshore profiles based on more recent offshore soundings. Where Johnson constructed only 3 projected profiles across the barrier islands on this coast (Johnson, 1919), 16 projected profiles were constructed for the present study in the same manner as by Johnson. These 5 mi wide projected profiles, therefore, gave almost 50 percent coverage of the entire barrier chain from Cape Henry, Virginia, to Bogue Inlet, North Carolina – a distance over 200 mi. In addition, profiles were also

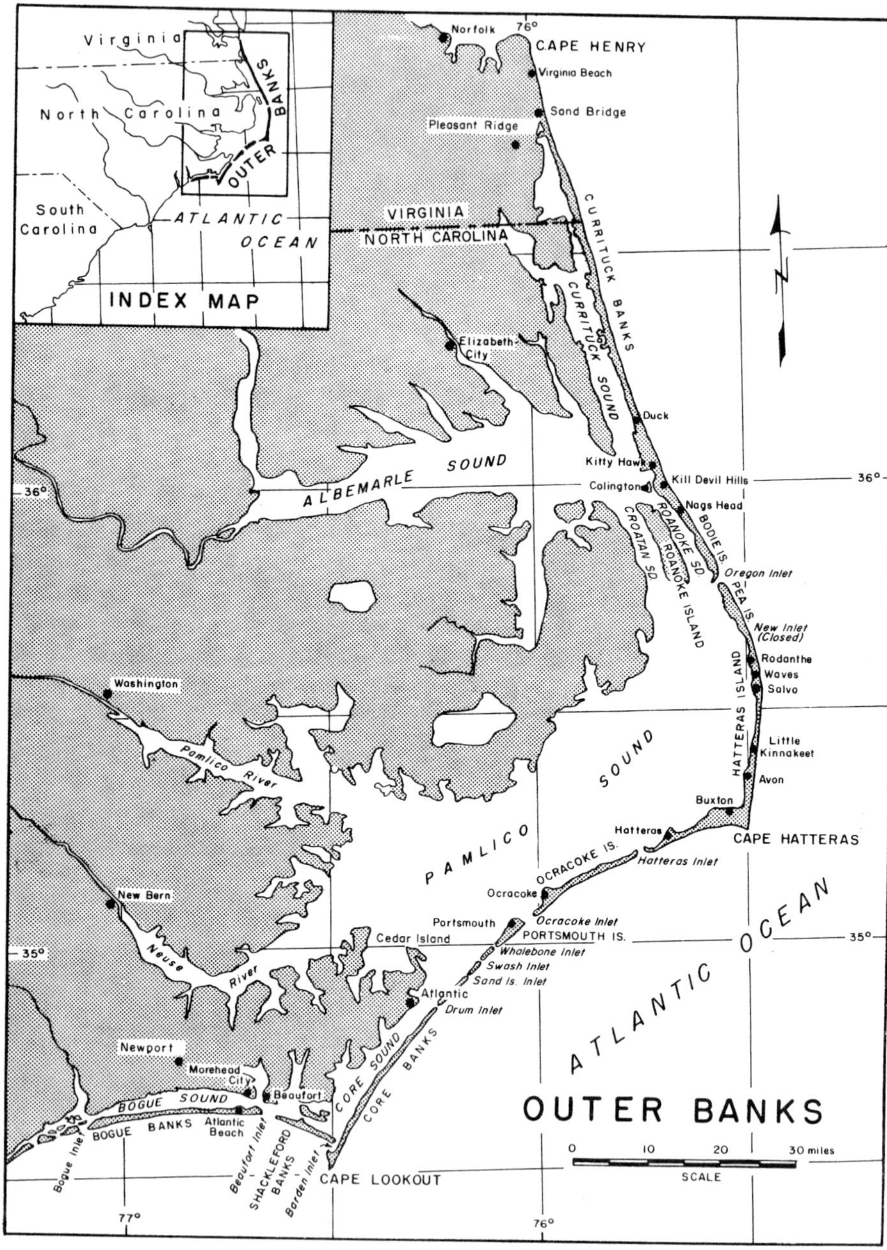

Figure 1. Location map of major coastline features of the Outer Banks barrier island chain, North Carolina—Virginia coast.

constructed across non-barrier shorelines along this coast, including cuspate spits and headland-mainland shorelines (Fig. 3). These bathymetric projected profiles were constructed using depth soundings on the National Ocean Survey (formerly U.S. Coast and Geodetic Survey) nautical charts numbers 1227, 1229, 1231, 1232, 1233, 1234, 1109, and 1110. For each projected profile, a 5 mi wide stretch of coastline was chosen and the profile extended offshore perpendicular to the coastline. All depth soundings within this 5 mi strip were projected to a common profile surface. This projection resulted in several points of about the same depth appearing at about the same distance from shore. The average slope was then drawn through the majority of these points and, in most cases, the line so generated was closer to the maximum depths than the minimum depths at a distance offshore. The vertical exaggeration for these bathymetric profiles was exactly the same as that used by Johnson in constructing his profiles for ease in comparison between the two studies. His published profiles have a horizontal scale of "one inch equal to about six miles" and a vertical scale of "one inch equal to a little less than 400 feet." He did not have a graphic scale for either the vertical depths or horizontal distances. Projection of the offshore slope was carried to an offshore distance where the slope of the nearshore continental shelf surface remained constant. In most cases for the Virginia-North Carolina coast this was at depths of 60-80 ft. This average continental shelf slope angle was then extended landward to the vicinity of the adjacent lagoon. Intersection or lack of intersection of the lagoon by this extension also is a factor in Johnson's analysis of projected bathymetric profiles as discussed in the next section.

JOHNSON'S DEDUCTIVE BATHYMETRIC ANALYSIS

Analysis of the bathymetric projected profiles, as developed by Johnson, is more complex than is first apparent, especially when detailed analysis is made of a regional coastline. For example, Shepard (1963) indicated that the basic feature, according to Johnson, was to determine whether the nearshore oceanic slope simply continued at the same angle landward and formed the bottom of the lagoon (Fig. 4A). If this were the case, then the barrier formed as a spit (Gilbert hypothesis as presented by Johnson). If the nearshore oceanic slope were at a lesser angle than the lagoon floor (Fig. 4B), then the material making up the barrier was eroded from the nearshore shelf (de Beaumont hypothesis as presented by Johnson). This erosion took place on an emerging shoreline as readjustment to an offshore profile of equilibrium. It should be pointed out that actual profiles constructed along barrier island shorelines do not show the geometry indicated by Shepard for either hypothesis. Actually the analysis by Johnson of these bathymetric projected profiles was more sophisticated than Shepard's idealized profiles would suggest.

Possible bathymetric profiles expected, if barrier islands developed on a stable shoreline by longshore deposition, were divided by Johnson (1919) into three groups (Fig. 5A). The first (A. 1) is of a coastal plain where close to shore a barrier has been built upon the uneroded nearshore bottom by longshore

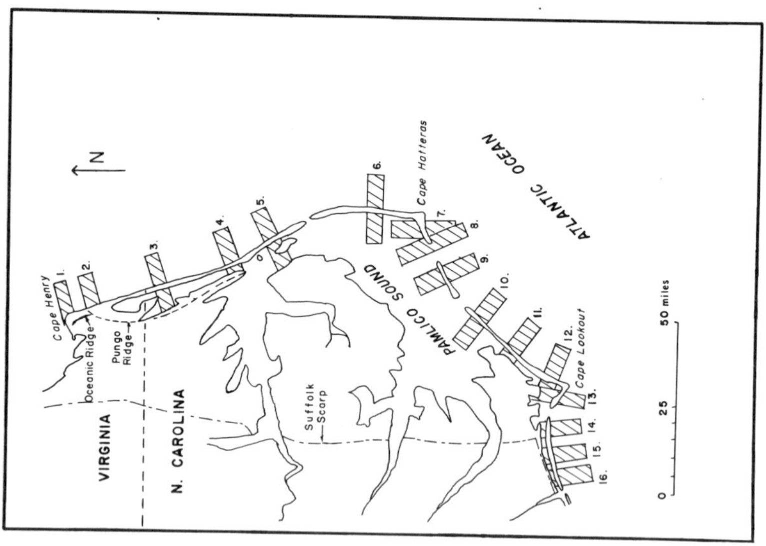

Figure 2. High altitude aerial photograph of Outer Banks coast. Taken on Apollo 9 space mission, March 12, 1969, 10:00 a.m. EST, altitude 120 mi. Ebb tide sediment plumes visible at most inlets. High tide at Cape Hatteras at about 8:00 a.m. Suffolk Scarp visible as light trace along left margin. Pungo Ridge visible as broad light trace in upper center of photograph.

Figure 3. Location of bathymetric projected profiles along the North Carolina–Virginia coast. Projected profiles of 5 mi width as indicated on map. Coastal plain topographic features, Suffolk Scarp, Pungo Ridge and Oceanic Ridge are indicated, (after Oaks and Coch, 1963).

deposition. Johnson points out that projection of the sea bottom slope in this case will intersect the sea level datum exactly at the inner edge of the lagoon. The second case (A. 2) is where there is additional deposition in the nearshore region, seaward of the barrier, but decreasing offshore. In this case, projection of the sea bottom slope will intersect the sea level datum within the lagoon. The third case presented by Johnson (A.3) is also that of a barrier built by longshore action but, in addition, deposition increases offshore. The projection of the sea bottom slope would now not intersect the sea level datum within the lagoon, but would instead intersect some distance landward at a projection of the sea level datum. Johnson did not think that this last case was probable since the barrier would be built in the zone of maximum breaking wave activity. Seaward from the barrier, breaking wave activity would be less and so should deposition. On the basis of this analysis, Johnson concluded (1919, p. 358) that if barriers formed as spits in response to longshore action, then in bathymetric profiles "the sea-bottom slope projected will intersect the sealevel datum at or seaward from the inner margin of the lagoon" (cases A.1 and A.2). Implied in both these cases was a stable sea level.

Johnson's analysis of the bathymetric profiles expected, if a barrier developed on an emergent shoreline, is more complicated than that expected on a spit barrier, and both are more complicated than that suggested by Shepard's

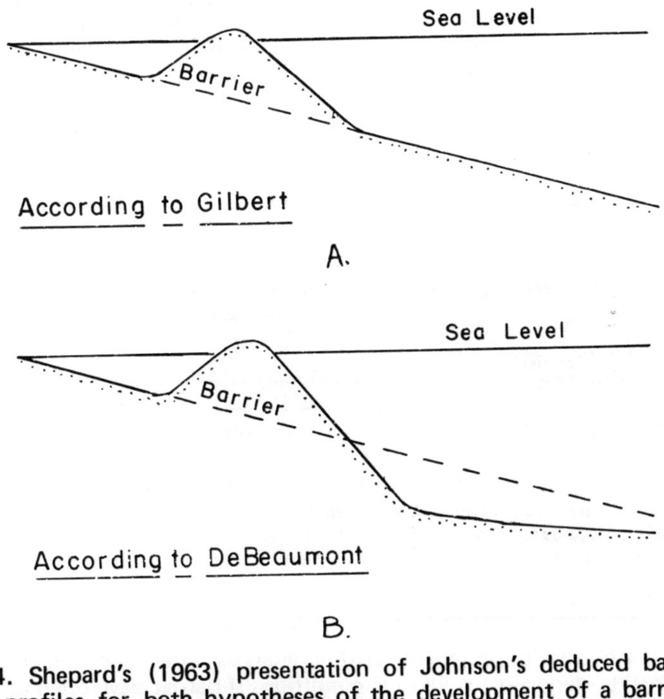

Figure 4. Shepard's (1963) presentation of Johnson's deduced bathymetric profiles for both hypotheses of the development of a barrier island, either as as a spit (according to Gilbert) or on an emergent shoreline (according to de Beaumont).

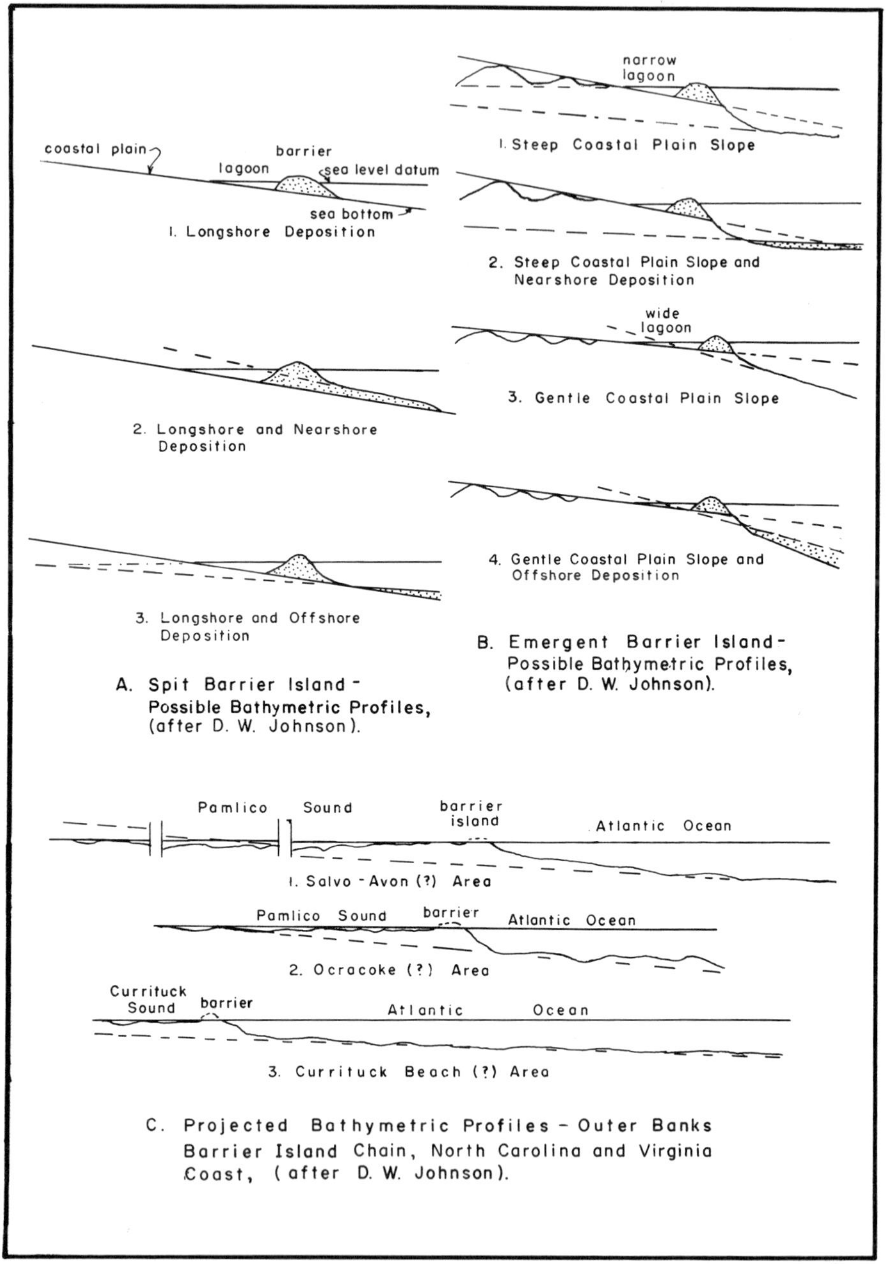

Figure 5. Bathymetric projected profile analysis after Johnson (1919). A. Profiles deduced for various conditions if barrier developed as a spit. B. Profiles deduced for various conditions if barrier developed on emergent shoreline. C. Actual bathymetric projected profiles presented by Johnson in support of his theory that barriers developed on a shoreline of emergence along the Outer Banks coast.

idealized explanation mentioned earlier. In addition, Johnson applied certain of the cases which follow specifically to his profile analysis of the North Carolina - Virginia coast and thus these additional principles should also be discussed here. The first case (B. 1) shows the expected bathymetric profile on an emerged coastal plain where the original coastal plain slope has been eroded by wave action to produce a new nearshore shelf slope. This latter slope is less than the original coastal plain slope and the material eroded forms the barrier. Johnson's analysis of the geometry of this case was that the projection of the sea bottom slope will intersect the sea level datum landward, away from the inner margin of the lagoon. The second case of an "emergent barrier" (B. 2) is similar to the first except Johnson thought it more probable because deposition occurs nearshore as well as forming the barrier. Also, in this second case, projection of the sea bottom slope intersects the sea level datum landward, at a greater distance away from the inner margin of the lagoon. Thus, in both these cases the slope of the initial plain is greater than the slope of the newly formed shelf surface. In contrast, the third and fourth emergent barrier cases (B. 3 and B. 4) are considered to develop along a coastline where the slope of the initial coastal plain is less than that of the new submarine slope developed during the building of the barrier from the adjacent sea floor. Thus, the third case (B. 3) profile bathymetry geometry indicates that the projection of the sea bottom slope, now greater than the coastal plain, will intersect the sea level datum seaward from the inner margin of the lagoon. In the case of offshore deposition along this particular coast (B. 4), the intersection of the slope with the sea level datum will again be somewhere in the middle of the lagoon surface. Johnson further pointed out that , for these last two cases, the initial gentle offshore slope will cause waves to break at a distance from the mainland shoreline, building a barrier at this point which would produce lagoons of greater width than in cases B. 1 and B.2.

Johnson summarized his deductive analysis of possible bathymetric profiles by developing three classes of barrier coastline geometry:

Class I. If the sea bottom slope projection intersects the sea level datum exactly at the inner margin of the lagoon, then the barrier formed as a spit (case A. 1).

Class II. If the sea bottom slope projection intersects the sea level datum *landward* from the inner margin of the lagoon then the barrier formed by emergence (cases B.1 and B.2).

Class III. If the sea bottom slope projection intersects the sea level datum *seaward* from the inner margin of the lagoon and the nearshore slope is steep, then the barrier formed as a spit (case A.2); but where the coastal plain slope is gentle and lagoons are wide, then the barrier formed by emergence (cases B.3 and B.4).

(Note: Johnson did not consider case A.3 to be an acceptable possibility.)

JOHNSON'S FIELD DATA ANALYSIS

Comparison was made by Johnson with the bathymetric profiles as deduced for the two hypotheses with actual field data along several barrier island coasts. Eighteen bathymetric projected profiles of American and European coasts were constructed: the Atlantic and Gulf coasts of the United States (13 profiles), the North Holland coast (3 profiles), the German and Italian Venetian coasts (1 profile each). Of these profiles 16 fell in Class II, 2 in Class III and 1 in Class I. From this tabulation, Johnson assumed that field analysis of bathymetric profiles proved his deductive reasoning that barriers formed on a shoreline of emergence. Of the two profiles falling into Class III, both are from the North Carolina coast, while another in Class II is also from the same coast. Figure 5C is an exact copy of Johnson's projected bathymetric profiles for the North Carolina coast. Unfortunately, Johnson did not indicate the location of these profiles, which may be a shortcoming of his presentation. In addition, in the profiles published, there is a vertical exaggeration of about 100:1 although the only scale given is a verbal scale in the figure caption. Profiles C.1 and C.2 are indicated as the "Cape Hatteras bar" using U.S. Coast and Geodetic Survey Charts Nos. 1232 and 1229. The Salvo-Avon area and the Ocracoke area are probable locations for two of Johnson's profiles as deduced from my analysis of his data (see Figs. 3, 5, and 6), as can be seen in Johnson's profiles in Figure 5 C.1 and 5 C.2, extension of the sea bottom intersects the sea level datum within the lagoon, locally called Pamlico Sound. Although this might indicate that the profiles belong in Class 1 (case A2) but because the coast is "an excessively flat plain and the lagoon exceptionally wide," both profiles belong in the special Class III. Of all 18 profiles constructed by Johnson, these two profiles along the North Carolina coast are the only ones that fell in Class III. The remaining bathymetric profile presented by Johnson for the North Carolina coast is shown as Figure 5 C.3 in this paper, the Currituck Beach area, just south of the Virginia-North Carolina stateline. Johnson points out that this part of the coast is less flat and the lagoon (Currituck Sound in Fig. 2) narrower than the area further south of the preceding two profiles. In Figure 5 C.3, extension of the sea bottom intersects the sea level datum well inland from the lagoon and this places the profile in Class II. The 15 profiles from other coastlines in the United States and Europe presented by Johnson that also fall in Class II are similar to this Currituck Beach profile.

ADDITIONAL BATHYMETRIC PROFILES, NORTH CAROLINA-VIRGINIA COAST

As mentioned earlier, my recent geomorphic studies indicate the barriers along the Outer Banks coast of North Carolina may have developed as spits. To restudy Johnson's bathymetric profiles of the barriers of the North Carolina coast, it was decided to reconstruct them using his technique of "projected profiles." These were reconstructed to investigate whether the more recent accurate depth soundings along this coast would materially change the geometry of his original profiles. In the present study, the profiles constructed off Currituck Beach (Profile 3, Fig. 3), the Salvo-Avon area (Profile 6, Fig. 3) and the Ocracoke Island area (Profile 9, Fig. 3) are thought to represent the same profile locations as Johnson's original work. In addition to reconstructing these 3 profiles of the original study, 13 additional bathymetric profiles were constructed, 11 of these are on barrier island coastlines, as might be expected, but 2 were constructed along non-barrier shorelines. Profile 1 was constructed off Cape Henry, a truncated cuspate foreland, while profile 2 was constructed off Sand Bridge, a mainland shoreline between Cape Henry and the remaining barrier island coastline. In addition, quantitative calculations were made of the slope of both the continental shelf and coastal plain along each projected profile.

These new profiles, together with slope calculations, indicate that certain assumptions Johnson made in his deductive reasoning for both hypotheses are open to question. Furthermore, certain conclusions he deduced from the proposed geometry of theoretical barrier island bathymetric profiles are not valid. Finally, the interpretations he made of the bathymetric profiles along the North Carolina coast are not correct when compared with the additional profiles made for the present study.

Examination of the additional profiles (Fig. 6 and 7) indicates they do fit into the various cases as deduced by Johnson for this shoreline using his reasoning. Profiles 3, 4, and 5 of barrier islands in the northern section of the Outer Banks, with a narrow lagoon and the projection of the shelf intersecting the sea level datum landward from the inner margin of the lagoon, fit Johnson's Class II (Fig. 5B1 and 5B2) for an emergent barrier. Profiles 6, 7, 8, 9, and 10 of barrier islands in the center section of the Outer Banks, with a wide lagoon and the projection of the shelf intersecting the sea level datum within the lagoon, fit Johnson's Class III (Fig 5B3 and 5B4) for an emergent barrier. Profiles 11, 13, 14, 15, and 16, also barrier islands with a narrow lagoon in the southern section of the Outer Banks, again fall into the emergent barrier of Class II of Johnson. Profile 12 would also fit into his class but, because of profile orientation, the apparent wide lagoon is actually a profile along the long axis of an east-west trending lagoon, rather than the short axis of Core Sound, the adjacent narrow lagoon.

Inherent in the above analysis, using Johnson deductive reasoning, is the geometry and the relationship of the slopes of the coastal plain and the continental shelf. Johnson, in his deductive reasoning of the various classes, tacitly assumed that the initial slope of the emerged sea bottom would be the same as the slope of the adjacent coastal plain mainland. There is no reason to believe that this would be the case. For an emerged shoreline the new sea bottom slope would simply be part of the former sea bottom slope and not part of the adjacent coastal plain nor have its slope. If Johnson were discussing a submerged shoreline, then it is possible that the slope of the adjacent coastal plain could be considered the new sea bottom slope. In addition, the immediate nearshore sea bottom slope on most shorelines is fairly steep down to depths of 30 to 60 ft, steeper than most adjacent coastal plain slopes. For example, along the Outer Banks coast, the immediate nearshore slope drops to depths of 30 to 40 ft in a distance of one or two miles and can be seen on the profiles (Fig. 6 and 7). This immediate nearshore slope change is simply the well-known parabolic-like profile of equilibrium that is found along most sandy shores. Its shape is related to present-day wave action. In contrast, the mainland coastal plain seldom rises more than 5 or 10 ft in the same distance inland in the northern section along Currituck Beach which Johnson considered as having a steep coastal plain slope. Actually it is only the gentle offshore continental shelf slope that Johnson considered in the development of a barrier on an emerging shoreline since he was concerned with reconstructing the original offshore profile before development of the barrier island. However, in Class III, the choice whether a barrier developed as a spit or emergent barrier depends on the relationship of coastal plain slope to continental shelf slope. Again this situation proposed by Johnson where the coastal plain slope is less than the offshore slope, is probably also not valid since it is not the coastal plain which emerges but the former sea bottom. Thus, whether a coastal plain slope is steep or gentle should have no relationship to the emerging sea bottom. The slope of the emerging sea bottom should be related to the dynamics of wave energy along a particular coast. It appears, therefore, that differences in the slope of the adjacent coastal plain should have little or no effect on the future profile of a barrier island developing on an emerged shoreline and Johnson's deductive hypotheses including this factor are not valid. The possibility of varying coastal plain slopes on the development of a barrier on a submerging shoreline, where it seems more logical, was not considered by Johnson.

It is important to recognize this ambiguity of the slope interpretation in Johnson's analyses because the only special profile cases requiring a gentle coastal plain slope occurred along the Outer Banks for profiles in the Pamlico Sound area. For 10 profiles constructed by Johnson of barriers along the remainder of the Atlantic coast and 6 along the Gulf coast, it was necessary that he consider the coastal plain slope to be steeper than the sea bottom slope, although no information is given to support either of these assumptions. The question therefore arises, just what are the quantitive values for the various land and sea bottom slopes? Table 1 gives these values as determined for the present study. Since all slopes are in the range of hundredth's of a degree, it was felt best to present the slopes as measured in changes of height in ft per unit statute mi (ft/mi). Values for the continental shelf were determined from the bathymetric profiles (Fig. 6 and 7) Values for the adjacent coastal plain were determined by measuring increased height (ft/mi) to the first major topographic break on the coastal plain from the inner margin of the lagoons. The Virginia - North Carolina coastal plain, as with all lower coastal plains along the Atlantic and Gulf coasts, is not a gentle sloping plain but is interrupted by various terrace levels and scarps. The presence of the Suffolk scarp with a crest from 50 to 90 ft and a toe

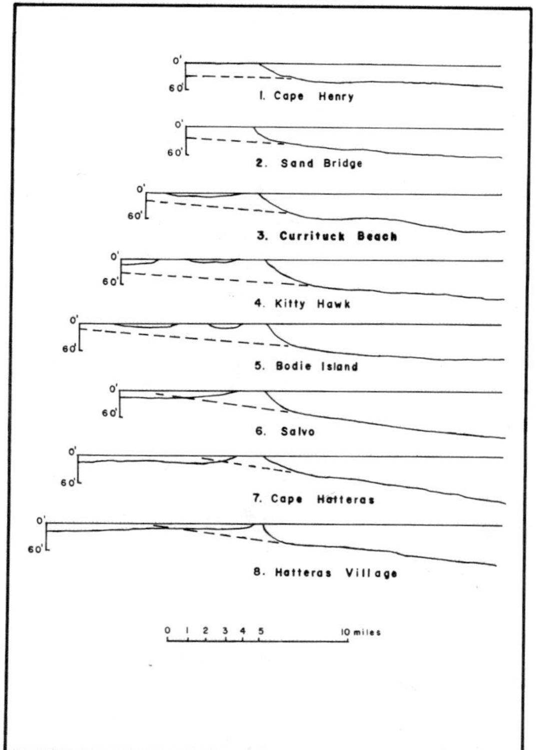

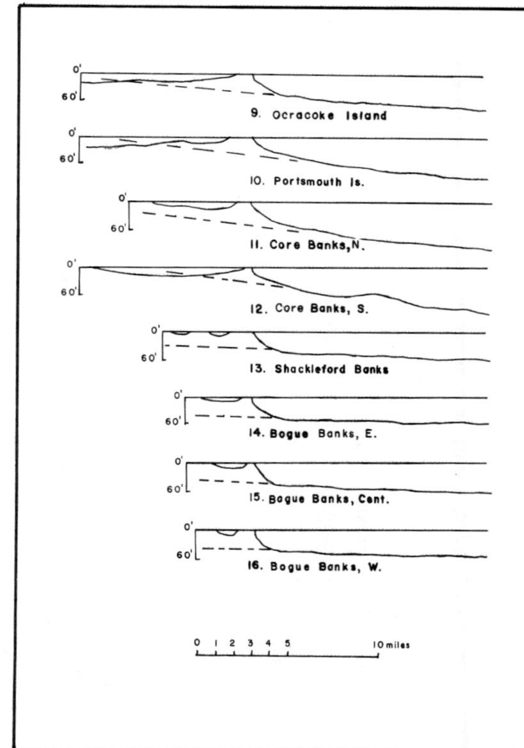

Figure 6. Bathymetric projected profiles along the Virginia–North Carolina coast, Cape Henry to Cape Hatteras area. Scale as indicated. Vertical exaggeration same as Johnson's original profiles. (Fig. 5C.).

Figure 7. Bathymetric projected profiles along the North Carolina barrier chain from just below Cape Hatteras to beyond Cape Lookout. Scale as indicated. Vertical exaggeration same as Johnson's original profiles (Fig. 5C.).

at 25 ft(Oaks and Coch, 1963) can be seen as a light linear tone extending south from the Dismal Swamp area along the western margin of the Apollo 9 Space Photograph (Fig. 1). The location of this scarp is also indicated on the profile index map (Fig. 3). South of Albermarle Sound this scarp represents the first major topographic change; however, to the north the Pungo Ridge west of Currituck Sound is perhaps the first topographic change. This ridge, also indicated in Figure 3, has a crest at 15 ft and a toe at only 5 ft (Oaks and Coch, 1963). Oceanic Ridge, also mapped by Oaks and Coch (1963), extends north of Pungo Ridge and crests to 25 ft with its buried toe extending perhaps below sea level datum. For the profiles surrounding Cape Hatteras (Profiles 6, 7, 8 and 9), an average coastal plain slope was determined along an axis that bisects the interior angle of the cape. These then were the geomorphic boundaries used in determining the lower coastal plain slopes for the bathymetric profiles (see Table 1).

Quantitative investigation of the actual slopes of coastal plains and continental shelves along the barrier shoreline of the Outer Banks indicates that the actual values often differ from those assumed by Johnson for this particular coastline. In general, it appears that the coastal plain slope is actually less than that of the adjacent continental shelf. This is in direct contradiction to the assumptions presented by Johnson in his deductive reasoning for an emergent

TABLE 1. CONTINENTAL SHELF AND COASTAL PLAIN SLOPES OF PROFILES

Profile	Shelf Slope, ft/mi	Coastal pl. Slope, ft/mi
1. Cape Henry	1.5	1.0 (3)
2. Sand Bridge	2.2	1.0 (2)
3. Currituck Beach	3.3	1.7 (2)
4. Kitty Hawk	2.6	2.5 (2)
5. Bodie Island	2.6	2.5 (2)
6. Salvo-Avon	5.5	0.6 (1)
7. Cape Hatteras	6.0	0.6 (1)
8. Hatteras Village	4.6	0.6 (1)
9. Ocracoke Is.	3.2	0.6 (1)
10. Portsmouth Is.	4.6	1.3 (1)
11. Core Banks N.	4.1	0.7 (1)
12. Core Banks S.	5.5	1.3 (1)
13. Shackleford Banks	2.0	3.0 (4)
14. Bogue Banks E.	1.2	2.0 (4)
15. Bogue Banks C.	2.1	25.0 (1)
16. Bogue Banks W.	1.9	25.0 (1)

Note: Coastal Plain Topographic Boundary Features:
(1) Suffolk Scarp: toe = 25 ft above sea level
(2) Pungo Ridge: toe = 5 ft above sea level
(3) Oceanic Ridge: toe = 5 ft above sea level
(4) Coastal plain below Suffolk Scarp

barrier. For example, Profiles 3 (Currituck Beach), 4 (Kitty Hawk) and 5 (Bodie Island) of Figure 6 are assumed by Johnson to fit the bathymetric profile requirements for a standard Class II emergent barrier (Fig. 5B1 and 5B2). This part of the barrier chain is adjacent to Currituck and Roanoke Sounds, both narrow lagoons. The projection of the continental shelf slope extends beneath the lagoon indicating, according to Johnson, that the coastal plain slope is steeper than the continental shelf slope. Actually along this shoreline the average of the three profiles indicates a coastal plain slope of 2.2ft/mi and a continental shelf slope of 2.8 ft/mi (Table 2). Thus, while the continental shelf slope is only about 20 percent steeper than the coastal plain slope, in no manner is the coastal plain steeper in slope than the continental shelf as assumed by Johnson in application of his deductive reasoning to his field data profile analysis for this section of a barrier shoreline.

The next section of the Outer Banks barrier chain is that part adjacent to Pamlico Sound, a very wide lagoon. Profiles 6 (Salvo-Avon), 7 (Cape Hatteras), 8 (Hatteras Village), 9 (Ocracoke Island) and 10 (Portsmouth Island) of Figures 6 and 7 are assumed by Johnson to fit the profile requirements for the special Class III emergent barrier (Figs. 5B3 and 5B4). This class, where the continental shelf slope projection intersects the sea level datum within the wide lagoon, according to Johnson occurs where the coastal plain slope is less than that of the continental shelf (Fig. 5B3 and 5B4). In this case Johnson's assumptions are borne out by the actual field values. The average coastal plain slope is 0.7 ft/mi while that of the continental shelf is 4.8 ft/mi (Table 2).

However, in the next section of the Outer Banks barrier chain, Johnson's assumptions for his field data fail again. Profiles 11 and 12 are along Core Banks, adjacent to the narrow lagoon Core Sound, and should fit into Johnson's standard Class II emergent barrier in a manner similar to the northern Currituck Beach section. With the continental shelf slope extending beneath the lagoon, the coastal plain slope should be steeper than the continental shelf slope if the barrier formed by emergence according to Johnson's deductive reasoning. The

TABLE 2. AVERAGE CONTINENTAL SHELF AND COASTAL PLAIN SLOPES OF DIFFERENT JOHNSON BARRIER CLASS SECTIONS

Profiles	Continental Shelf(1)	Coastal Plain(1)
1,2	1.9	1.0
3-5	2.8	2.2
6-10	4.8	0.7
11,12	4.8	1.0
13-16	1.8	2.5-25.0

(1) slope in ft/mi

actual slope data from this section of the Outer Banks coastline do not support his assumptions for the slopes. For Profiles 11 and 12, the average coastal plain slope is only 1.0 ft/mi while the continental shelf slope is 4.8 ft/mi (Table 2). It is also interesting to point out that in this section the average continental shelf slope is the same as that of the preceding Cape Hatteras section, adjacent to a wide lagoon.

The final section of the Outer Banks extends westward from Cape Lookout, with bathymetric profiles constructed at Shackleford Banks (Profile 13) and at three points along Bogue Banks (Profiles 14, 15, and 16). These "Banks" are typical barrier islands with narrow lagoons and the projection of the continental shelf slope for these profiles does extend beneath the lagoons as assumed by Johnson for an emergent barrier of Class II. Unlike the Core Banks section mentioned above, here the relationship of the coastal plain and continental shelf slopes are as assumed by Johnson for a narrow lagoon. The average continental shelf slope is 1.8 ft/mi while the average coastal plain slope is 2.5 to 25.0 ft/mi. This range in coastal plain slope is because, for the first two profiles (13 and 14), the immediate coastal plain is extremely low, is below the Suffolk Scarp and is similar to the coastal plain adjacent to Pamlico Sound. The remaining 2 profiles (15 and 16), although no different in their continental shelf slope from the previous two, have a high coastal plain slope because it incorporates the Suffolk Scarp, almost immediately at the edge of the lagoon.

Bathymetric projected profiles 1 (Cape Henry) and 2 (Sand Bridge) were constructed to determine what the continental shelf slope was along non-barrier coastlines. Both these areas are north of the Currituck Beach barrier island section of the Outer Banks barrier chain. Cape Henry is a truncated cuspate spit shoreline, while Sand Bridge is a mainland coastal plain shoreline. The average continental shelf slope for these two shorelines is 1.9 ft/mi and is similar to the slope offshore from the Outer Banks southern barrier island section discussed above. The average coastal plain slope is 1.0 ft/mi and is similar to that of the barrier island chain adjacent to Core Sound. It is not known what assumptions Johnson would have made for coastal plain and continental shelf slopes along non-barrier emergent shorelines. His Class II emergent barrier shoreline, which he considered typical of most barrier shorelines, has the coastal plain slope steeper than the continental shelf slope. Since these slope relationships are not influenced by the presence of the barrier, it is likely he might have considered that an emergent non-barrier shoreline, if such a thing could exist in Johnson's theory, would have a coastal plain slope steeper than the continental shelf slope. However, along this Virginia shoreline north of the Outer Banks barrier chain, the coastal plain slope is less than the continental shelf slope. The presence of the Oceanic and Pungo Ridges of the coastal plain, so near the shoreline in this region, influences measurements of the coastal plain slope. In any case, the measurements of the continental shelf slopes are not open to the same questions and it is interesting to point out that there is a regular increase in continental shelf slope from the Cape Henry mainland south towards Currituck Beach, a barrier shoreline. In addition, the bathymetric continental shelf slopes along this non-barrier shoreline (1.5 and 2.2 ft/mi) are duplicated along the barrier shoreline of Shackleford and Bogue Banks (2.0, 1.2, 2.1 and 1.9) with averages of 1.9 and 1.8 ft/mi respectively (Table 2).

CONCLUSIONS

1. D. W. Johnson's bathymetric projected profiles proof that barrier islands develop on shorelines of emergence was re-investigated and expanded on the basis of more recent offshore soundings along the North Carolina–Virginia coast. Johnson's deductive reasoning as to the profile geometry of a barrier formed on a shoreline of emergence ("an emergent barrier") was found to still apply to the 13 new bathymetric profiles including the special class for barriers adjacent to wide lagoons, but my quantitative studies cast doubt on the base premise of his reasoning.

2. Inherent in Johnson's deductive reasoning of the profile geometry are certain relationships between the slope of the continental shelf and that of the adjacent coastal plain. Assumptions for these slopes, made by Johnson in his field testing of bathymetric profiles along the Outer Banks barrier coast, were found not to be completely valid based on actual quantitative slope calculations. For Johnson's special case, of a barrier adjacent to a wide lagoon and considered by him to exist only along the Outer Banks, the coastal plain slope was less than that of the continental shelf as assumed by Johnson. However, for Johnson's typical case, of a barrier adjacent to a narrow lagoon, the coastal plain slope should be steeper than the adjacent continental slope. In two out of the three barrier sections of the Outer Banks coast where this situation occurs, the actual slope relationship was found to be reversed and the continental shelf slope was steeper than the adjacent coastal plain. In one of these reverse slope cases, the slope of the continental shelf off the narrow lagoon barrier was exactly the same as that off the special case of the barrier bounding the wide lagoon. In addition, bathymetric projected profiles constructed along the neighboring Virginia mainland non-barrier shoreline exhibited exactly the same averaged bathymetric continental shelf slope as along the one typical barrier that did fit Johnson's slope relationships.

3. Regional analysis of the mean gradient of bathymetric projected profiles along the Outer Banks, barrier and non-barrier coast, indicates there is a regular increase in the slope of the inner continental shelf slope. This increase extends from the northern Cape Henry truncated cuspate spit and mainland coast, south along the barrier island shoreline to Cape Hatteras. The maximum inner continental shelf slope occurs at this major cuspate feature, Cape Hatteras, and then again regularly decreases southward along the barrier chain. A second lesser increase in gradient occurs off the smaller cuspate feature of Cape Lookout. The remainder of the barrier chain, as it approaches close to the southern mainland, shows a regular decrease in continental shelf slope with values similar to those

along the northern section. The regional coastal plain slope decreases in gradient from the northern mainland to a point inland from Cape Hatteras and then increases again as it approaches the southern section, but it is complicated by several topographic features (terrace scarps and ridges), relicts of higher Pleistocene sea levels.

4. Since Johnson considered only a stable or falling sea level in his deductive models, the possibility of applying his geometric profile analysis to a submerging shoreline is not valid unless one assumes what reasoning Johnson might have developed for such a model. In any case, in view of the results of the present study, it might be questionable whether Johnson's theoretical model is valid for either stable, emergent or submergent shorelines. This present study was not done as an attempt to apply Johnson's projected profile technique to any particular theory that some barrier islands may develop on shorelines of submergence, but instead this study is an application of his technique and analysis, applied quantitatively and in greater detail to a regional barrier island chain coastline. In general, however, the changes in the offshore inner shelf profile gradient of different segments of the North Carolina-Virginia barrier chain, related by Johnson to "typical" and "special cases" in the original coastal plain gradient before emergence, might be related in a regular fashion to a changing coastline orientation and perhaps dominant wave action. It is more probable, therefore, that the mean gradient of the offshore slope does not have the relationship to barrier island development that Johnson envisioned, whether for an emergent barrier or a spit barrier. Thus, the entire premise on which Johnson "proved" the relationship of barrier islands to a shoreline of emergence is not valid along the North Carolina–Virginia coast, one of the "type locality" coastal areas he studied for his original theory.

REFERENCES

Abbe, C., Jr. 1895, Remarks on the cuspate capes of the Carolina coast: Boston Soc. Nat. Hist. Proc., v. 26, p. 489-506.

Davis, W. M. 1896, The outline of Cape Cod: Proc. Am. Acad. Arts and Sci., v. 31, p. 303-332.

Fisher, J. J. 1967, Origin of barrier island chain shorelines, Middle Atlantic States: Geol. Soc. Amer. Spec. Paper, v. 115, p. 66-67.

Fisher, J.J. 1968. Barrier island formation, discussion: Geol. Soc. Amer. Bull., v. 79, p. 1421-1426.

Feltner, C. E. 1948, Beach erosion in North Carolina: Bull. No. 37, Dept. of Engineering Research, N. C. State College, Raleigh, N. C.

Fenneman, N. M. 1938, *Physiography of the Eastern United States:* McGraw-Hill Co., Inc., New York, N. Y., 714 p.

Hoyt, J. H. 1967, Barrier island formation: Geol. Soc. Amer. Bull., v. 78, p. 1125-1136.

Johnson, D. W. 1919, *Shore processes and shoreline development:* John Wiley and Sons, Inc., New York, N. Y., 584 p.

Oaks, R. Q. and Coch, N. R. 1963, Pleistocene sea levels, southeastern Virginia: Science, v. 140, p. 979-983.

Otvos, E. G. 1970, Development and migration of barrier islands northern Gulf of Mexico: Geol. Soc. Amer. Bull., v. 81, p. 241-246.

Pierce, J. W. and Colquhoun, D. J. 1970, Configuration of Holocene primary barrier chain, Outer Banks, North Carolina: Southeastern Geology, v. 11, p. 231-236.

Schwartz, M. P. 1970, Multiple casuality of barrier islands: Jour. Geol., v. 79, p. 91-94.

Shaler, N. S. 1895, Beaches and tidal marshes of the Atlantic coast: Natl. Geog. Monog., v. 1, n. 5.

Shepard, F. P. 1963, *Submarine Geology:* Harper and Row, New York, 557 p.

Swift, D. J. P.,Sanford, R. B., Dill, C. Z., Jr. and Avignone, N. F. 1971, Textural differentiation on the shore face during erosional retreat of an unconsolidated coast, Cape Henry to Cape Hatteras, western North Atlantic Shelf: Sedimentology, v. 16, p. 221-250.

Zenkovitch, V. P. 1957, The origin of barrier and lagoon coasts: Trudy Inst. Okeanol. Akad. Nauk. SSSR, (Works of the Oceanological Institute), No. 21.

Zenkovitch, V. P. 1967. *Processes of Coastal Development:* John Wiley, Inc., New York, N. Y., 738 p.

Chapter 9

BARRIER AND LAGOON SYSTEMS ALONG THE ZULULAND COAST, SOUTH AFRICA

Antony R. Orme

ABSTRACT

The Zululand coast extends 320 km SSW from the Mozambique border to the Tugela River, and is backed by a tapering coastal plain which, though 80 km wide in the north, pinches out 40 km north of the Tugela. The coast consists of long barrier beaches, surmounted by sand dunes up to 180 m high, behind which lie extensive swamps, lagoons, and old dune ridges. The present barrier has grown in response to high energy and material transfers during and since the Holocene transgression, Nevertheless, older nearshore sediments, dune sands, and swamp deposits indicate Pleistocene sea-level oscillations across this coast, and show that the present barrier is superimposed upon remnants of a major Pleistocene barrier system represented by Port Durnford Beds.

Shorezone dynamics are conditioned by prevailing northeasterly and southwesterly winds, by complex nearshore and coastal circulation systems, and by seasonally high sediment discharges from neighboring watersheds. Nearshore circulation patterns, consisting mainly of asymmetric cells, generate longshore currents exceeding 1 m/sec. Inner and outer breaker zones reflect offshore bars. Beyond the nearshore circulation, reversing coastal currents related to the Agulhas Current system flow parallel to the shore with surface velocities of 0.2 to 0.8 m/sec, reduced by a factor of 0.4 near the seabed. During the winter dry season, strong littoral drift creates barrier beaches across estuaries. During the summer rainy season, these barriers are breached and vast sediment plumes extend seaward and alongshore. The Tugela brings 10.5×10^6 tons of sediment to the coast annually.

Behind the coastal barrier lie Kosi Bay, Lake Sibayi, Lake St. Lucia, and Richards Bay, remnants of once extensive lagoons reduced by infilling and segmentation. As Holocene barriers became more complete, drainage outlets were plugged with sediments and lagoons became vast sediment traps. Holocene sedimentation patterns are complex and impressive. Furthermore, present sedimentation rates for Lake St. Lucia are three times the mean rate for the past

Department of Geography, University of California at Los Angeles, Los Angeles, California 90024

5000 yrs, largely due to accelerated erosion from mis-used Zululand watersheds. Reedswamp encroachment, spit growth, lagoon segmentation, and high salinities now pose serious threats to the survival of aquatic life.

INTRODUCTION

The Zululand coast extends for 320 km SSW from Ponta do Ouro on the Mozambique border to the mouth of the Tugela river (Fig. 1). The coast consists of long sandy barrier beaches of Holocene age, interspersed with crumbling cliffs cut into remnants of poorly consolidated barrier and lagoon deposits of Pleistocene age. This complex is surmounted by massive sand dunes of Pleistocene and Holocene age that commonly rise over 100 m above present sea level and reach 188 m near Cape St. Lucia. These dunes are mantled with luxuriant scrub and forest vegetation, modified outliers of the evergreen subtropical forest that once covered much of the Zululand coastal plain. This composite coastal barrier rises everywhere higher than the terrain lying immediately inland.

Behind the coastal barrier complex lie several impressive lagoons, the largest of which are, from north to south, Kosi Lake, Lake Sibayi, Lake St. Lucia, and Richards Bay. These lagoons are remnants of once extensive lagoon systems that were brought into existence by the Holocene transgression. These systems have since experienced appreciable modification owing to sedimentation, segmentation, and swamp encroachment, processes that appear to be accelerating at the present time. Beach spit and foreland growth has divided the formerly linear lagoon systems into a series of smaller, more nearly circular segments. These remnant lagoons are thus margined by extensive reedswamps, dominated by *Phragmites communis* and *Cyperus papyrus,* and locally by swamp forest and mangroves, all of which form habitats for a rich and varied fauna.

The coastal barriers and lagoons are backed by a tapering coastal plain which, though 80 km wide in the north, narrows southward to pinch out at Mtunzini, 40 km northeast of the Tugela estuary. This coastal plain, which is everywhere below 200 m, is the southward termination of the much more extensive Mozambique coastal plain with which it shares many features in common. The Zululand coastal plain is thickly mantled with Pleistocene sands, mainly in the form of north-south oriented dune ridges, the magnitude of which decreases inland. Weathering has accentuated the clay content and reduced the permeability of these deposits so that interdune swamps are now common, particularly with higher water tables consequent upon the Holocene marine transgression. Toward the inner margins of the coastal plain, the Pleistocene sands thin out across Cretaceous marine and nearshore sediments. These Cretaceous materials in turn pinch out against tilted Karroo rhyolites, which rise to form the Lebombo and Ubombo mountains in the north, and against basalt and older rocks farther south. Cenozoic nearshore and marine sediments crop out locally within the Zululand coastal plain.

From Mtunzini to the Tugela estuary, the coastal plain is absent and a series

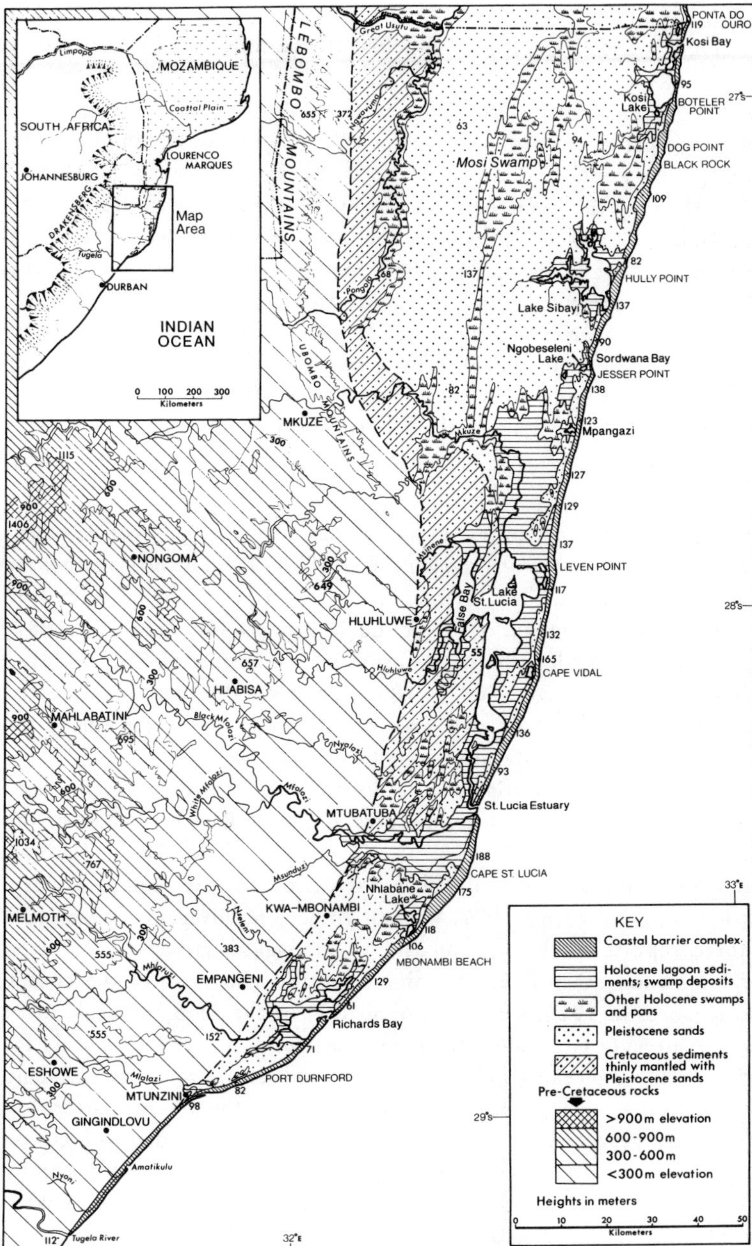

Figure 1. Geomorphology of the Zululand coastal plain.

of Holocene beach ridges forms a prograding shore at the margins of subdued hills. This area is transitional between the linear barriers and lagoon systems that parallel the Zululand coast to the north, and the massive spits that extend across the blind estuaries aligned perpendicular to the hilly Natal coast farther south.

Offshore, a comparatively shallow continental shelf terminates in a pronounced slope change at depths of between 90 and 135 m. Off the northern Zululand coast, from Ponta do Ouro to Cape St. Lucia, this shelf is only 2 to 7 km wide, thus allowing the main Agulhas Current to flow close to shore. From Cape St. Lucia the shelf widens southward, reaching a maximum width of 45 km southeast of the Tugela estuary before narrowing again to the south. Thus, the coastal plain narrows southward and the continental shelf widens. Seemingly, therefore, the coastal barrier complex trends diagonally across a continental shelf which slopes ESE and is underlain by a wedge of Cretaceous and younger sediments. This wedge begins at the western margins of the coastal plain and thickens eastward, reaching 2500 m thick in two borings made at Inharrime on the Mozambique coast.

Background to research

Though the approaches to Durban and Lourenco Marques had been surveyed in the early 19th century, not until the Zulu War of 1879 and its aftermath were concerted efforts made to map the intervening Zululand coast. Such efforts were associated with Britain's annexation of St. Lucia Bay (1884), Zululand (1887) and Tongaland (1895), and with the eventual incorporation of Zululand and Tongaland into the Colony of Natal in 1897. Even so, the maps constructed during this period provide scant and grossly inaccurate data concerning the lagoons and other geomorphic features along this coast.

In 1902, C. W. Methven investigated the possibilities for harbor development at St. Lucia Bay and Umhlatuzi Lagoon (now Richards Bay) on behalf of the Natal colonial government. In 1905, C. J. Crofts of the Natal Harbour Department completed the first detailed survey of the St. Lucia estuary. Meanwhile, between 1902 and 1907, William Anderson, government geologist, published three reports on his reconnaissance geological survey of Zululand which contain, among other items, the first scientific discussion of the coastal barriers and lagoons. In 1910, with the publication of the 1:63,360 scale maps of Natal by the Surveyor General's Office, the coast was portrayed accurately for the first time. The directions taken by spits and barriers across Zululand estuaries shown on these maps conform in all instances to the present pattern.

Since these early years, research pertinent to the coast has tended to focus on three principal themes. First, following the leads provided by Anderson and by du Toit (1922), geologists have examined the scattered and somewhat ambiguous evidence for the evolution of the Zululand coastal plain since Cretaceous times. Problems raised by this evidence are by no means resolved. Second, ecologists have examined some aspects of the rich flora and fauna of Zululand. Nevertheless, a definitive flora for the Zululand coast has yet to be written. Third, since 1945, the South African Council for Scientific and Industrial Research (CSIR) has pursued oceanographic research within 300 km of the coast, giving special attention to offshore circulation patterns. Despite this work, however, coastal barrier and lagoon dynamics have received little attention. This is due in part to the relative isolation of the Zululand coast, in part to the lack of research personnel, and in part to the dangerous research environment provided by the high energy shore-zone and by the very real menace posed by sharks and crocodiles.

I am presently engaged upon a geomorphic and ecological analysis of the entire Natal coast, and this paper draws upon field data from these investigations. This paper will outline the salient elements in the evolution of the Zululand barrier coast, and discuss in greater detail the present system of barriers and lagoons within a process-response framework.

EVOLUTION OF THE ZULULAND COAST

The present coastal barrier complex is the most recent in a series of coastal barriers that have been constructed in an approximately north-south direction across the Zululand plain and continental shelf since Cretaceous times. The coast and continental margins of southeast Africa were blocked out during the disruption of Gondwanaland by late Jurassic diastrophism. During and shortly after the rupturing of Gondwanaland, flexuring and faulting further modified the continental margins of Zululand, establishing ideal environments for marine deposition across the present coastline. Cretaceous shorelines lapped against the inner margins of the Zululand coastal plain while 1800 to 2500 m of fossiliferous marine deposits accumulated offshore. Cretaceous and older rocks in turn provided a shallow shelf across which Cenozoic marine oscillations occurred, leaving a legacy of relatively thin nearshore and deeper water deposits of Miocene and younger age. These were then mantled with a variety of Pleistocene and Holocene material. In general, Cretaceous, Miocene, Pleistocene and present shorelines converge southward toward Durban and run closely parallel to the present south coast of Natal and the neighboring Transkei coast.

In a recent analysis of late Cenozoic sediments from the Durban region, McCarthy (1967) has shown that the widespread Berea Red Sand and the underlying calcareous sandstones forming the Bluff Beds, previously regarded as separate formations, are respectively the weathered and unweathered remnants of extensive cemented coastal dunes. These dunes overlie basal beach deposits which, in turn, rest upon a step-like sequence of narrow rock benches descending from 90 m above sea level near Durban to 90 m below sea level in the Aliwal Shoal, forming an offshore reef of eolinite some 65 km south of Durban. Though McCarthy favors a Pliocene age for these deposits, foraminifera evidence is also consistent with a Pleistocene age. Certainly, the sequence accords readily with our knowledge of Quaternary sea-level changes, as demonstrated by Maud (1968) and Davies (1970) for Natal. Borehole evidence for low Pleistocene sea levels and for sedimentation patterns consequent upon the Holocene transgression are presently being investigated by this author.

Unequivocal evidence for multiple Pleistocene shorelines is less readily discerned in Zululand, owing mainly to the widespread mantle of Plesitocene and Holocene sands. Though this evidence will not be detailed here, certain parts of the Pleistocene record are pertinent to this discussion, namely the Port Durnford Beds, the Pleistocene sands, and the submerged river valleys. These Pleistocene features provided the environnent across which Holocene barriers and lagoons have developed.

Port Durnford Beds

In 1907, Anderson described a sequence of shales, lignites, and sands from coastal cliffs near Port Durnford, southeast Zululand. A sandy mudstone toward the base of this sequence yielded a fossil mammalian fauna that included an extinct elephant *Paleoloxodon zulu* and an extinct buffalo *Bubalus andersoni* (Scott, 1907). Both these species are now known from the younger Vaal river terraces, which suggests a middle to upper Pleistocene age for the Port Durnford Beds.

I have recently re-examined these deposits, tracing them 45 km along the southern Zululand coast from just south of Port Durnford to north of Mbonambi Beach, though interrupted at Richards Bay. At the type locality, the Port Durnford Beds comprise the following stratigraphic sequence (Fig. 2):

9.	Red and brown sand	2 m
8.	Light yellow sand, crossbedded	5-15 m
7.	Orange sand, crossbedded and flatbedded	1-4 m
6.	Peat and/or spore bed	0.5 to 2.5m
5.	Orange sand	2.5 m
4.	Blue-gray sandy mudstone, fish and molluscan fauna	3-4 m
3.	Brown sand, Foraminifera and Ostrea 2.	
2.	Blue-green shale and sandy mudstone, mammalian and fish fauna, wood remains, comminuted shells	
1.	Red sandstone, unfossiliferous	

Unit 4 forms the present wave-cut platform and is only exposed at low water. The underlying beds have not been observed since Anderson's time. The base of unit 6, the peat, lies between 1 and 3 m above present mean sea level. The overlying sands are exposed along the coast in crumbling cliffs up to 60 m high, and are covered with up to 15 m of Holocene dune sands. Coloration of the Pleistocene sands is due mainly to weathering and ferruginization, and has little diagnostic significance.

The Port Durnford Beds seemingly represent a major barrier-lagoon complex of middle to upper Pleistocene age, related to a shoreline configuration and sea level not greatly different from the present. The sequence is apparently comformable and unwarped. Units 1 thru 4 suggest a transgressive or fluctuating nearshore environment such as may be associated with a large embayment periodically open both to the sea and to debris discharges from nearby estuaries. As beach barriers and sedimentation gradually transformed this embayment into a more restricted lagoon, swamp encroachment gave rise to extensive peat accumulations. Pollen and macrofossils indicate a mainly freshwater environment at this stage characterized by Potamogetonaceae, Cyperaceae, some ferns and grasses, together with swamp forest trees such as *Syzygium* (van Zinderen Bakker, personal communication). These swamps were in turn invaded by coarse, medium and fine sands, with crossbedded sets dipping landward at 10°, suggesting a major barrier beach and dune complex sweeping inland across the infilled lagoon. The sea subsequently withdrew below the present level, allowing the weathering, induration, and thus survival of these deposits. This simple model thus attributes most of the stratigraphy to shore-zone dynamics during a single stillstand little above present sea level, and avoids the need for several shoreline fluctuations. The stratigraphy revealed in these cliffs is duplicated in many modern barrier-lagoon environments along the Natal coast.

Traced northward along the Zululand coast, the Port Durnford Beds commonly form a subdued and intermittent ridge, generally occurring as indurated red sand hills and eroding bluffs protruding through recent dunes. Initially at least, this Pleistocene barrier system may have extended some 350

Figure 2. The Port Durnford Beds at their type locality, showing the upper portions of the Pleistocene barrier-lagoon complex. The peat bed, here 1.5 m thick, lies about 2 m above the cliff base and is overlain by beach and dune sands.

km from Mtunzini to Delagoa Bay.

Fossiliferous nearshore marine deposits from Picnic Point, 3-6 m above sea level along the western shore of False Bay, Lake St. Lucia, have been dated as older that 50,000 yr (UCLA-1830). These deposits apparently accumulated in False Bay when the open coast lay along the western shore of Lake St. Lucia. From their comparable elevation, it is probable that barrier islands composed of Port Durnford materials may have been forming offshore at this time. Both the Picnic Point deposits and the Port Durnford Beds may be correlated with an Eem (Sangamon) Interglacial sea level a few meters above the present.

Pleistocene sands

Pleistocene sands occur in a variety of textures, colors, consolidation phases, and topographic situations across the Zululand coastal plain and continental shelf. Sands at or near the surface are mostly eolian in origin, but marine and fluvial sands also occur. Of particular pertinence to this discussion are the subdued ridges of the coastal plain, the foraminferal sandstone ledges at the coast, and the eolianite reefs offshore.

Several episodes of dune formation are represented in the Zululand coastal plain but erosion, weathering, and eluviation have blurred such differences as may once have existed. In general, the subdued dune ridges of northern Zululand are oriented to within 10°of true north, and the relative relief between the dunes and their adjacent hollows increases significantly from west to east. Closed depressions are common and interdune areas often contain linear swamps. The trend surface exhibited by the dune crests slopes eastward appreciably below the crest of the coastal barrier complex. These relationships are less clear in the more dissected and weathered dunes farther south, where red and grey clayey sands are the residual product of pronounced weathering and eluviation. With favorable slope and hydrology, shell fragments and other calcareous matter have long since been removed from these dunes. Weathering of magnetite, ilmenite, and other ferromagnesian minerals has provided a strong red coloration, while the weathering of feldspars and other unstable minerals has emphasized the clay content. Iron and clay pans have in turn impeded drainage.

At the coast, low reefs of cemented calcareous sandstones create and reinforce the many small headlands along the otherwise straight coastline. These sandstones are composed mainly of quartz but feldspars, ferromagnesian minerals, shell fragments and foraminifera also occur, cemented within a matrix of crystalline calcium carbonate. Textural data and strong crossbeds, with individual sets dipping up to 30°,suggest an eolian origin for these deposits, in which case they are the unweathered analogies of certain red clayey sands farther west. Some textural and structural data, however, support an origin for these deposits within a beach and nearshore environment. Eolianite and/or beach rock: the problems are by no means resolved. These calcareous sandstones have been planed by wave action, both at present sea level and at levels some 5 to 12 m higher, for example at Mission Rocks.

Eolianites probably form many of the pronounced rock reefs that parallel the Zululand coast from 1 to 3 km offshore. An eolian composition has already been proved for the Aliwal Shoal and other submerged reefs off Durban (McCarthy, 1967), and eolianites seemingly extended to more than 100 m below sea level beneath Durban Bluff (Anderson, 1906). Off the Zululand coast, a similar composition may be suggested for Glenton Reef (-2m below low water) off Mtunzini, Tenedos Shoal (-2m) off Port Durnford, Nhlabane Rock (-5m) off Mbonambi, Leven Shoal (-3m) and Leadsman Shoal (-5m) off Leven Point, and many other shallow reefs. Certainly, eolianite fragments are frequently washed onshore by wave action, presumably from nearby reefs. These reefs are probably the wave-planed counterparts of the cemented foraminiferal sandstones outcropping along the shore. At greater depth, a discontinuous submarine ridge follows the edge of the continental shelf off much of the Zululand coast. This ridge which rises 9-18 m from depths of 60-75 m, could conceivably represent a late Pleistocene barrier beach and eolianite complex.

Submerged river valleys

The deeply entrenched rock channels beneath many Zululand estuaries indicate appreciable downcutting during later Pleistocene times followed by erratic Holocene sedimentation. Borehole data indicate a buried rock channel in Dwyka tillite at -52 m below the banks of the Tugela river, 5 km from the sea. Boreholes failed to reach Cretaceous bedrock at -55 m in Richards Bay but the buried Mhlatuzi channel is defined by neighboring shallower boreholes (Fig. 3). Boreholes also failed to reach bedrock at -40 m below sea level in the Mfolozi valley some 16 km from the present coast. In the St. Lucia estuary, Holocene sediments have not been penetrated below -21 m. Isotopic dating of peat from a depth of -22 m beneath Durban has yielded a date of 24,950±950 yr B.P., while wood from -29 m in the nearby Umgeni valley, where deep scouring persisted well into the Holocene, has been dated at 8420±140 yr B.P. (Maud, 1968).

Off the northern Zululand coast, a number of steep-sided submarine canyons begin at depths of -30 to -40 m within 1 or 2 km of shore and plunge to over -650 m within 8 km of shore. Though the origin of these canyons is complex and clearly unrelated to present-day river mouths, their persistence may conceivably be linked with former outlets of the Pongola and Mkuze drainage systems. East of the Lebombo-Ubombo mountains, these rivers have been deflected parallel to the present coast by Pleistocene barriers.

Quaternary shoreline model

A general model may be proposed from the above evidence to provide the framework for the ensuing discussion of shore-zone dynamics.

During Pleistocene times, the sea withdrew intermittently across the Zululand coastal plain, leaving a residue of marine and nearshore sediments, and a mantle of sand dunes oriented north-south. A number of transgressions were superimposed on this regressive sequence. In particular, a middle or upper Pleistocene transgression initiated a major barrier-lagoon complex along the present coastline, now preserved as the Port Durnford Beds. During later Pleistocene times, certainly during the last glaciation, the sea fell at least 100 m below present sea level, rivers entrenched themselves seawards through earlier deposits, and a blanket of marine, fluvial, and eolian sands was plastered across the continental shelf. Onshore, pronounced weathering and eluviation transformed the Pleistocene sands into red and gray clayey sands, but at lower levels calcareous cementation transformed similar sands into eolianite and beach rock. Erosion reduced the Port Durnford Beds into a number of discrete blocks. With the Holocene transgression, the sea transgressed across the offshore eolianites and invaded these blocks. Remnants of a Pleistocene barrier-lagoon complex have thus provided the framework around which Holocene barrier and lagoon development has taken place.

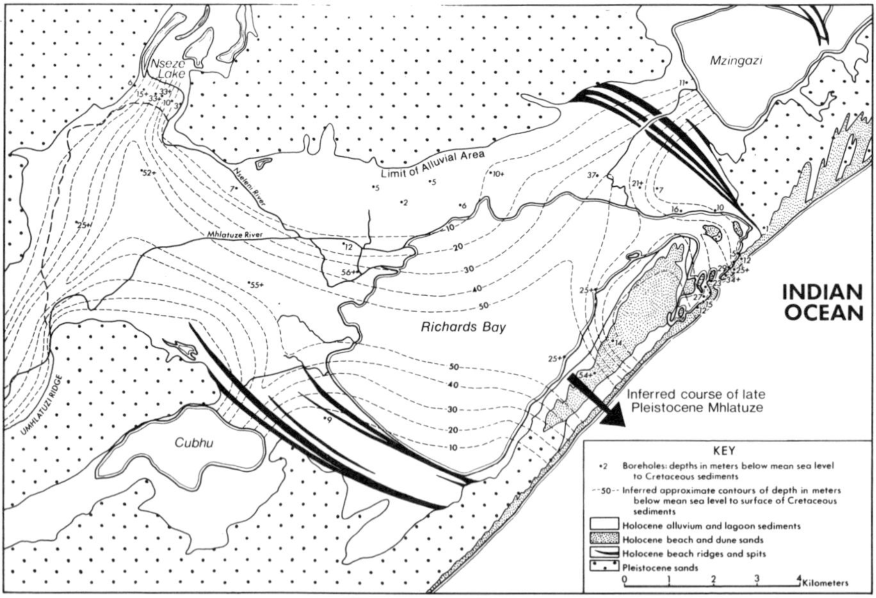

Figure 3. Geomorphology of Richards Bay, Zululand. The approximate form of the late Pleistocene drainage system is shown. Borehold data are provided by courtesy of D.L. Webb and Associates, M.S. Zakrzewski and Partners, Natal Roads Department, and South African Railways.

COASTAL PROCESSES

The principal energy factors affecting modern barrier and lagoon dynamics along the Zululand coast are surface winds, waves, and wave-generated currents within the surf zone. Coastal currents related to the Agulhas Current system are of secondary importance. The semidiurnal tides are a relatively insignificant factor. The hydrologic budget of the lagoons exerts strong influence over physical, chemical, and biological responses within the lagoons. The materials contributing to barrier growth and lagoon sedimentation come primarily from the high sediment loads brought down seasonally by Zululand rivers. Cliff erosion is a minor contributor of materials. Let us therefore examine each of these energy and material components, before analyzing the response of the barriers and lagoons to these processes.

Surface winds

The prevailing surface winds along the Natal coast are northeasterly and southwesterly. These winds flow roughly parallel with the coastline and are almost equally divided in frequency and velocity. They are spread fairly evenly over the 12 months of the year. Wind velocities are subject to strong diurnal variations, daytime winds being significantly stronger than winds at night.

At Durban in January, 32 percent of winds flow from between S and SW, and 29 percent flow from between NNE and ENE (Fig. 4). In July, 20 percent of winds flow from between SSW and WSW, and 17 percent from between N and NE. Based on hourly means, winds from these directions often exceed 25 knots and sometimes reach 40 knots, with gusts of up to 140 knots during spring and summer. Similar directional frequencies and velocities prevail along the Zululand coast, with a tendency for winds to flow more nearly north and south in the far north. Winds from other directions are transient and rarely above moderate velocity, though occasional fresh to strong southeasterlies generate high seas.

The prevailing surface winds are closely related to the movement and intensity of high and low pressure systems advancing along the coast from SW to NE. During the passage of low pressure systems, moderate to strong northeasterly winds may be succeeded almost instantaneously by strong to gale force southwesterlies. Not only may such winds have considerable effect on nearshore surface waters, but they are the principal forces governing coastal dune development and lagoon hydrodynamics.

Superimposed upon those wind components flowing parallel with the shore are significant onshore-offshore components related to diurnal variations in the temperature gradient between land and sea. Maximum onshore wind velocities occur just after midday. The daily onshore wind component is greater than the daily offshore component, thus net transport of surface water due to wind effects along the coast is onshore. By vectorially adding the prevailing wind components that flow parallel with the shore to these onshore-offshore components, the resultant wind may be calculated. Despite the paucity of wind

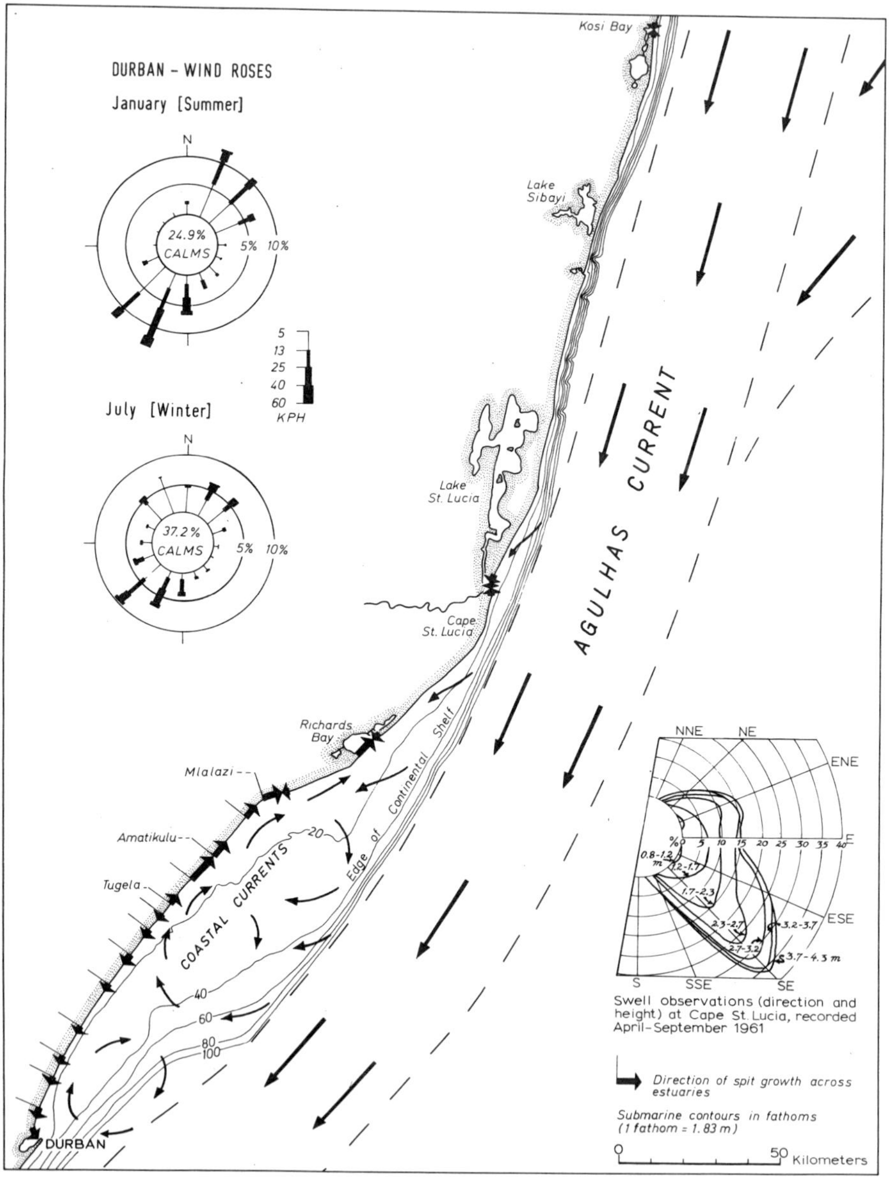

Figure 4. Zululand coast: selected energy components. Submarine contours are not shown below 100 fathoms (-183 m).

data north of Durban, the orientation of coastal dunes is seemingly correlated with the direction of the prevailing rather that the resultant wind.

Nearshore circulation systems

Measurement of the vigorous nearshore circulation systems along the Zululand coast has begun only recently. In general, the nature and intensity of these circulations depend primarily on the direction of wave approach and the shape of the nearshore bottom profile, and to a lesser extent on wave height and wave period.

Swell observations by lighthouse keepers at Cape St. Lucia show that swell approach varies from 30°to 170°, the median direction being 115° or ESE (Fig. 4). Some 72 percent of the swells come from directions south of east. These findings accord well with records from Durban (Oliff, 1969). However, the dominant direction of swell approach is 15°farther north at Cape St. Lucia than at Durban. Wave heights at Cape St. Lucia vary between 0.9 and 4.0 m, with a median height of 2.1 m or 0.6 m higher than at Durban. The distribution of wave periods at Cape St. Lucia falls between 6 and 18 sec, with a median wave period of 9.7 sec, or 1 sec smaller than at Durban. Longer wave periods and higher waves occur during winter at both Cape St. Lucia and Durban. Waves originating from more southerly directions have on average greater significant wave periods and heights than those originating from more northerly directions. These data refer to deep-water waves.

On nearing the shore, waves are affected by shoreline configuration and bottom topography. Because the Zululand coast lacks major headlands, wave refraction due to shoreline configuration is found mainly around the localized sandstone ledges, notably north of Cape St. Lucia. Elsewhere, swells approach the shore with little directional modification and the effects of an irregular bottom topography are more readily discerned. Submerged sand bars and troughs are persistent features of the nearshore bottom, reflecting an abundance of available materials and their continued reworking as deep-water waves feel bottom. Submerged alongshore bars are particularly notable some 100 to 200 m offshore. Waves generally break on these bars, then reform and continue unbroken toward the shore, to break again close inshore. Extremely high energy and material transfers are commonly involved in these breakers. These inner and outer breaker zones are expressions of two interrelated circulation sub-systems coupled by wave-mass transport, longshore currents, and rip currents.

Nearshore circulation patterns exhibit considerable variety in time and space. At one time or another, circulation systems embrace a broad spectrum from symmetrical cells with strong onshore-offshore components to unidirectional alongshore flows. When wave approach is more or less normal to the shore, a series of symmetrical circulation cells is established in which modest longshore currents and strong seaward return flows feed rip currents at mean intervals of 600 m. Off the Natal coast farther south, rip currents are 30 to 60 m wide, with velocities of 0.3 to 1.0 m/sec and a discharge of about 28 m^3/sec, and may extend up to 0.8 km offshore (Oliff, 1969). Off Zululand, rip currents may extend at least 300 m offshore, through the outer breaker zone and into the coastal current circulation.

As wave approach becomes more oblique, circulation cells become increasingly asymmetric, rip currents diminish in magnitude and frequency, and alongshore flows become increasingly important. Asymmetric cells with a significant longshore component occur on about half the observed occasions. Eventually, the longshore components may become sufficiently strong to impose a continuous unidirectional flow on the circulation, flow broken only by headlands and discharging estuaries. Under such conditions, longshore currents attain velocities exceeding 1 m/sec, forming a virtual river flowing along the shore with a discharge of several score m^3/sec, probably greater than the mean annual discharge of most Zululand rivers. Velocities of 1 m/sec may of course readily erode and transport the coarsest sands and small pebbles. Though unidirectional alongshore flows are clearly related to strongly oblique wave approach, such flows are accentuated when wind stress and coastal currents move in the same direction.

Interaction between the nearshore circulation and coastal currents is a significant feature of the Zululand coast, resulting in appreciable transfers of water and sediments into and out of the surf zone. With symmetric and asymmetric circulation cells, interaction may occur at frequent time and space intervals throughout the coastal zone. With unidirectional flows, interaction is more localized but is observable in the vicinity of most major estuaries.

The Agulhas Current and coastal currents

Because the southeasterly trade wind circulation of the central Indian Ocean produces a major hydraulic head against the east African coast, the South Equatorial Current diverges and major components flow southward both east and west of Madagascar. These currents converge off northern Zululand to form the warm Agulhas Current, the direction and position of which are subsequently determined largely by the shape of the coastline and the continental shelf. The Agulhas Current flows southward parallel to the coast throughout the year as a comparatively well defined stream, 50-75 km wide, with velocities of 1.0 to 2.5 m/sec, and a volume of flow which varies from 25-60 million m^3/sec (Fig. 4). The western edge of the Agulhas Current generally flows near the edge of the continental shelf, the distance of the current offshore correlating significantly with the atmospheric pressure gradient.

Wave motion along the western edge of the Agulhas Current generates an eddy circulation between the current and the coast, which appears offshore as a system of reversing coastal currents. Though coastal currents have long been recognized (viz. *Africa Pilot*, v. III, 12th ed., 1967), only recently have their complexities been examined and explanations sought.

The coastal currents flow parallel to the shore with near-surface velocities of 0.2-0.8 m/sec, reduced by a factor of 0.3-0.4 near the seabed (Oliff, 1969; Oliff and Addison, 1970; Anderson et al, 1970). Off southern Zululand, these currents flow alternately NE and SW, reversing direction periodically at 1-3 day intervals. These reversals appear linked with changes in atmospheric pressure, allowing for a variable lag of 3-36 hours. Off Richards Bay and Port Durnford, the overall exchange of water is slow, northerly, and offshore. At Richards Bay, mean monthly velocities average 0.3 m/sec and onshore currents occur on 10-60 percent of occasions. At Port Durnford and farther south, current reversals are common in both summer and winter, velocities for north-flowing and south-flowing currents average 0.2 m/sec, and onshore components occur on 10-30 percent of occasions, Summer reversals seem to follow short-term changes in atmospheric pressure. Winter reversals are random.

Coastal currents are much stronger and more persistently south-flowing off northern Zululand. The mean monthly velocity at Cape St. Lucia in summer is 0.4 m/sec south-flowing, and the extreme range is 0.6-0.8 m/sec. Winter measurements indicate more north-flowing currents than in summer. Nevertheless, the overall current is strong and south-flowing 80 percent of the time, with onshore components on 10-50 percent of occasions. This strong flow suggests that the main Agulhas Current impinges on the coast as far south as Cape St. Lucia, then diverges southward allowing eddies to develop on the widening continental shelf. Off the St. Lucia Estuary, the main current is 5 km offshore, south-flowing 50 percent of the time, with onshore components on 30 percent of all occasions.

These coastal currents constitute a basically simple system responding to energy changes imposed by atmospheric pressure variations. Their driving energy presumably comes from the main Agulhas Current and is consistent with a large-scale periodic wave motion induced upon a steady south-flowing system. This snake-like motion seemingly generates a series of clockwise eddies that move down the coast with the wave form. The flows within these eddies, together with eddy dimensions, are probably a function of the magnitude of the disturbances within the Agulhas Current.

Coastal currents are strongest in summer, impinging directly on Cape St. Lucia and breaking into eddies farther south. In winter, the currents are weaker or farther offshore, and eddies form throughout the offshore zone. Observations by the present author show that these strong coastal currents are a locally significant component of the offshore-nearshore circulation complex off the Zululand coast.

Tides

The mean range of spring tides along the Zululand coast is 1.8 m, and 0.5 m for neap tides. Thus, along the open coast, tidal influences are very modest compared with the energy generated by waves and currents. Tidal scour helps to maintain the mouths of larger estuaries but tidal influences diminish rapidly into the lagoons. Maximum tidal ranges diminish to 1.5 m about 1 km inside the St. Lucia Estuary, decreasing up The Narrows until tidal influences die out near the Mpate river mouth, 14 km from the sea and well short of Lake St. Lucia proper. Exchange of water between the lagoons and the sea depends primarily on water levels in the lagoons, the greatest inflow of sea water occurring when lake levels are lowest after prolonged drought. When Lake St. Lucia falls 1 m below mean sea level, an inflow of about 7 m^3/sec of seawater floods into the estuary.

Material components

Barrier and lagoon dynamics along the Zululand coast are greatly influenced by the discharge of water and sediments into the coastal zone from neighboring watersheds. Flowing from a summer rainfall area, Zululand rivers reach high flow rates between September and April. Over 80 percent of the mean annual discharge and sediment yields of perennial rivers occurs during these summer months. During major flood discharges, sediment plumes may extend several kilometers seaward beyond the mouths of major rivers while large quantities of coarse sediments are introduced to the nearshore zone. During winter months (May thru August) when discharge is low and sediment loads negligible, the mouths of many lesser rivers are readily closed by littoral drift. These sand barriers are breached by the first major spring floods.

Despite the paucity of reliable hydrologic data for Zululand rivers, a positive linear relationship appears to exist between drainage basin size, mean annual discharge, and mean annual sediment yields. Ranked according to these criteria, the principal rivers of the Zululand coastal plain are the Tugela, Mfolozi, Mkuze, and Mhlatuzi. Farther north, the Pongola and Ngwavuma breach the Lebombo Mountains in impressive 'poorts,' but then turn north to join the Great Usutu river in forming the Maputo river system which drains into Delagoa Bay. They thus make no contribution to the present Zululand coast.

The Tugela is the only river for which reasonable hydrologic data have been collected from a gaging station near the coast. At Mandini gaging station, 22 km above the mouth, the Tugela drains a 28,000 km^2 watershed and records a mean annual discharge of 226 m^3/sec. Discharge rates vary from 480 m^3/sec during the summer rainy season to 74 m^3/sec during the dry season. During the period 1955-58, the Tugela carried a mean annual suspended sediment load of 10.5 x 10^6 tons (Middleton and Oliff, 1961). Though this record is regrettably short, the suspended sediment data may be compared with mean annual values of 4.0 x 10^6 for the St. Lawrence, 10.5 x 10^6 for the Yenisei, and 18.0 x 10^6 for the Po (Kukal, 1971).

Of greater significance to the Natal environment is the fact that this suspended sediment yield represents mechanical denudation of the Tugela watershed in the order of 375 t/km^2/yr. This is substantially greater than values for the Zambezi watershed (75 t/km^2/yr), the Mississippi (154-230 t/km^2/yr), and the Colorado (271 t/km^2/yr), and approaches values for the Indus (420 t/km^2/yr).

The mean annual concentration of suspended sediment by weight in the Tugela is 0.23 percent (Middleton and Oliff, 1961). This value is similar to the 0.22 percent concentration recorded for the upper Vaal River between 1929 and 1940 (Mackenzie, 1949), but substantially higher than the 0.10 percent concentration estimated for the Pongola near Gollel, just west of the Lebombo Mountains. The Pongola drains a somewhat drier and geologically distinct watershed.

Northward through Zululand, rainfall decreases while evaporation increases. Mean annual rainfall exceeds 1250 mm along the southern Zululand coast north to the St. Lucia Estuary, ranges between 1000 and 1250 mm along the coast north to Kosi Bay, but falls to below 700 mm in central Tongland and west of the Lebombo Mountains. Potential evaporation rates are everywhere higher than mean annual rainfall. Furthermore the low slopes and extensive swamps of the widening coastal plain reduce stream velocities and encourage sediment deposition. Thus discharge and sediment yields from other Zululand rivers are predictably less impressive than for the Tugela. In the low sandy plains of northern Zululand, less than 5 percent of the mean annual rainfall contributes to stream discharge.

Extrapolating from meager data and allowing for the nature and size of the contributing watersheds, the mean annual discharge for the Mfolozi mouth (10,680 km^2 watershed) is probably about 33 m^3/sec, for the Mkuze (5,380 km^2 watershed) about 9 m^3/sec, for the Hluhluwe (960 km^2 watershed) about 3 m^3 and for the Nyalazi (670 km^2 watershed) about 2 m^3/sec. Though these values are low compared with the Tugela, discharge may be temporarily very large following high intensity summer thunderstorms over the inland watersheds. It is unlikely that the concentration of suspended sediments in these rivers exceeds, or even approaches, the value recorded for the Tugela. But even allowing for a comparable 0.23 percent concentration and accepting the above discharge estimates, the mean annual load of suspended sediments would only reach about 1.5×10^6 tons for the Mfolozi and 0.4×10^6 tons for the Mkuze.

Thus, whereas the discharge and sediment loads of Zululand rivers are important factors in coastal geomorphology, their effect on the open coast is likely to diminish from south to north. It will be shown that the introduction of Tugela materials produces a dramatic response along the open coast of southern Zululand. Farther north, as discharge and sediment yields diminish, low gradients and lagoons encourage deposition behind the open coast.

COASTAL RESPONSES

Barrier beaches and dunes

Open sandy beaches extend almost unbroken from Ponta do Ouro to the Tugela estuary, their continuity interrupted only by a few larger river mouths and then mainly during the rainy season. Indeed, at low tide levels during exceptionally dry periods, it is sometimes possible to walk or even drive along the beach over virtually the entire 320 km of coastline. The continuity of the beaches attests to the abundance of beach-forming materials and to the powerful onshore and alongshore movement of these materials throughout the coastal zone.

Zululand beaches are composed predominantly of materials in the medium to coarse sand range, median diameters commonly range 0.7-1.0 mm. Coarser materials are often found near major estuaries while finer sands occur increasingly away from such estuaries. These materials support foreshore slopes of 7-12°, with steeper slopes correlating with increasing moisture content. Most beaches are 50-150 m wide at low water, with berms reaching elevations of 2-5 m above mean sea level. A close interdependence between grain size, foreshore slope, moisture content, and wave action is readily observable. Behind the berm, surficial materials are finer grained and spits slope inland towards lagoons at up to 4°. Beaches are composed predominantly of quartz sands, but natural concentrates of ilmenite, rutile, zircon, and other heavy minerals are also significant.

Modern beaches occur either as spits and barriers extending across estuaries and lagoons, or as fringing beaches deposited over and against the remnants of Pleistocene barriers. The continuing accumulation and redistribution of wind-blown sands everywhere augment the height of this barrier complex. Under favorable conditions, Holocene dunes have climbed readily over Pleistocene barrier deposits to reach elevations frequently in excess of 100 m. Recent dune sands vary in thickness from a few meters to as much as 60 m. Where broad exposed beaches occur, as at Cape St. Lucia and Cape Vidal, mobile dune fields may occupy a considerable width of backshore, migrating along the coast in belts up to 500 m wide. Where mobility is less pronounced, the backshore and dunes are readily colonized by a succession of sand-tolerant plants. These plants are mostly species typical of tropical and subtropical sandy shores, and many find their most southerly expression around the Indian Ocean in the warm permissive environment afforded by the Natal coast.

The most important pioneer species of the backshore and foredunes are *Scaevola thunbergii,* which generally grows nearest to the sea, *Ipomoea biloba,* and the grass *Digitaria eriantha.* These species flourish in shifting sand and help to promote foredune accumulation. Other pioneers include *Mesembryanthemum edule, Tephrosia canescens, Gazania uniflora,* and *Osteospermum moniliferum.*

Farther back from the shore, the pioneer colonies are invaded by *Othonna sarnosa, Gloriosa virescens, Passerina rigida, Anthospermum littoreum,* and various grasses. The mat provided by these plants more or less stabilizes the backshore and enables a low dense scrub to become established. Under this scrub, which grows to about 2 m high, the outer dunes attain complete stability.

Away from the extreme effects of wind, shifting sand, and salt spray, several species of the dune scrub develop into a complex and luxuriant coast dune forest composed of trees 10-25 m in height. Dominant species include *Mimusops caffra,* south of Lake Sibayi, *Diospyros rotundifolia*, north of Lake Sibayi *Brachylaena discolor,* and *Euclea natalensis.* The coast dune forest does not differ significantly in physiognomy and composition from the evergreen subtropical forest that may once have covered much of the Zululand coastal plain but which is now found only in a few localities, such as the Dukuduku Forest west of the St. Lucia Estuary (Bayer, 1938).

For further analysis, the Zululand coast may be conveniently divided into four segments From south to north, these are: the Tugela-Mlalazi segments; the Richards Bay segment between Mlalazi and Cape St. Lucia; the St. Lucia Estuary segment between Cape St. Lucia and Cape Vidal; and the northern Zululand coast from Cape Vidal to Ponta do Ouro (Fig. 1).

Tugela – Mlalazi segment

Between the Tugela estuary and Siyana mouth south of Mtuzini, the straight coastline trends NE for 40 km. From the Siyana mouth the coast trends ENE an additional 5 km to the Mlalzi estuary, and continues this trend beyond Port Durnford. Though deep-water waves approach more or less normal to this coast, the continental shelf shallows more rapidly to the north than around the Tugela estuary. This helps explain the wave refraction commonly observed along this coast, refraction which allows waves to approach the surf zone from more southerly directions and causes wave crests to make a 5-20° angle with the shore. Furthermore, waves approaching from SSE are frequently reinforced by south to southwesterly winds associated with the passage of low pressure systems along the coast. Pronounced wave obliquity in turn promotes a strong unidirectional longshore current flowing NE within the surf zone. In addition, the coastal current off this section of coast is predominantly NE and is observed to play a significant though secondary role in transporting sediments up the coast beyond the surf zone, and in redistributing materials across the inner parts of the continental shelf. Onshore components of the coastal current system are also important in moving sediments back towards the nearshore circulation system.

Thus both coastal current and nearshore circulation systems favor a northeastward movement of materials along the southern Zululand coast, with strong onshore components added. The materials are made abundantly available by the Tugela river, and to a lesser extent by neighboring streams. Following summer storms, suspended sediment plumes may extend 3-5 km to sea off the Tugela estuary and stretch for at least 15 km along the coast under the influence of nearshore and offshore circulation systems.

In response to strong unidirectional energy flows and heavy material inputs, an impressive series of beach ridges and barriers extends northeast for 45 km along the coast, diverting the mouths of the Nyoni, Amatikulu, Siyani, and Mlalazi rivers parallel to the shore for varying distances. These beach ridges are oriented approximately parallel with the coast throughout but suprajacent eolian sands trend 15°-20° farther inland around the Mlalazi estuary in response to prevailing winds. The beach ridges occupy a zone with a maximum width of 1 km but within which 7-15 ridges are present. Except where disturbed by stream flushouts, agriculture, or afforestation, these ridges exhibit a spatial and temporal succession of coastal strand and dune scrub plants, culminating in coast dune forest on the innermost ridges.

A negative correlation exists between stream discharge and the distance over which stream mouths have been diverted by littoral drift, but this relationship weakens with increasing distance from the Tugela. The Tugela floodplain is 1500 m wide at its mouth, 700 m of which is occupied by a stable spit carrying dune forest which extends northeastward into the estuary from the south bank. This spit is extended by foredune vegetation for a further 70 m, from which an unstable sand spit periodically builds out across the estuary, restricting outflow to a deep narrow channel less than 50 m wide (Fig. 5). This spit is often destroyed by peak discharges.

On reaching the beach ridges farther north, the Nyoni river is diverted for 10 km NE into the larger Amatikulu river, which is in turn deflected 4.5 km NE, though peak flows sometimes temporarily breach the single unstable barrier at the diversion point. The Siyani is diverted NE for 2.5 km, while the Mlalazi is diverted ENE for 3.5 km. About 1.5 km of the Mlalazi barrier carries dune forest, a further 1.5 km carries foredune scrub, and the outer 0.5 km forms an unstable spit which seasonally blocks the river mouth (Fig. 6). Net accretion of

Figure 5. Tugela estuary, December 1971, showing sediment discharge following summer storms farther inland. Predominant swell is from the SSE, and the northward dispersion of the sediment plume by both nearshore and offshore circulations is evident. Note tha stable and unstable portions of the sand spit across the river mouth.

beach sediments ceases just beyond the Mlalazi estuary.

Richards Bay segment,

For 20 km from the Mlalazi estuary to beyond Port Durnford, the coast trends ENE and is fully exposed to dominant wave action. Over the remaining 60 km north to Cape St. Lucia, the coast trends NE and experiences similar nearshore and offshore circulation patterns to the Tugela-Mlalazi coast but with a much reduced input of materials. Owing to exposure and/or the lack of a sediment buffer, the poorly consolidated Pleistocene barrier-lagoon complex forming much of this coast is undergoing rapid erosion. Seacliffs 30-60 m high, topped by recent dune sands, are being cut back along a broad front north and south of Richards Bay. Pleistocene and Holocene dune ridges run inland from these cliff tops, trending between NE and NNE at 5°-25° angles to the shore, and reaching elevations of 60-188 m.

The only major Holocene barriers along this coast are those enclosing Richards Bay (Figs. 3 and 7). A massive barrier complex, 8 km long, up to 2 km wide, and as much as 60 m high, is primarily responsible for enclosing Richards Bay. This complex extends NE from the Pleistocene barrier deposits at Port Durnford and the higher inner barrier is composed in part of similar ferruginous sands over which recent dune sands have accumulated. The narrower outer barrier is composed entirely of Holocene materials. A similar Pleistocene barrier complex extends SW from the north side of the outlet channel and is continued by a highly unstable Holocene spit. Evaluating the Pleistocene and Holocene deposits of this region indicates that Richards Bay has reoccupied the site of a much larger Pleistocene lagoon system, portions of whose enclosing barriers have survived to provide foundations for the present barrier complex. Morphological and sedimentological evidence indicates that the curving 6 km long Mhlatuze Ridge, rising to 90 m west of Richards Bay, is a still earlier Pleistocene spit built across an earlier Mhlatuze estuary.

Littoral drift across the present outlet from Richards Bay is extremely variable, reflecting both wind and wave activity and the velocity and direction of waters discharging from the Bay. With winds and waves approaching from the south, the movement of materials NE across the outlet forms a 2 km long spit that ultimately pushes outflow against the northern bluffs. When this spit is breached by flood discharge followed by NE wind and wave approaches, a spit of comparable length builds SW across the outlet. This spit becomes increasingly unstable as it grows and is normally breached by meanders undercutting its inner face. Farther north, Nhlabane Lake also occupies the site of a larger Pleistocene lagoon, maintaining an outlet to the sea through a corridor between Pleistocene barrier remnants. Littoral drift across this outlet is predominantly toward the SW, and forms a 650 m long spit which has been largely stabilized with scrub vegetation. Thus, littoral drift along this coast is by no means as consistent as in the Tugela-Mlalazi segment.

HOMERTON
COLLEGE
LIBRARY

Figure 6. Mlalazi estuary and spit, near Mtunzini, Zululand. Unstable foredune scrub characterizes outer portions of the spit. On the inner side of the estuary, stable vegetated dunes reach 42 m. Mlalazi spit is 3.5 km long and reaches 18 m high.

Figure 7. Richards Bay, Zululand. On the far side of the outlet channel the inner barrier, composed in part of Pleistocene sands, and the outer barrier can be seen.

St. Lucia Estuary

Between Cape St. Lucia and Cape Vidal, the 45 km of coast trends NNE but is indented 7°-10° towards the St. Lucia Estuary. With the median direction of wave approach at Cape St. Lucia being from the ESE and the strongest swells coming from the SE, the predominant longshore current sets strongly toward the north. Thus, under natural conditions, water and sediments discharged by the Mfolozi are deflected north into the St. Lucia Estuary by a sand barrier that grows north from the sandy Pleistocene bluffs at Maphelana and Sugar Loaf Hill. Under low discharge conditions, the waters from both the Mfolozi and St. Lucia drainage systems are held back by a wave-built barrier augmented by abundant eolian sands. Though natural conditions no longer prevail in the St. Lucia Estuary, the record is sufficiently complete to indicate the problems posed to these drainage systems by strong littoral drift.

Since navigators first visited this coast in the 16th century, the Estuary has been known to close periodically as a result of sand drifting, commonly between August and November when river discharge is at a minimum. When the St. Lucia Estuary was first mapped by C.J. Crofts in 1905, distributaries of the Mfolozi entered the St. Lucia drainage some 0.5 to 1.5 km inland from the coast (Fig.8). The combined waters then breached the 7 km long barrier extending across the Estuary, but with difficulty. When this outlet was closed by littoral drift, river waters spread along the coast northward behind the barrier. In 1903, C.W. Methven noted that: *"The whole condition of the mouth of the Lagoon [Estuary] is very unstable and liable to constant change. It is evident from the debris along the shores, and the muddy character of the Lagoon that the Umvolosi [Mfolozi], when in flood, brings down and deposits in it large quantities of alluvial matter and floating debris, such as reeds, timber, etc., and that much of it remains in the Lagoon and is gradually silting it up."* Subsequent agricultural development and increased soil erosion in the contribution watersheds soon led to higher sedimentation rates in the Estuary, while drainage of the Mfolozi swamps, in particular the completion of Warner's Drain in 1936, effectively bypassed a major sediment trap. The changing morphology of the Estuary between 1905 and 1951 is shown in Figure 9.

By 1951, the serious threat posed to the physical and ecological well-being of the Lake St. Lucia system as a whole by the continuing deposition of Mfolozi sediments was realized. Accordingly, in 1952, the Mfolozi was diverted to a new mouth near Maphelana, a confining levee was built, and a dredging program was initiated within the St. Lucia Estuary in an attempt to enlarge the tidal basin and maintain an outlet through the coastal barrier. These measures reduced siltation but, without the flushing action of the Mfolozi discharge, closure of the Estuary became increasingly persistent (Fig. 9, 1960). In recent years, retaining walls have been constructed through the barrier in a further attempt to maintain an outlet, and some effort is being made to divert some Mfolozi discharge, but not sediments, into the Estuary (Fig. 9, 1972; Fig. 10). It seems likely that these efforts can only postpone the inevitable siltation and decay of the Lake St.

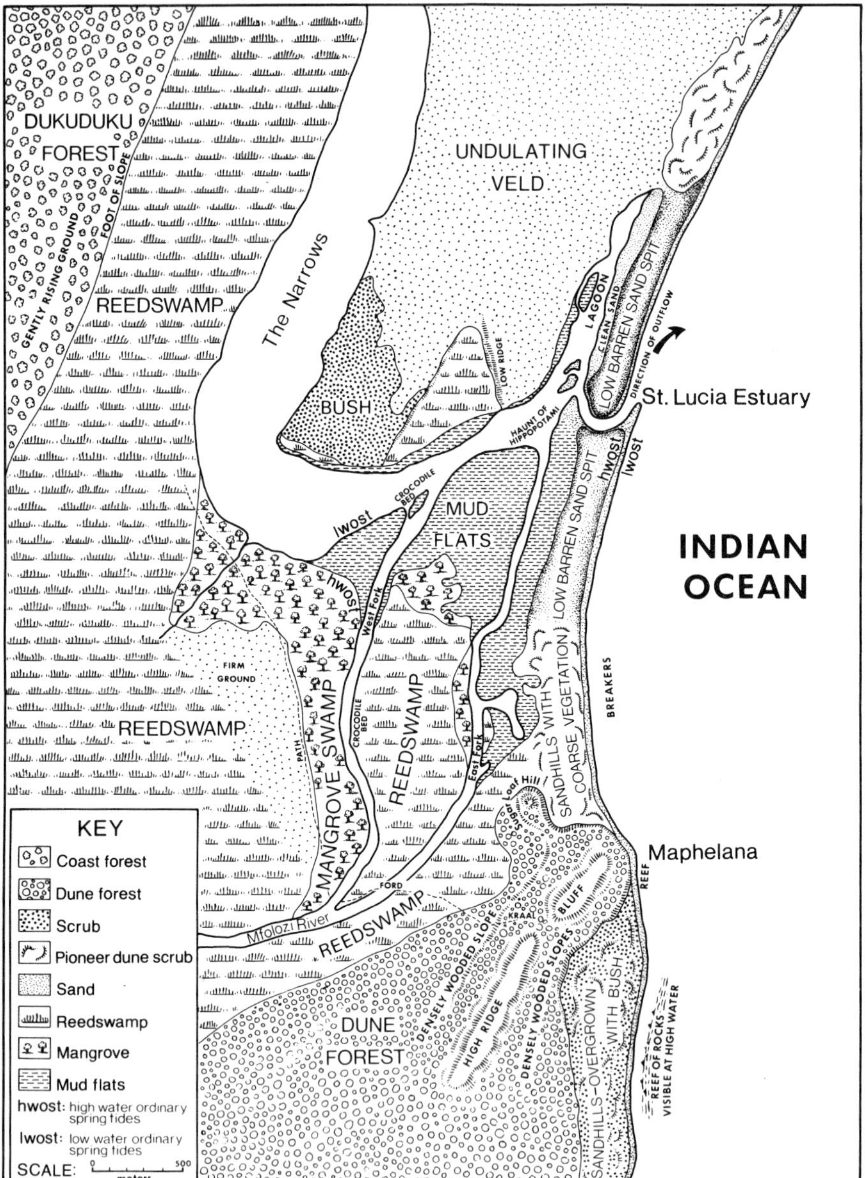

Figure 8. St. Lucia Estuary, 1905. [Redrawn with minor adjustments from C.J. Crofts' *Survey of mouth of Umvolosi River and entrance to St. Lucia Lake,* Natal Harbour Department, Zululand Coast Surveys, 1905.]

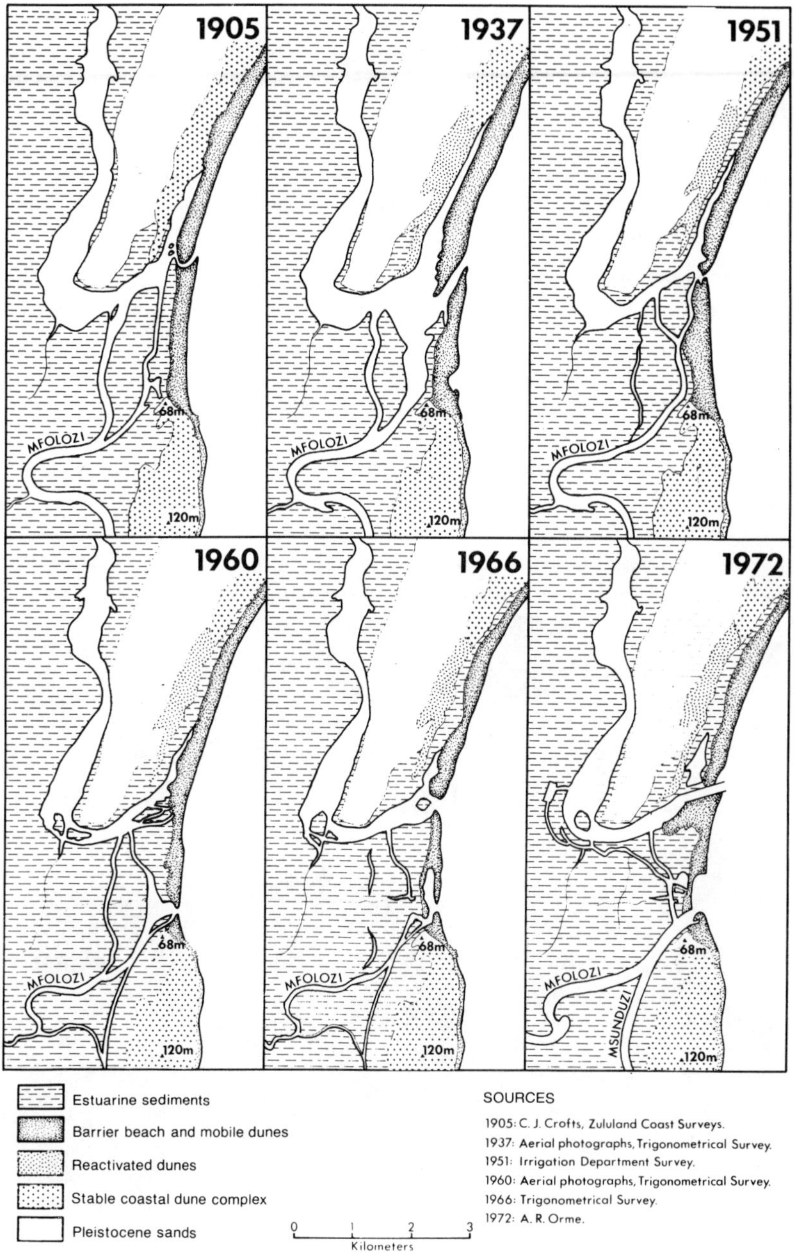

Figure 9. Natural and man-made changes in the St. Lucia Estuary, 1905-1972.

Lucia system.

The threat to the survival of plant and animal life as presently known within the Lake St. Lucia system is compounded by the problem of high salinities well above sea-water values. Increased salinities have resulted from increasing use of watershed waters for irrigation, from afforestation practices which have drawn off appreciable water through transpiration, from decreasing water volumes consequent upon lake sedimentation and, somewhat ironically, from the loss of periodic incursions of Mfolozi waters which helped to dilute high lake salinities.

Northern Zululand coast

Between Cape Vidal and Ponta do Ouro, the 150 km long coast is oriented N 10° E-N 20° E and, except at Sordwana Bay and Kosi Bay, the coastal barrier complex is continuous and commonly over 100 m high. For most of this distance, Holocene dunes are plastered over remnants of Pleistocene dune, beach, and nearshore deposits which crop out locally along the beach. At several localities however, Holocene dunes abut onto the sediments of Holocene lagoons, indicating that former connections between the lagoons and the sea have been closed only during comparatively recent Holocene times.

Foremost among these localities is a 20 km stretch of coast around Leven Point where Holocene dunes, no more than 1 km wide but now stabilized

Figure 10. St. Lucia Estuary, 1971, during construction of retaining walls designed to maintain outlet channel through the sand barrier. Inner and outer breaker zones are readily visible. Holocene sands are partially stabilized by pine plantations.

beneath dune forest, abut against Selley's Lakes and the North Eastern Shallows of Lake St. Lucia. Farther north, Mpangazi is closed off from the sea by a 1.5 km stretch of Holocene dunes. Ngobeseleni Lake maintains an intermittent outlet to the sea at Sordwana Bay through the south end of a 3.6 km long barrier that extends south toward Jesser Point. Lake Sibayi now has no direct outlet to the sea but a narrow Holocene beach and dune barrier abuts directly against the southernmost part of the present lake system. A similar closure has apparently taken place at the southern end of the Kosi Lake system where Holocene dunes, only 37 m high and 100 m wide at their narrowest point, abut directly onto the lagoon along a 2 km front. The one surviving outlet for these lagoons through Kosi Bay is maintained only with difficulty through highly mobile beach spits, mainly with the aid of tidal scour (Fig. 11).

Between Cape Vidal and Jesser Point the coast is oriented N 10° E. From Jesser Point northward, the coast is oriented N 20° E. This subtle change of direction is seemingly responsible for a change in the effective wind direction responsible for sand transport over the backshore. South of Jesser Point, the effective wind direction is from the north and dunes move inland to make a 10°-30° angle with the coast, or otherwise migrate southwards along the backshore in belts up to 600 m wide. North of Jesser Point, the effective wind direction is from the south and Holocene dune orientation makes a similar 10°-30° angle with the coast in moving N and NNW. Around Kosi Bay, the mean orientation of Holocene dunes is toward N 16° W, whereas Pleistocene dunes trend due N. Whether this significant deflection of the Holocene dunes is due to the influence of coastal terrain and thermal controls causing winds to be drawn across the present coast, or whether it reflects a major departure from the Pleistocene wind circulation in this part of southeast Africa, is not at present clear. Littoral drift around Kosi Bay is predominantly towards the north and, farther south, sandstone ledges serve to trap beach sediments on their southern or updrift side while allowing erosion on the downdrift side, especially with wind and wave action from the north.

Coastal lagoons

Since the close of the Holocene transgression, Zululand lagoons have been much reduced in size by infilling and segmentation. Table 1 compares the former surface area of the lagoons, as deduced from degraded cliffs and old beach deposits, with their present mean surface area. The percent decrease offers an areal measure of Holocene sedimentation. In practice, however, the present water area depends on several variables, primarily stream inflow, evaporation, and neighboring ocean level. For example, with mean sea level in the Estuary, the Charters Creek gauge records +0.58 m and Lake St. Lucia covers 310.8 km^{2}. With high water springs in the Estuary and flood discharge from inland, Charters Creek records +1.52 m and Lake St. Lucia covers 417 km^{2}, so shallow and gently shelving are the lake shores. In contrast, were sea level to fall to 2 m below present mean datum, lake levels would fall to -1.5 m and Lake St. Lucia would

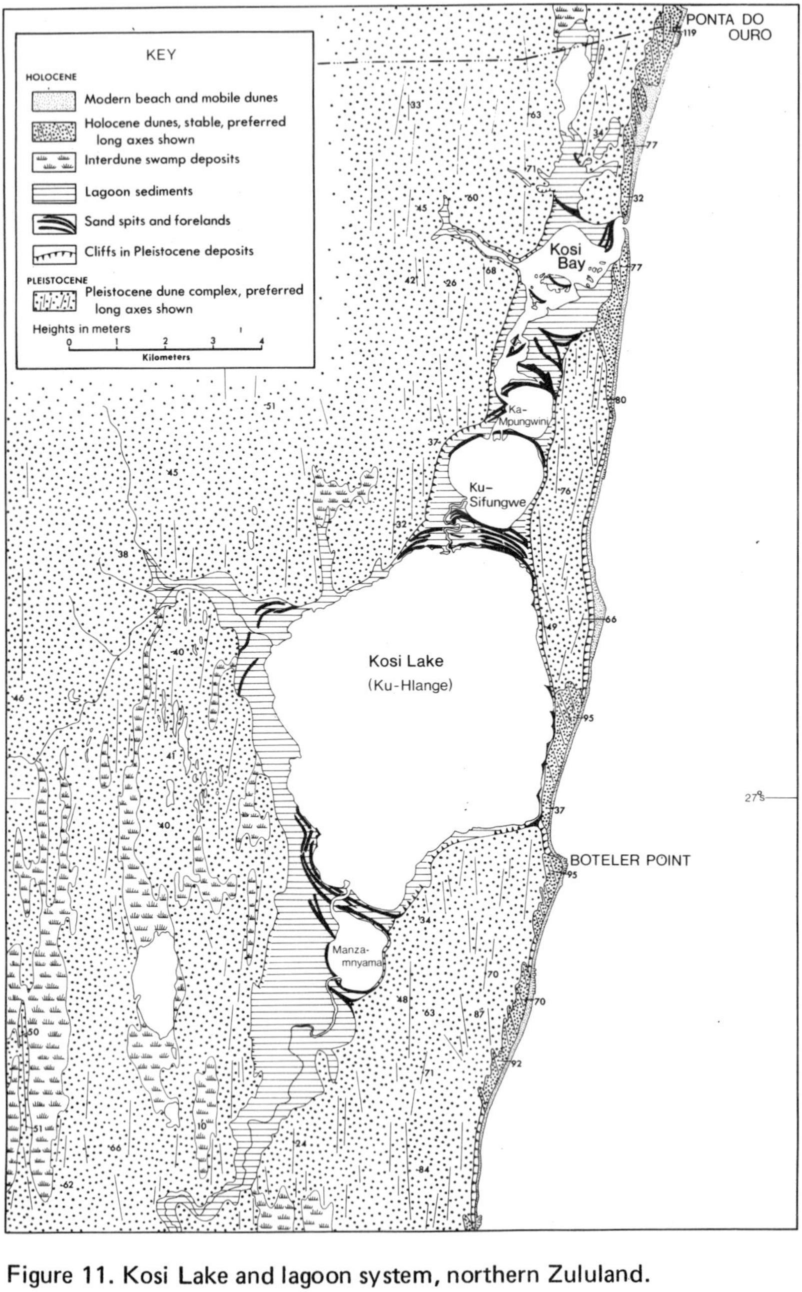

Figure 11. Kosi Lake and lagoon system, northern Zululand.

be virtually dry. Furthermore, the position of the lake shoreline at any one time also depends on the prevailing wind. Such are the effects of wind stress and seiche oscillations that the northeast shoreline may fluctuate as much as 5 km laterally, depending on lake stage and whether onshore or offshore winds are blowing. Onshore winds and related transgressive wave action cause appreciable disturbance of nearshore sediments, leading to the construction of long low beach ridges. Thus, the areas presented in Table 1 should be considered approximate values.

Nevertheless, Zululand lagoons have all experienced at least a 60 percent reduction in area during Holocene times. The greater Lake St. Lucia system formerly extended for 112 km parallel with the open coast from the Mkuze swamps to the Mfolozi swamps, embracing an area of 1165 km^2. This has now been reduced to about 312 km^2, and Lake St. Lucia now measures only 40 km long and 1-20 km wide. The Mfolozi lagoon has been more or less extinguished, the final stages having been accomplished by swamp reclamation and levee construction beginning in 1927. The Kosi lagoon complex and greater Richards Bay both formerly stretched over 25 km but have since been greatly modified. The smaller lagoons have been reduced to little more than perennial swamps whose volumes and salinities are subject to wide variations.

Three interrelated processes are responsible for the gradual extinction of

TABLE 1. PAST AND PRESENT SURFACE WATER AREAS OF HOLOCENE LAGOONS IN ZULULAND

Lagoon System*	Initial area (km^2)	Present area (km^2)	Percent decrease
Kosi	104.8	33.9	67.7
Lake Sibayi	145.2	56.5	61.1
Ngobeseleni	40.3	3.0	92.6
Mpangazi	8.5	2.1	75.3
Lake St. Lucia**	911.3	310.8	65.9
Mfolozi**	253.2	1.5	99.5
Nhlabane	24.2	4.0	83.5
Richards Bay	153.2	51.6	66.3
Mlalazi	12.9	1.0	92.2

* Refer to Figure 1 for locations

** Greater Lake St. Lucia discussed in the text is the formerly continuous lagoon formed by Lake St. Lucia and the Mfolozi lagoon.

these lagoons, namely sedimentation, segmentation, and reedswamp encroachment. Sedimentation and reedswamp encroachment are most evident near stream inlets, whereas segmentation occurs throughout the lagoon system. These processes will be discussed with reference primarily to Lake St. Lucia and the Kosi lagoon system.

Sedimentation

Sediments are carried into Zululand lagoons primarily by inflowing rivers, to a lesser extent by winds, and by tidal currents only into the mouth of Kosi Bay, the St. Lucia Estuary, and Richards Bay. Cliff erosion occurs locally, mainly along the western shores of Lake St. Lucia proper.

Lake St. Lucia receives the drainage of watersheds covering 8900 km^2. Of the mean annual runoff reaching the lake, the Mkuze system probably contributes about 45 percent, the Hluhluwe about 15 percent, the Nyalazi about 10 percent, and the Msinene about 3 percent. The remaining runoff is contributed by lakeside watersheds but these supply very little sediment. Except when high floods overflow the levee, the 10,680 km^2 Mfolozi drainage system no longer empties sediments towards the lake although the Narrows, the 20 km long constriction between Lake St. Lucia and the sea, were probably plugged mainly with Mfolozi debris. The Narrows now inhibit free exchange of water between the lake and the sea.

The Mkuze, Hluhluwe, Nyalazi, Msinene, and Mfolozi have supplied the vast bulk of sediments that have infilled Lake St. Lucia since the Holocene transgression. Borehole data from bridges on the new National Road west of the lake reveal the magnitude of Holocene alluviation well inland. The Msinene, 40 stream km above False Bay, has a late Pleistocene channel cut in basalt some 2.5 m below present mean sea level, covered by 10 m of Holocene debris. The Hluhluwe, 18 stream km above False Bay, has a Pleistocene channel cut in Cretaceous limestone at mean sea level overlain by 20 m of Holocene sediments. The Nyalazi, 24 stream km above False Bay, has a similar channel 1 m above mean sea level overlain by 11 m of Holocene sediments. Farther south, the Mfolozi is choked with at least 40 m of Holocene debris 38 km from the sea. The neighboring Msunduzi, 45 km from the sea above the now freshwater Eteza Lake, has a late Pleistocene channel in decomposed granite at present sea level overlain by 15 m of alluvium (Fig. 12). The Nseleni, above Nseze Lake 27 stream km from the sea at Richards Bay, is choked with 20 m of Holocene material. The burial of these late Pleistocene stream channels by Holocene alluvium reflects the high base levels provided by the lagoons over the past 5000 yr.

Except during extreme floods, the coarser materials derived from Lake St. Lucia watersheds are commonly deposited as stream gradients and velocities decrease on entering the coastal plain. Finer sands and silts are trapped in the extensive swamps farther downstream. Very fine silts and clays tend to flocculate with high lake salinities and to settle readily within the lake. The Narrows now inhibit sediment losses seawards, and such losses as may have occurred in the past were more than compensated by inflowing Mfolozi sediments. Thus, Lake St. Lucia serves as a very efficient sediment trap and has probably done so for most of its Holocene existence. Soft muds extend down to at least 16 m below mean sea level in the northern parts of the lake, and to at least 20 m in the southern Narrows.

Using steel probes, Kriel (1965) has estimated that Lake St. Lucia contains 2,376 x 10^6 m^3 of soft muds and that a further 739 x 10^6 m^3 lie beneath the nearby swamps, making a total of 3,115 x 10^6 m^3. If, as seems likely from radiocarbon dates, sedimentation began here about 5000 yr B.P. then these sediments represent an accumulation of 623,000 m^3/yr.

Radiocarbon dating or organic material from these sediments off Picnic Point in False Bay has yielded an age of 3960$\pm$60 yr B.P. for materials at 8 m below water level, and 2380$\pm$80 yr B.P. for materials from a 4 m depth (E.M. van Zinderen Bakker, in Kriel, 1965). An estimated 2,115 x 10^6 m^3 of sediments occupy the lake and swamps beneath the 4 m level, while the remaining 1,000 x 10^6 m^3 lie above that level. Thus, between 5000 and 2380 yr B.P., sediments accumulated at a mean rate of about 807,000 m^3/yr, or higher than the average for the past 5000 yr. During the period since 2380 yr B.P., sediments have accumulated at a mean rate of about 420,000 m^3/yr, or below the average for

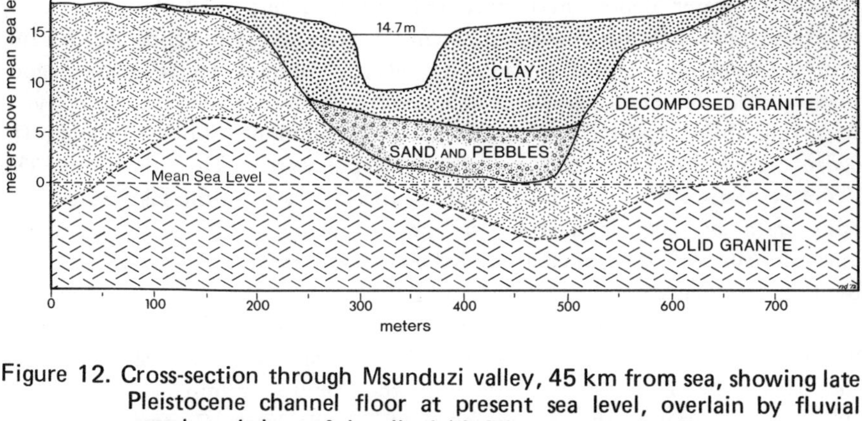

Figure 12. Cross-section through Msunduzi valley, 45 km from sea, showing late Pleistocene channel floor at present sea level, overlain by fluvial gravels and clays of the alluvial infill.

the past 5000 yr. However, by multiplying the estimated mean annual runoff into Lake St. Lucia by a sediment concentration factor of 0.50 percent, Kriel has suggested that sediments are presently being delivered into the lake at a mean rate of over 2,100,000 m^3/yr, well over three times the sedimentation rate for the past 5000 yr. Even adopting a sediment concentration factor of 0.23 percent, comparable with the Tugela, yields a rate of 980,000 m^3/yr, well over twice the rate for the past 2380 yr. These estimates are of course very crude, and fraught with all the imponderables implicit with estimating past and present sediment loads from ungaged watersheds. Nevertheless, the trend may be realistic, namely an initially high sedimentation rate after the lagoon and the higher base level were initiated, followed by a declining rate as stream gradients developed some dynamic equilibrium and watersheds became adjusted to the Holocene environment, followed in turn by greatly accelerated sedimentation as Zululand watersheds adjusted to increasingly exhaustive landuse practices. Certainly, from other data presently being developed by this author, sediment yields from Natal watersheds during recent decades are significantly higher than the values from earlier periods.

Reedswamp encroachment

Zululand lagoons harbor a subtropical vegetation which, under favorable conditions, greatly accentuates sedimentation and segmentation processes. Three communities are potentially important in this context, namely reedswamp, swamp forest, and mangrove forest. Theoretically, the less saline inner margins of lagoons should be dominated by reedswamp interspersed with riverine swamp forest, while the more saline outer reaches of lagoons should support mangrove forest. In reality reedswamp is well developed throughout most Zululand lagoons but suffers dieback in the more saline upper reaches of Lake St. Lucia. Mangrove forests have been greatly reduced by excessive siltation and human interference. Of the aquatic weeks, *Potamogeton* is found in less saline areas, giving way to *Ruppia* and *Zostera* towards estuaries.

Once water depths have been reduced to about 1.0-1.5 m reedswamp vegetation becomes of major significance in trapping sediments. Reedswamp encroachment and sedimentation henceforth interact to prograde lagoon margins. The dominant plants are *Pharagmites communis* and *Cyperus papyrus*, occurring either separately or in mixed stands commonly about 3 m high. The sedge *Scirpus littoralis* often grows into deeper waters below the reeds. This dense vegetation filters sediments and floating debris from the lake waters, and adds decaying organic matter to the sedimentary fill. Reedswamps are exceptionally well developed around the northern and eastern margins of Lake St. Lucia proper, around streams entering False Bay, in the Mfolozi swamps, and in Richards Bay. Evergreen swamp forests grow locally up to 25 m along watercourses through the reedswamps, the best surviving examples lying along the Mbazwane that flows south into the Mkuze swamp, and along the Inkhazamu swamp on the eastern shores of Lake St. Lucia. Common species are *Ficus hippopotamus, Ficus sycamorus, Syzygium cordatum, Cussonia umbellifera,* and *Barringtonia racemosa.* These swamp forests are the most southerly outliers of the gallery forests of tropical Africa.

In more estuarine environments, the pioneer herbs and grasses colonize mudflats and prepare the ground for mangrove forest dominated by *Bruguieria gymnorhiza, Avicennia marina,* and *Rhizophora mucronata,* growing 2-5 m out of the water. *Bruguieria* once grew extensively through the St. Lucia Estuary and The Narrows, but siltation and human interference have caused its recent rapid decline. *Avicennia* still survives in Richards Bay.

Reedswamp, swamp forest, mangrove forest, and aquatic plants all suffer periodic stress from excessive siltation and high salinities. Too much sediment tends to exclude salt water from around mangrove roots, while high turbidities reduce light intensity and cause aquatics to die out. Lake St. Lucia has recorded salinities 7-53 ppt (parts per thousand), compared with estuarine salinities of 35 ppt. Fluctuations are closely correlated with the seasons. During summer, salinity decreases from around 35 ppt in the Estuary to less than 15 near river mouths, but this gradient may be reversed after prolonged drought when stream inflow is negligible and evaporation has concentrated dissolved salts within the lake. Owing to the constriction imposed by The Narrows, even sea water may not pass readily inland to dilute these high salinities. With such fluctuations, the flora and fauna of the lake are highly dynamic. When salinity increases in the lake, marine species such as *Zostera* may migrate inland to supersede *Potamogeton,* while reedswamp may suffer appreciable dieback leading to erosion of unprotected lagoonal sediments and even to some cliff erosion. Conversely, during freshwater flushouts, a flourishing aquatic and reedswamp vegetation is soon re-established. The efficacy of these plants as sediment traps thus depends to a large extent on water chemistry, which in turn reflects several hydrologic variables. Without disturbance or vegetative recession imposed by salinity, lagoonal vegetation will eventually build the mudflats into firmer ground, and a succession toward evergreen forest may occur.

Segmentation,

Segmentation is a process whereby a large lagoon is converted into a series of smaller, round or oval lagoons by means of spits, barriers, and cuspate forelands building out at intervals from the initial lagoon margins. The efficiency of this process depends on three sets of variables: (1) the geometry of the initial lagoon (2) the magnitude and location of water and sediment discharge into the lagoons, and (3) the direction and strength of winds and related waves and wave-generated currents operating within the lagoon. Segmentation occurs more readily in tideless lagoons than in lagoons where tidal currents may disrupt nearshore circulations and oppose the coalescence of outbuilding spits. Whereas some marginal spit and foreland growth may occur soon after the initiation of most lagoons, segmentation accelerates rapidly as sediments build up extensive shallows. Sedimentation is in turn affected by segmentation which creates semi-independent basins for sediment deposition.

Segmentation has occurred to some extent in all the larger Zululand lagoons, but is best developed in the Kosi lagoons and in Lake St. Lucia. Both these lagoon systems have their long axes aligned subparallel with the open coast and with the prevailing winds. Strong prevailing winds thus often generate moderately high wave energy within these lagoons, which in turn create nearshore circulations capable of moving appreciable sediment loads along the shore. Furthermore, both systems receive most of their water and sediment input at their upper ends, and thus experience no major disturbance to their shore-building processes farther into the lake. Erosion and deposition within the lagoons operate in such a way as to align the shoreline at right angles to the wave resultant at a given point, with spit and barrier growth being further favored by constrictions in the intial shoreline.

Segmentation is admirably illustrated from the Kosi lagoons (Fig. 11). The initial Holocene lagoon ran subparallel with the open coast for 25 km and was up to 7 km wide. Segmentation has transformed this lagoon into six surviving water bodies separated by low beach barriers. South of Kosi Bay, four lagoon segments have achieved considerable circularity. Using the diameters of the inscribed and circumscribed circles yields circularity ratios for these lagoons of, from north to south, 0.74, 0.72, 0.69, and 0.62. These relatively high values may reflect the opposing wave energies generated by the northerly and southerly prevailing winds. A close correlation between the geometry of these lagoon segments and lagoon hydrodynamics undoubtedly exists and it is likely that, with adequate hydrodynamic data, prediction of the ultimate size and shape of lagoon segments for a given lagoon configuration would be possible. I am now investigating this problem with reference to a larger lagoon population.

Lake St. Lucia has also experienced appreciable segmentation, with several beach ridges rising to 2 m above lake level between segments. Excluding False Bay, the lake is now divided into six reasonably distinct segments, each forming a small sedimentary basin. From north to south, these segments have circularity ratios of 0.44, 0.52, 0.64, 0.64, 0.36, and 0.52. Compared with the Kosi lagoons, the greater variance and lesser overall circularity of these values may reflect, among other things, the less advanced stage of segmentation in Lake St. Lucia.

In Lake Sibayi, segmentation is mainly confined to the eastern shore against the coastal barrier from which sands are more readily available. With its deep and steeply shelving floor, its long axis aligned normal to the coast. and its limited input of sediments, segmentation is restricted. Furthermore, several Holocene strandlines up to 5 m above the lake margins bear witness to recent fluctuations of lake level which would also inhibit barrier stability. Such fluctuations have presumably occurred since the lake was cut off from the sea by the coastal barrier, and probably reflect the dependence of the now freshwater lake on the highly variable runoff and oscillating watertables of its small 536 km^2 watershed. Segmentation in Richards Bay has been restricted by the initial depth of the lagoon and by periodically disruptive water and sediment discharges through the central area (Fig. 3). Nevertheless, segmentation has occurred in the more tranquil portions of the initial lagoon that paralleled the coast just inside the barrier, thereby separating Mzingazi and Cubhu from the main lagoon.

CONCLUSION

The Zululand coast is particularly noteworthy in that persistently high energy components combine with seasonally high material inputs to generate a coastal system that is extremely dynamic. The main contributing energy factors are the strong northeasterly and southwesterly winds that flow more or less parallel with the coast, the complex nearshore circulation patterns dominated by strong surf and powerful longshore currents, and the reversing currents that occur farther offshore. Materials are introduced to this coastal system mainly in the form of fluvial sediments derived from eroding watersheds farther inland. The system responds by establishing sandy barrier beaches surmounted by massive sand dunes more or less continuously from the Tugela estuary to Delagoa Bay, a distance of over 400 km. Behind this coastal barrier, Holocene lagoons are infilling rapidly with fluvial materials, aided by segmentation and reedswamp encroachment. The system has added complexity, however, because this Holocene coastal zone is essentially superimposed upon the remnants of a major Pleistocene barrier and lagoon complex, represented by the Port Durnford Beds, while still older Pleistocene barriers survive farther inland. The individual components of the Zululand coastal zone are by no means unique, but the way in which these components have been assembled and continue to be manipulated affords a wealth of interest for coastal scientists.

ACKNOWLEDGMENTS

Research for this paper was supported by the Geography Programs, Office of Naval Research, United States Department of the Navy, under Project NR 388-102, Contract No. N00014-69-A-0200-4035.

The author gratefully acknowledges the following: The Natal Parks, Game, and Fish Preservation Board for field and data assistance in the St. Lucia region and in the Pongola swamps; the Natal Roads Department for permitting access to original borehole data; D.K. Hobday for companionship and discussion in the field; L.C. King, M.J.McCarthy, and R.R. Maud for discussion of Pleistocene problems; F.P. Anderson and personnel of the Council for Scientific and Industrial Research, Durban, for discussion concerning coastal currents; and the Natal Mercury for Figures 6, 7, and 10. C.R. Berger analyzed the Picnic Point material in the UCLA Radiocarbon Laboratory.

REFERENCES

Anderson, F.P., Sharp, S.O., and Oliff, W.D. 1970. The reaction of coastal waters off Durban to changes in atmospheric pressure: Paper H2 Oceanography in South Africa–1970: SANCOR, Pretoria, p. 1-22.

Anderson, W. 1902. Report on a reconnaissance geological survey of the eastern half of Zululand, with a geological sketch map of the country traversed: First Report of the Geological Survey of Natal and Zululand, Pietermaritzburg, p. 39-50.

Anderson, W. 1904. Furthur notes on the reconnaissance geological survey of Zululand: Second Report of the Geological Survey of Natal and Zululand, London, p. 39-52.

Anderson, W. 1906 . On the geology of the Bluff Bore, Durban, Natal: Trans. Geol. Soc. South Africa, v. 9, p. 111-116.

Anderson, W. 1907. On the discovery in Zululand of marine fossiliferous rocks of Tertiray age, containing mammalian remains: Third and Final Report of the Geological Survey of Natal and Zululand, London, p. 121-127. Bayer,

Bayer, A.W. 1938. An account of the plant ecology of the coastbelt and midlands of Zululand: Ann. Natal Mus., v.8, pt.3, p. 371-454.

Crofts, C.J. 1905. Survey of mouth of Umvolosi River and entrance to St. Lucia Lake: Zululand Coast Surveys, Natal Harbour Department.

Davies, O. 1970. Pleistocene beaches of Natal: Ann. Natal Mus., v. 20, n. 2, p. 403-442.

du Toit, A.L. 1922. The evolution of the South African coastline: South African Geog. J., v. 8, p. 1-9. Hydrographer to the Navy, 1967. *Africa Pilot, Volume III*, 12th edit

Hydrographer to the Navy. 1967.*African Pilot, Volume III,* 12th edition: London, Hydrographer to the Navy, 530p.

Jordaan, J.M. (ed.) 1968. Monthly flow records of gauging stations up to September 1960: Pretoria, Government Printer, Hydrographic Survey Publication No. 8.

Kriel, J.P. 1965. Report on the hydrology of the St. Lucia Lake system: Appendix No. 5, in Report of the Commission on Inquiry into the alleged threat to animal and plant life in St. Lucia Lake 1964-1966: Pretoria, p. 228-303.

Kukal, Z. 1971. *Geology of recent sediments:* Prague, Czechoslovak Academy of Sciences, 490 p.

Mackenzie, L.A. 1949. Silt in South African rivers: Irrig. Dept. Prof. Pap., No. 16.

McCarthy, M.J. 1967. Stratigraphical and sedimentological evidence from the Durban region of major sea-level movements since the late Tertiary: Trans. Geol. Soc. South Africa, v. 70, p. 135-165.

Maud, R.R. 1968. Quaternary geomorphology and soil formation in coastal Natal: Zeitschrift fur Geomorphologie, n.f., s. 7, p. 155-199.

Methven, C.W. 1903. Umhlatuze Lagoon: Report to the Prime Minister of Natal, Durban, 31p.

Methven, C.W. 1903. St. Lucia Bay: Report to the Prime Minister of Natal, Durban, 11 p.

Middleton, E.A., and Oliff, W.D. 1961. Suspended silt loads in the Tugela River: Trans. South African Inst. Civ. Eng., v.3, n.12 p.237-244 .

Oliff, W.D., ed, 1969. The disposal of effluents into the sea off the Natal coast: Natal Town and Regional Planning Report, v. 14, 139 p.

Oliff, W.D. and Addison, B.R. 1970. Currents on the Natal coast: Pretoria, CSIR Contract Report No. C Wat 23, 15 p.

Scott, W.B. 1907. A collection of fossil mammals from the coast of Zululand: Third and Final Report of the Geological Survey of Natal and Zululand, W. Anderson (ed), London, p. 253-262.

Weather Bureau, Department of Transport. 1960. Climate of South Africa, Part 6, Surface Winds (WB 26): Pretoria, Government Printer.

Chapter 10

EFFECTS OF EROSION ON BARRIER–ISLAND MORPHOLOGY FIRE ISLAND, NEW YORK

Kenneth Ruzyla

ABSTRACT

Erosion on the order of 3 to 5 ft per year occurred at Fire Island, New York, from 1959 to 1969. Surveys made during 1970 show that a maximum in excess of 65 ft of landward erosion occurred from July to December within a part of the study area. These same 1970 surveys show approximately 56,800 yd^3 of material was eroded from the dune system and approximately 18,000 yd^3 of material was eroded from the berm along a 1660 ft section of beach.

Wave action is most directly responsible for the erosion. Longshore transport of beach sediment is an important process on eastern Fire Island which moves beach sediment out of the area after it has been eroded by wave action. Wind action does not directly influence erosion, and most dunes are stabilized by vegetation. Neither berm, dunes, nor dune vegetation within the study area were altered by humans or beach vehicles during 1970.

Major factors which may be controlling erosion at Fire Island and other east coast beaches are: (1) number of hurricanes and other severe storms (2) eustatic rise in sea level (3) interference of natural shore processes by man, and (4) lack of available beach material due to natural causes.

Dunes, berms, washover fans, beach cusps, and beach scarps occur on Fire Island and each responds to erosion. Sedimentary structures of particular geomorphic features (dunes, berms, washover fans) are diagnostic of their depositional environment.

Erosion is variable, depending upon crescentic shoreline form position and probably offshore topographic configuration. Crescentic shoreline forms are relatively stable under conditions which normally result in erosion to straight portions of the beach.

Fire Island is currently undergoing ravinement. The shallow marine facies is transgressing landward while the lagoonal and dune facies are being reworked by wave action.

Marathon Oil Company, P.O. Box 120, Casper, Wyoming 82601

INTRODUCTION

According to El-Ashry (1971), U.S. east coast shorelines during the past 100 yr have been eroded several hundred feet. Fire Island (Fig. 1), a barrier beach located off the south shore of Long Island, New York, is also being eroded at the present.

The quantitative estimate of erosion at Fire Island and analysis of erosion effects on the land forms was accomplished by systematic measurements and observations taken over a period of one year (1970). Determination of erosion and other beach changes occurring from 1959 to 1969 were made from air photos. Effects of physical processes on Fire Island (wind, wave, and longshore current action) were studied so that process could be related to erosion. Geomorphic features of Fire Island, which include berms, dunes, washover fans, beach cusps, and beach scarps, were also studied to determine how erosion is manifest in Fire Island landforms.

Fire Island, one of a series of barrier beaches which parallel the south shore of Long Island (Fig. 1), is separated from Jones Beach, to the west, by Fire Island Inlet. Moriches Inlet separates Fire Island from Westhampton Beach on the east. Fire Island is approximately 32 mi in length and varies in width from 0.5 mi to less that 0.25 mi. The study area (Fig. 2) covers approximately 1.5 mi of beach.

PROCESSES ACTING ON FIRE ISLAND

There are basically three natural processes which act upon Fire Island: wave action, longshore transport of beach sediment, and wind. Erosion of the Fire Island dunes and berm by wave action was witnessed during several visits to the study area. Effects of wave action were most profound during the late fall and winter months.

Longshore transport of beach sediment is an important process since it controls the amount of sediment which is available to the beach and effectively moves sediment along the beach. The dominant direction of littoral transport of Fire Island beach sediment is westward. Estimates of the rate of littoral transport differ widely, but Taney (1961) gives a rate of 450,000 yd^3 per year at Fire Island Inlet.

Most dunes on Fire Island, with the exception of those at Democrat Point, are largely stabilized by vegetation. Thus the dunes spanning most of the island, including those within the study area, are not migrating. Visual observations and comparative photographs failed to reveal any construction of new dunes nor destruction of old dunes, by wind, during the interval from June, 1970, to June, 1971. However, comparison of air photos taken in 1966 and 1969 showed that some new dunes had formed during this period.

Visual observations and photographs also revealed no significant alteration of the berm, dunes, or vegetation which stabilizes the dunes, as a result of human

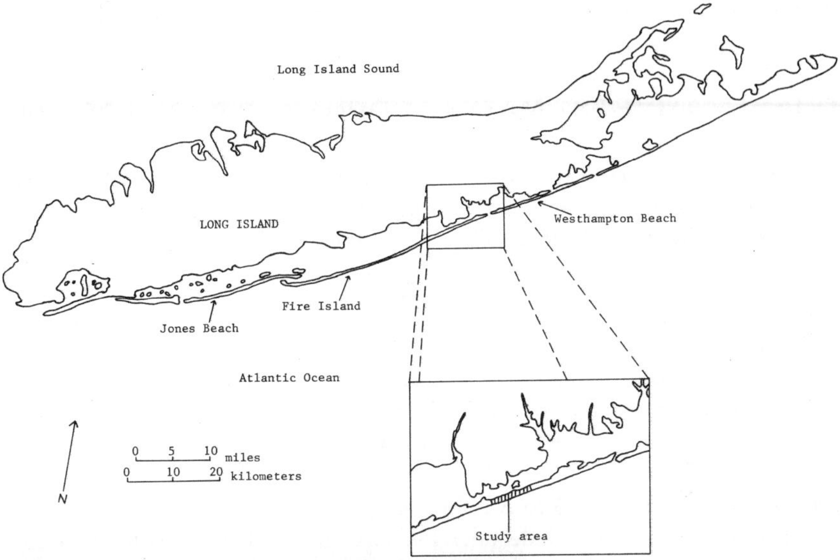

Figure 1. Location map of study area.

Figure 2. Aerial view study area looking east toward Moriches Inlet. Atlantic Ocean lies in background. A washover fan lies among dunes on right side of photograph.

activity. Dune vegetation was not appreciably damaged or destroyed by humans or beach vehicles.

QUANTITATIVE ESTIMATES OF DUNE AND BERM EROSION

Quantitative estimates of the volume of material eroded from both a section of dune ridge and berm within the study area from July, 1970, to December 1970, were computed using the following equation:

$$V = \frac{A^1 + A^2}{2} L$$

where:

V = volume of eroded dune (berm material)

A^1 = cross-sectional area of dune (berm) material eroded at cross section A-B

A^2 = cross-sectional area of dune (berm) material eroded at cross section E-F.

L = distance between cross sections A-B and E-F

From Figure 3, V was 56,800 yd^3 from A-B to E-F for the dune ridge. Likewise, V was 18,000 yd^3 from A-B to E-F for the berm.

The above estimates include only the material eroded from the dune ridge and berm for a part of the year and therefore are less than the annual amount of erosion. Observations and comparative photographs revealed that significant amounts of erosion did occur on the dune system and berm during December, 1970, to July, 1971, although of a much smaller magnitude than the erosion which occurred during July to December, 1970.

Observations and comparative photographs also revealed that much of the material which had been eroded from the berm was returned to the berm during periods of moderate wave activity in late spring and early summer of 1971. This was not the case in regard to the dune system. No significant accretion occurred to the dune system following the erosion.

RESPONSE OF THE DUNE SYSTEM AND BERM TO EROSION

Comparison of two surveys made in July and December, 1970, reveals that a considerable amount of erosion had occurred on the dune system during this inverval (Fig. 3). The dune system had been cut back shoreward as much as 70 ft, causing serious breaching of the dune ridge and complete destruction of some isolated dunes. Cross sections at three locations were constructed from the July and December survey maps to further illustrate dune ridge erosion (Fig. 4). Cross-section locations for July correspond to those for December. It is apparent from the cross sections that erosion of the dune system did not occur in equal amounts along the beach. Most erosion occurred at the location of cross section E-F, where the entire dune ridge was destroyed. Considerable erosion is indicated by the A-B and C-D cross sections although less severe than E-F.

A record of erosion and accretion on the dune system from 1959 to 1969 was compiled from air photo measurements (Table 1). The measurements were made

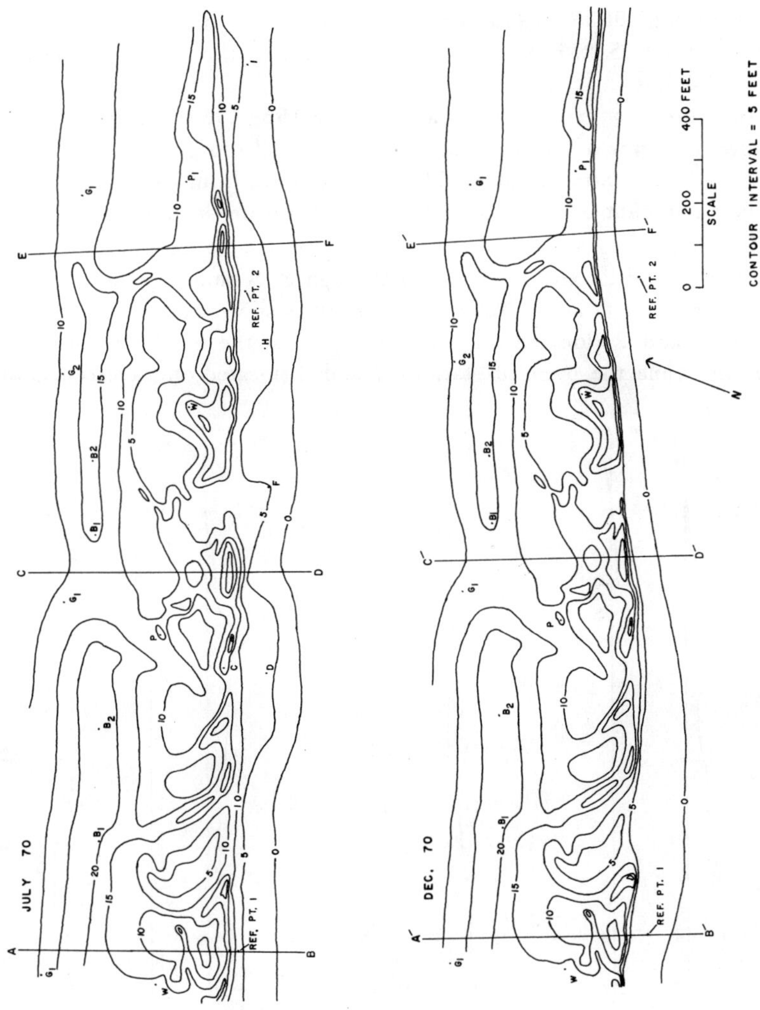

Figure 3. July and December 1970 survey maps. Locations of reference points 1 and 2 are indicated. Stations Y and Z are located approximately 1 mi and 1.2 mi east of reference point 2 respectively.

at the same locations as the cross sections indicated on the July and December 1970 survey maps. Years represented by available photos include 1959, 1962, 1966, and 1969.

The dune system has experienced episodes of both landward erosion and seaward accretion since 1959 (see Table 1). Maximum erosion was from 1959 to 1962; accretion occurred from 1966 to 1969; and erosion predominated during the 1959 to 1969 period. The amount of erosion has not been uniform along the length of the beach. For example, the dune system at location E-F was eroded 41 ft during 1962 to 1966, while the dune system at location A-B was eroded landward only 5 ft. However, the erosion rates for the entire 1959-1969 interval are within 2 ft.

In comparison with erosion rates for the 1959-1969 interval, the amount of landward erosion of the dune system occurring from July 1970, to December, 1970, is of an extreme nature. Maximum erosion from 1959 to 1969 was 47 ft, while an amount in excess of 65 ft was eroded from the dune ridge during 1970.

Beach profiles at four different locations (reference points 1 and 2, and stations Y and Z) are in Figures 5 through 8. A summary and comparison of beach profile data are in Table 2. The profiles show that the berm at reference points 1 and 2 (located on a straight part of the beach) experienced much erosion, while the berm at stations Y and Z (located on a crescentic shoreline

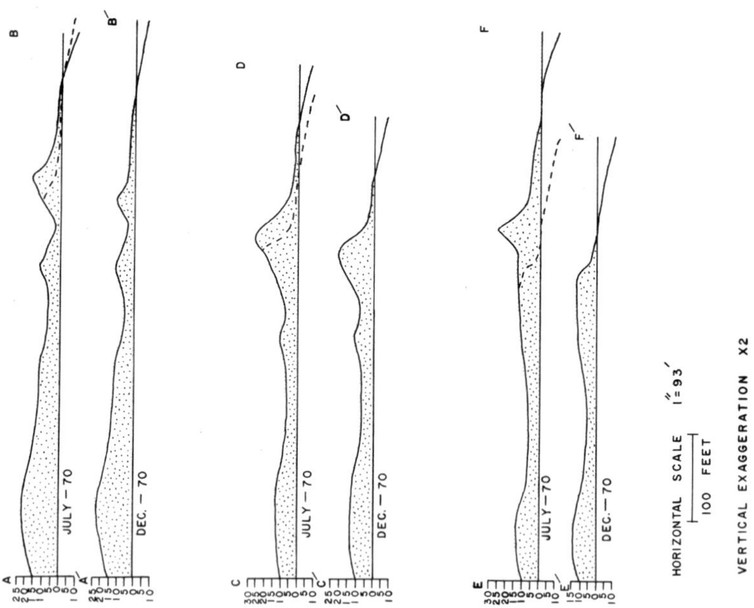

Figure 4. Study area cross sections. Locations of cross sections are indicated on Figure 3. Locations for July correspond to those for December. Dashed lines on July cross sections represent the superposition of December cross sections.

form) had comparatively little erosion. The original berm at reference point 2 was completely destroyed while the berm at station Z had accretion. Beach scarps were a common feature of the reference points 1 and 2 berms; tidal channels (runnels) were a common feature of stations Y and Z berms.

GEOMORPHIC FEATURES OF FIRE ISLAND

Berms, dunes, washover fans, beach cusps, and beach scarps occur on Fire Island; their characteristics are summarized in Table 3. The relative occurrences of sedimentary structures in berms, dunes, and washover fans are in Table 4.

Berms

The berm at a particular location varies greatly according to time of year (see Figs. 5-8). Berms are subdivided by formation of beach scarps. During the summer of 1970, two distinct berms were created by formation of two beach scarps (Fig. 9), each scarp forming the seaward termination of each berm. Two runnels (tidal channels) formed on the berm in response to storms and abnormally high tides on August 16, 1970. The runnels were normal to the shoreline and covered with interference ripple marks. Beach strata consisting of sub-horizontal layers of light and dark minerals characterize the berm. These layers range from about 0.1 in to about 6.0 in thick; the dark layers contain up to 98 percent heavy minerals by weight.

Dunes

The Fire Island dune ridge is continuous, but some irregularities exist where erosion has occurred (Fig. 2). Clay lenses occur at several locations beneath the dune ridge. The lenses are generally as much as several inches thick and pinch out laterally. The largest clay lens was 50 ft wide and about 6 in thick (Fig. 10). Plant fragments (mainly thin stems) are common within the clay lenses. In all cases, dune sand overlies the clay lenses. Large pebbles and shell fragments within the material underlying some of the clay lenses suggests lagoonal or marine deposition of the underlying material. In other cases, the origin of sediment immediately below the clay lenses could not be definitely established.

Inclined stratification characterizes the dunes. Dune strata are composed of alternating layers of light and dark minerals, as in the case of berm strata, but dune strata are generally much thinner, varying from 0.2 to 2.0 in thick. However, some dune strata are as much as 3-4 in thick. Dunes are completely stratified except where the stratification has been destroyed by roots.

Most dune strata are even-bedded, or cross-bedded, but disturbed and contorted bedding can occur. The dominant cross-bedding consists of the large-scale planar type. Some trough cross-bedding occurs, but is rare.

Washover Fans

Washover fans are easily identified on aerial photographs (Pierce, 1970) and

TABLE 1. AIR PHOTO MEASUREMENTS OF EROSION AND ACCRETION ON DUNES

Location	1959-1962	1962-1966	1966-1969	Net gain or loss	1959-1969 Erosion rate	July 1970 to* Dec., 1970
A-B	-46	-5	+4	-47	-5 ft/yr	-42
C-D	-32	-25	+24	-33	-3 ft/yr	-28
E-F	-40	-41	+49	-32	-3 ft/yr	-65

*July 1970 to Dec., 1970 measurements taken from survey data
- sign indicates erosion, + sign indicates accretion

TABLE 2. SUMMARY AND COMPARISON OF BEACH-PROFILE DATA FOR 1970

Profile locations	Berm width	Berm elevation	Features present	Comments
Reference point 1	Decreases from Aug. 25 to Dec.29 in general; temporarily shortened on Aug. 25 due to stormy surf	Lowered by 5 ft at datum on Dec. 29 profile	Scarps occur on Aug. 25 through Oct. 3 profiles	Most erosion occurred between Oct. 25 and Dec. 29; berm experienced net erosion from Aug. 25 to Dec. 29
Reference point 2	Decreases from Aug. 25 to Dec. 29 in general	Lowered in excess of 9 ft at datum on Dec. 29 profile	Scarps occur on Aug. 25, Oct. 1, and Oct. 3 profiles	Changes are similar to those of ref. pt. 1 profiles; original berm was completely destroyed between Oct. 25 and Dec. 29
Station Y	Decreases only slightly from Sept. 8 to Nov. 27	Not appreciably affected	Tidal channel occurs on Nov. 27 profile; scarps not prominent	Profiles are quite similar; berm experienced very slight net erosion from Sept. 8 to Nov. 27
Station Z	Increased from Sept. 8 to Dec. 27 in general; greatest width occurred on Nov. 27	Not appreciably affected	Tidal channel occurs on Nov. 27 profile; scarps not prominent	Berm experienced net accretion from Sept. 8 to Dec. 27

TABLE 3. CHARACTERISTICS OF GEOMORPHIC FEATURES OCCURRING ON FIRE ISLAND

Feature	Definition	Occurrence (time of year)	Occurrence (portion of beach)
Berm	A nearly horizontal surface of the beach formed by the deposition of material by wave action	Are present year round; can be absent in some localities during winter; most prominent from late spring through early fall	Includes portion of beach extending from dune ridge base out to swash zone
Dunes	Consist of ridges or mounds of loose, wind-blown sand	Present throughout year	Generally limited to seaward side of island; most of the larger dunes form part of a dune ridge parallel to the beach
Washover fans	Deposits formed when the sea breaches the dune ridge and invades the landward side of a beach or barrier island during periods of abnormally high tides and intense wave activity	Present throughout year	Occur within dune system; extend normal to shoreline
Beach cusps	Consist of one of a series of low mounds of beach material separated by crescent-shaped troughs spaced at more or less regular intervals along the beach face	Most prevalent during summer months; none observed during winter	Restricted to beach face
Beach scarps	Consist of an almost vertical slope extending along the beach roughly parallel to the strand line	Most prevalent during summer months; rare in winter	Can occur anywhere on berm

Feature	Dimensions	Sedimentary structures	Comments
Berm	From 0 to over 300 ft in width; height varies up to about 10 ft above sea level	Beach stratification, ripple marks (current and wind), swash marks, backwash marks, small scale low angle planar cross stratification, shell fragments, pebbles	Main berm may become subdivided by erosion into a series of berms; tidal channels may occur following storm activity
Dunes	Height varies, maximum of 30 ft	Laminations, contorted laminations, cross laminations, wind ripple marks, bioturbation by plant roots	Dunes within study area are relict and stabilized by vegetation; blowouts common
Beach cusps	Width (measured parallel to shore) about 100-300 ft; length indeterminable	Stratification, cross bedding, graded bedding, pebbles, shell fragments	Appear as broad, low level, slightly undulating surfaces which abruptly interrupt the dune ridge
Beach cusps	Cusp spacing varies from about 50 ft to just over 100 ft during summer months; will vary according to surf conditions	Same as for berm	Two distinct series of cusps can occur along the beach; shape of cusps was consistently symmetrical; fall cusps not as well developed as summer cusps
Beach scarps	Height varies from a few inches to 4 or 5 ft; length generally thousands of feet	Not applicable	Beach scarps are transitory features lasting several days at most; more than one scarp may occur at any one time and place

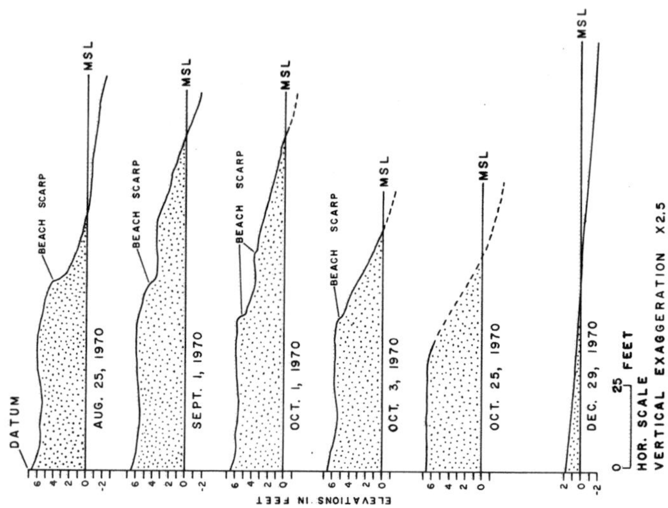

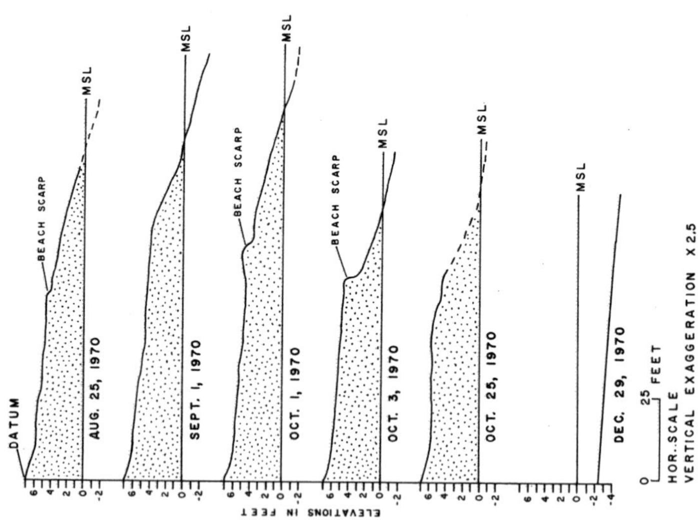

Figure 5. Reference point 1 profiles. Dashed lines represent extrapolation of profile.

Figure 6. Reference point 2 profiles. Dashed lines represent extrapolation of profile.

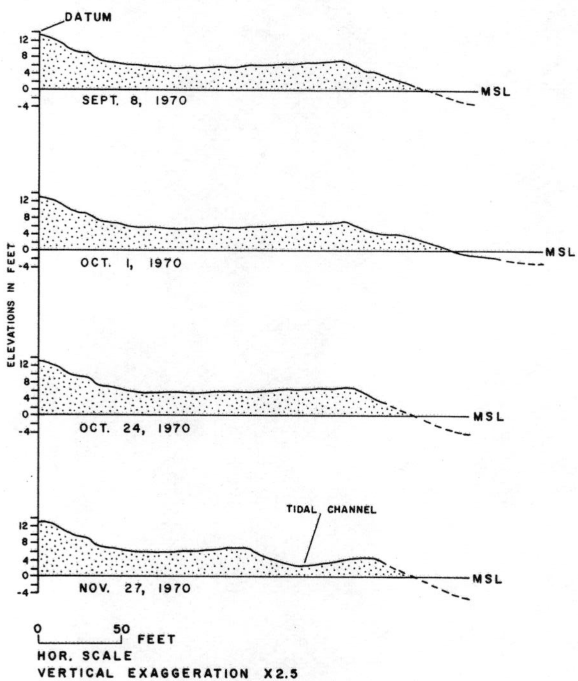

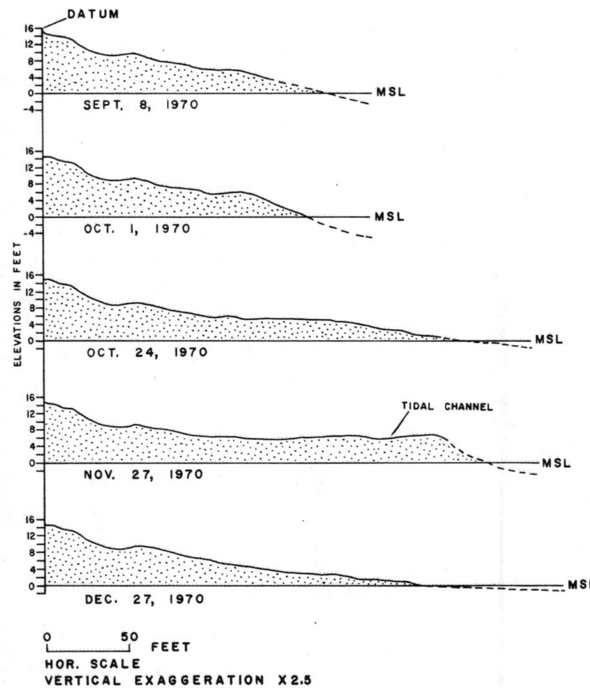

Figure 7. Station Y profiles. Dashed lines represent extrapolation of profile.

Figure 8. Station Z profiles. Dashed lines represent extrapolation of profile.

appear to be rather common features of Fire Island. A washover fan occurring in the study area (Fig. 2) is capped by wind-deposited sands which have since become stabilized by vegetation. The stratification of the washover fan possesses features that identify it (Fig. 11).

The suites of sedimentary structures which are characteristic of the dunes, berms, and washover fans of Fire Island, can be used to differentiate them in the field (see Table 4). Such distinction is important to interpret correctly the depositional history of recent or ancient barrier beaches.

Beach cusps and beach scarps

Beach cusps occur commonly on the Fire Island beaches, especially during the summer months. On one occasion the cusps disappeared following a storm on August 23, 1970. By August 25, 1970, an entirely new series of cusps had formed.

Beach scarps are less stable than cusps. Form and dimensions of scarps depend in part upon beach configuration prior to their formation. As a result, the height of a scarp may be either uniform or variable.

Scarps may occur adjacent to beach cusps. In one such case, beach cusps terminated abruptly along the beach face. Beyond this cusp termination point, a beach scarp originated and continued along the beach, following the perimeter

Figure 9. Berm subdivided by two beach scarps.

Figure 10. Clay lens (at base of exposure) overlain by dune sands.

Figure 11. Exposure of a washover fan occurring within the dune system. Cross-bedded dune sands (containing no pebbles) lie at the base of the washover fan. A thin veneer of wind-blown sand (containing plant roots) lies above the washover fan.

of a small protrusion of the berm. On the opposite side of the protrusion, the beach scarp terminated and a series of cusps began again.

THE RELATIONSHIP AMONG PROCESS, EROSION, AND MORPHOLOGY AT FIRE ISLAND

Wave action is the process which contributes most to the erosion of Fire Island. During winter months, waves undercut the dunes and redistribute the dune sediment upon the berm. Longshore transport contributes to erosion indirectly by removing material, eroded from the beach by wave action, from the area (westward). Neither wind nor human activity contribute directly to the erosion of Fire Island, or appreciably alter the berm, dunes, or stabilizing vegetation.

The morphology of Fire Island is affected most profoundly by wave action. Beach profiles surveyed along straight parts of the beach illustrate how the berm responds to changes in wave action. Increase in wave activity, which accompanies changing weather conditions, causes the berm to erode on both daily and yearly basis. A storm on October 2, 1970, generated waves which caused erosion shown on the reference points 1 and 2 October 3 profiles (Figs. 5 and 6), and resulted in beach scarp formation.

Yearly erosion of the berm is documented by decrease in berm width from summer to winter (Figs. 5 and 6). Such erosion is the result of severe and relatively frequent wave activity during fall and winter. Extent of yearly erosion is illustrated at reference point 2 (Fig. 6), where the entire berm eroded prior to December 29, 1970.

Segments of the Fire Island berm do not respond similarly to the same wave conditions, as revealed by comparison of reference points 1 and 2 profiles (Figs. 5 and 6) with stations Y and Z profiles (Figs. 7 and 8). The berm at reference points 1 and 2 had severe erosion from August 25 to December 29, 1970. During this same period, the berm at stations Y and Z had little if any net erosion. Since stations Y and Z were located on a crescentic shoreline form, the data suggest crescentic shoreline forms are rather stable when straight parts of the beach are being eroded. Beach scarps were not found to be prominent features in the vicinity of stations Y and Z, further suggesting the relatively stable nature of crescentic shoreline forms.

The dunes at Fire Island are no longer influenced directly by the wind, but now respond to wave action. The data of Table 1 indicate that the Fire Island dune system has experienced erosion since 1959. Because the dunes are stabilized and are no longer able to migrate landward, they can no longer escape marine erosion. Although the dunes are now relict, their superposition on clay clay lenses indicates they once actively migrated toward the mainland. The clay lenses which in places underlie eolian sand, represent material deposited in brackish-water marshes. Erosion of the dune ridge does not occur in equal amounts along the shoreline (Figs. 3 and 4). It is possible for such variations of erosion to be controlled by crescentic shoreline form position (Dolan, 1971), but no such forms were present within the area surveyed. The configuration of offshore topography may be responsible for the observed distribution of dune erosion. Zeigler, et al, (1959) suggest that configuration of offshore topography is a factor influencing erosion on Cape Cod, Mass.

Washover fans form under violent current conditions in a relatively short time interval, perhaps a few hours. Sedimentary structures of washover fans attest to a high energy depositional environment. Once having formed, washover fans become relict geomorphic features. Wind-blown sand accumulates above the fans and conceals them. This cap of eolian sand forms an undulating surface which may become stabilized by vegetation. In this manner washover fans become integrated into the dune system. Washover fans of Fire Island, barring a recurrence of conditions of formation, are now subject only to erosive wave action, as are the dunes.

The occurrence of beach cusps depends mainly upon wave conditions. Cusps were always present on the Fire Island beach during the summer, and were absent only on the few occasions when they had been destroyed by stormy surf. On similar occasions, the berm was also partially eroded, and a beach scarp was formed. Thus cusps may be completely or only partially destroyed by erosion. In the case of complete destruction, an entirely new set of cusps forms in response to the onset of normal surf conditions following storm activity. In the case of partial erosion, a knife-edge truncation abruptly terminates each apex of the old cusps and, following storm activity, a new set of cusps forms seaward of the old set. Thus, the partial destruction of the old cusps makes it possible for two distinct sets of cusps to co-exist. Formation of new cusps is independent of those formed earlier. The spacing of beach cusps depends on wave height at the time of cusp formation (Shepard, 1963). The mean spacing of study area cusps on July 28, 1970, was 72.4 ft, whereas mean spacing on October 25, 1970, was 107.8 ft. Therefore the summer and fall cusps apparently originate under different surf conditions.

Beach scarps develop from wave erosion. Conditions favorable to beach-scarp development exist during unusually intense wave action which normally accompanies storms. Two or more scarps may occur at the same time (Fig. 9). In such cases, the first-formed scarp originates during a higher tide than the second. The second-formed scarp originates within a few days of the first. Scarps adjacent to a series of beach cusps indicates that scarps and cusps can coexist under the same wave conditions.

CAUSES OF FIRE ISLAND EROSION

El-Ashry (1971) lists three major factors which may be responsible for erosion of east coast beaches: (1) increase in the number of hurricanes and other severe storms (2) recent eustatic rise in sea level, and (3) the interference of natural shore processes by man. Dunn and Miller (1960) state that an average of 8 hurricanes per year have occurred during the past 70 years. In the past 40 years, this average has increased to 9 per year; and during the past 20 years the average has increased to 10 per year.

Zeigler, et al, (1959) have discussed the effects of specific storms and hurricanes upon the beaches of Cape Cod. They concluded that it is sometimes impossible to correlate beach changes with specific wind conditions. They also state that waves generated by storms some distance away from a coast may have no relationship to local wind conditions. It therefore may not be sufficient to say that the increase in the number of severe storms is the principal factor responsible for the recent increase in erosion of the U.S. east coast.

El-Ashry (1971) concludes that east coast shore recession can be explained in terms of eustatic rise in sea level. Other workers also support the hypothesis that submergence and recession of the east coast is a result of eustatic rise of sea level. For example, Bloom and Stuiver (1963) have shown that the Connecticut coast has been submerged about 9 ft in the last 3000 yr.

Man-made structures, such as jetties and groins, interfere with the longshore transport of beach sediment. In effect, jetties and groins cause the up-current beaches to accrete, while the down-current beaches erode. Jetties and groins, as well as numerous other man-made structures, can cause serious erosion problems if they are not used wisely.

An additional factor which may be related to the recent increase of shoreline erosion of Fire Island and other east coast beaches is the possible lack of sediment available to beaches as a result of natural causes. For example, Taney (1961) states that a large difference exists between estimates of material being moved along the south shore Long Island beaches and estimates of material being supplied. It may be that additional sources of sediment were available in the past and that erosion has since depleted this surplus sediment.

To summarize, at least four major factors control beach erosion. These major factors influence the three natural processes (wind, wave, and current action) which act upon Fire Island. The natural processes can be considered as the direct erosion factors, since they are the actual mechanisms or erosion. The three direct and the four major controlling erosion factors, along with some possible effects of each, are itemized in Table 5.

FUTURE OUTLOOK FOR FIRE ISLAND

According to Swift (1968), the surf zone of a transgressing sea destroys part or all of its own marginal record of low-energy marsh and lagoon deposits and higher-energy barrier-island deposits. The resulting disconformity is called a ravinement (Stamp, 1922). Swift refers to this transgressive process as the ravinement process. He states that the ravinement mechanism is a result of the shoreward progradation of various marginal-marine facies. Swift reports the following aspects of the ravinement process which are displayed in the Bay of Fundy: (1) In the case where wave energy dominates over sediment supply, the spits (barriers) of the area move shoreward over the marsh deposits by means of shoreface erosion and washover. (2) The marsh deposits, which are overridden by the transgressing beach deposits, are exposed and eroded at the beach face. Additional erosion may occur by means of scour by offshore currents. (3) In the case where sediment supply exceeds wave energy, the barriers do not prograde shoreward, but build upward instead until shifts in sea level cause them to be abandoned.

Fischer (1961) has shown that the coast of New Jersey is undergoing ravinement as a result of post-Pleistocene sea-level rise. Evidence is provided by the remains of mollusks found on the beaches of the barrier islands. Marsh and lagoonal species are exhumed on the beaches by wave erosion as the barrier islands transgress over the lagoonal deposits.

This same ravinement process is occurring at Fire Island. The clay lenses (which represent a lagoonal marsh facies) underlying some of the Fire Island dunes are being eroded by wave action along with the dunes. As erosion progresses, shallow marine sediments will transgress the lagoonal sediments, and the dune and lagoonal sediments will continue to be reworked by wave action. Wave action thus predominates over sediment supply at Fire Island. If the erosion of Fire Island continues at the present rate, it seems rather unlikely that any part of the beach will avoid being reworked. Ultimately, all of the geomorphic features described in this study (berms, dunes, washover fans, beach cusps and beach scarps) will be destroyed by erosion.

CONCLUSIONS

The net effect of active processes operating within the Fire Island study area from 1959 to 1970 has been erosion. Wave action is the force most responsible for this erosion. Longshore currents contribute indirectly to the erosion by removing wave-eroded material from the area and transporting it westward. Wind action and human activity do not contribute directly to the erosion of Fire Island. It is possible that all four major erosion-control factors (number of hurricanes, eustatic rise in sea level, interference of natural shore processes by man, and lack of sediment supply) may be influencing the erosion of Fire Island.

Measurements on air photos show that the rate of landward erosion of the study area from 1959 to 1969 was on the order of 3 to 5 ft per year. Surveys and air photos indicate the amount of erosion occurring at a particular time is not the same for all parts of the Fire Island beach. The amount of erosion depends upon crescentic shoreline-form position. The configuration of offshore topography probably also governs erosion.

Fire Island is currently undergoing ravinement. The dune and lagoonal sediments are being eroded by wave action, and the shallow marine sediments are transgressing the lagoonal sediments. If the ravinement process and accompanying erosion of Fire Island continue, it is unlikely any part of the island will avoid reworking. The geomorphic features described in this chapter (dunes, berms, washover fans, beach cusps and beach scarps) will probably be destroyed by wave action and will not be preserved in the geologic record.

ACKNOWLEDGMENTS

This paper is a condensed version of a thesis submitted in partial fulfillment of the requirements for the degree of Master of Arts in State University of New York at Binghamton, 1972. The author wishes to thank the following individuals: Dr. Marie Morisawa, thesis advisor, Dr. Donald R. Coates and Dr. Leonard Vasher, members of the thesis examining committee, for their suggestions, comments, and evaluation of the original manuscript; the faculty and staff of S.U.N.Y. at Stony Brook Department of Marine Sciences, for providing living accommodations and assistance; Mr. L. E. Koppelman, and the Suffolk County Police Department for arranging a helicoptor flight over the study area; and the Smith Point County Park rangers for their cooperation.

REFERENCES

Bloom, A.L. and Stuiver, M. 1963. Submergence of the Connecticut coast: Science, v. 139, p. 332-334.

Bruun, P. 1962. Sea-level rise as a cause of shore erosion:Amer. Soc. Civil Eng. Proc., Jour. Waterways and Harbors Div., v. 88, p. 117-130.

Dolan, R. 1971. Coastal landforms:crescentic and rhythmic: Geol. Soc. Amer. Bull., v. 82, p. 177-180.

Dunn, G. and Miller, B. 1960. *Atlantic Hurricanes* : Louisiana State University Press, Baton Rouge, 326 p.

El-Ashry, M.T. 1971. Causes of recent increased erosion along United States shorelines: Geol. Soc. Amer.Bull., v. 82, p. 2033-2038.

Fischer, A.G. 1961. Stratigraphic record of transgressing seas in light of sedimentation on the Atlantic coast of New Jersey: Amer. Assoc. Petr. Geol. Bull, v. 45, p. 1656-1666.

Pierce, J.W. 1970. Tidal inlets and washover fans: Jour. Geol., v. 78, p. 230-234.

Shepard, F.P. 1963. *Submarine Geology:* Harper and Row, Inc., New York, 557 p.

Stamp, L.D. 1922. An outline of the Tertiary geology of Burma: Geol. Magazine, v. 59, p. 481-501.

Swift, D.J.P. 1968. Coastal erosion and transgressive stratigraphy : Jour. Geol., v. 76, p. 444-456.

Taney, N.E. 1961. Geomorphology of the south shore of Long Island, New York: U.S. Army Coastal Eng. Research Center, Tech. Memo. 128, 50 p.

Zeigler, J.M., Hayes, C.R. and Tuttle, S.D. 1959. Beach changes during storms on outer Cape Cod, Massachusetts: Jour. Geol., v. 67, p. 318-336.

TABLE 4. OCCURRENCE OF SEDIMENTARY STRUCTURES IN BERMS, DUNES, AND WASHOVER FANS

Sedimentary structure	Berms	Dunes	Washover fans
Cross stratification			
planar (large scale)	C	D	R
trough (large scale)	R	R	R
trough (small scale)	R	R	D
Laminations			
thick, greater than 0.5 in	C	C	C
thin, less than 0.5 in	C	D	C
disturbed	R	C	R
Heavy-mineral accumulations	A	C	R
Graded bedding	R	R	A
Large shell fragments	C	R	A
Pebbles	C	R	A

D - dominant, A - abundant, C - common, R - rare

TABLE 5. DIRECT AND MAJOR BEACH-EROSION FACTORS AND THEIR EFFECTS

Direct erosion factors	Effects of direct erosion factors	Major erosion-control factors	Effects of major erosion-control factors
1. Wind action	Movement of beach sand and unstabilized dune sand	1. Number of hurricanes and other severe storms	Increase in occurrence and intensity of wave and wind activity
2. Wave action	Movement of berm material seaward; undercutting of dunes and movement of dune material seaward	2. Eustatic rise in sea level plus local tectonic and oceanographic rise in sea level.	Bottom of nearshore littoral drift zone raises until covered with same depth of water at same distance from new shoreline as it was befoe the rise; material needed to raise bottom comes from corresponding shore area (Bruun, 1962)
3. Current action	Movement of beach material alongshore by longshore current action; movement of beach material offshore by rip currents	3. Interference of natural shore processes by man	Decrease in amount of material available to beach
		4. Lack of available beach material due to natural causes	Recession of shoreline

COMPARISON OF ECOLOGICAL AND GEOMORPHIC INTERACTIONS BETWEEN ALTERED AND UNALTERED BARRIER ISLAND SYSTEMS IN NORTH CAROLINA

Paul J. Godfrey and Melinda M. Godfrey

INTRODUCTION

The Outer Banks of North Carolina, a chain of narrow, sandy islands with a fringe of salt marshes and lagoons behind them, are one of the world's best examples of a continuous offshore barrier system (Fig. 1). First seen by Eruopeans during the explorations of Verrazano in 1524, the Outer Banks have long been used and abused by people. Today they are faced with increasing demands from recreation and development programs which threaten these islands more seriously then have all the years of use since white men first settled the region. Because of shortsighted development schemes combined with the belief among many coastal managers that stabilization of dunes and beaches is essential to preserve the shoreline, undesirable changes are being made in ecological and geomorphic processes which create and maintain these islands.

While such development pressure has been deplored by conservationists, there has been little research on the effects of the stabilization that necessarily accompanies development on the barrier island system. Fortunately, certain sections of the Outer Banks are still available in a natural condition to provide comparisons between developed and undeveloped barrier islands. Our research on the ecological and geomorphic changes of a natural barrier system (Cape Lookout National Seashore) has paralleled the geomorphic research of Dr. Robert Dolan (Chapter 12) on a developed beach (Cape Hatteras National Seashore), and we have come to the same general conclusions regarding the present condition of the Outer Banks independently and by different routes. These studies were initiated by the National Park Service to avoid the mistakes made at Cape Hatteras as new seashores became established, particularly Cape Lookout, and to plan development programs for barrier islands more in keeping with natural processes. Our goal is to provide guidelines for the recreational use of National Seashores that will fit into these processes instead of conflicting with them.

Department of Botany/National Park Service
University of Massachusetts, Amherst, Mass. 01002

CONTRASTS BETWEEN CAPE HATTERAS AND CAPE LOOKOUT

The northern section of the Outer Banks from Nags Head to Ocracoke Inlet comprises Cape Hatteras National Seashore, established in 1953, and the islands south of Ocracoke to Beaufort Inlet are authorized to become Cape Lookout National Seashore (Fig. 1). All these islands once looked much the same, judging from historical records, but since the 1930's the northern section has been progressively stabilized and developed. Much of the early stabilization efforts were made in the belief that the generally low islands with migrating dunes were in a deteriorating condition due to past misuse, although little evidence supports that view, except in localized places (Dunbar, 1958). Programs were instituted to plant beach grass, erect fences, and stabilize as much moving sand as possible with the general blessing of the scientific community, up to the present time.

The southern part of the Outer Banks, however, has largely been left alone except for grazing, which in places was apparently quite heavy. Livestock were removed from the Banks in the late 1950's with the exception of Shackleford Banks, where today herds of horses, cattle, goats, and sheep still hark back to the day when all of the Banks were used for grazing. While the signs of grazing pressure on Shackleford are quite evident, there does not seem to be a wholesale general deterioration of the island due to the livestock as some have maintained. What changes have occurred in the past century, such as migrating dunes covering forests, we believe are primarily due to hurricanes and natural recession. Grassland cover has increased on Shackleford since the early part of this century despite heavy grazing pressure.

Today the two sets of islands present a strikingly different appearance. The Cape Hatteras sections show a narrow and irregular berm, and continuous, artificial dune lines that are eroding in numerous places on every island (Fig. 2). Some of this erosion is serious enough to have prompted massive and expensive control projects, usually in front of some threatened facility. Nearly $½ million was spent during the early part of 1972 for beach nourishment at Buxton, only to have it all disappear 10 months later (Fig. 3). A $2 million project has been approved and is now in the planning stages to restore this same section of the beach. In spite of endless expensive erosion control efforts (between 1957 and 1968 the Park Service spent an average of $420,000 per yr on dunes, with additional sums of $1,200,000 to repair storm damage), buildings and roads close to the barrier dune are constantly vulnerable to destruction by the sea. Despite the increasing threat of eroding dunes and storm damage, buildings are now under construction just behind the barrier dune in those interior areas not owned by the Park Service. Unfortunately, the Service is committed by law to maintain the dune in front of these inholdings at public expense, Nearly all of Cape Hatteras National Seashore is an artificial system with a high, continuous barrier dune protecting the highway and towns and halting the natural process of barrier island recession, but not the erosion. Instead of "saving" the barrier islands, these programs seem to be threatening their very existence as natural

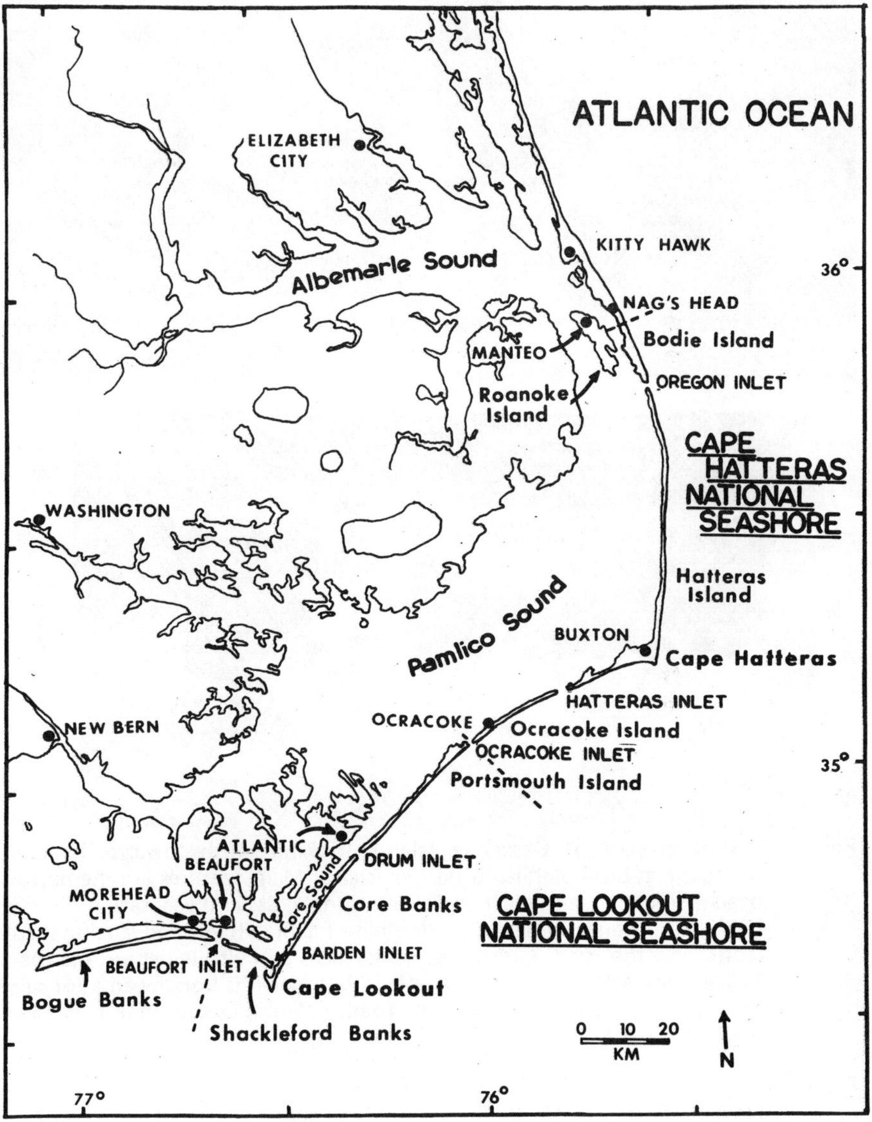

Figure 1. The Outer Banks of North Carolina, showing the locations of the Cape Hatteras and Cape Lookout National Seashores.

Figure 2. Eastern part of Ocracoke Island, looking south toward Ocracoke Village; a typical stabilized barrier island. Main features are the narrow beach, single, man-made, continuous dune; grassland vegetation directly behind the dune (in dense bands where fertilized) with advancing shrub lands; and the road. Old overwash terraces still visible behind the road. Beyond are salt marshes. Large pit in salt marsh is borrow area for sand to build secondary dike along the road. Atlantic Ocean on left, Pamlico Sound on right.

Figure 3. Buxton and the Cape Hatteras Lighthouse during a winter storm 17 November 1971. Dune line has been nearly destroyed in front of motels. Overwashes behind dune are plainly visible. Plastic sand bags to protect dune are on beach. Groins built opposite the lighthouse created the bulge in the beach line. By 17 November 1972, all of this beach nourishment material was gone and the beach is now essentially the same as this photograph taken a year previously. View is south towards Cape Hatteras. (Photo courtesy of Cape Hatteras National Seashore.)

features.

On the Cape Lookout islands, however, the berm is consistently wide, and the dunes are in a zone, rather than line, which is low and broken by frequent overwash passes (Fig. 4) and erosion occurs only in areas which are undergoing natural geomorphic change. The few permanent buildings are well back from the ocean (Fig. 5) and no money has been spent on erosion control in recent years. During major storms, ocean water floods across these islands without doing any real damage. Based on our research we believe these islands represent the natural condition of low barrier islands undergoing recession, and that they show a type of dynanic stability that has allowed their survival through time.

Sea Level and Storms

Both sets of islands are exposed to approximately the same environmental conditions. Of paramount importance are rising sea level and oceanic storms. Sea level has been rising about 30 cm per century during the past 5,000 years (Milliman and Emery, 1968), with a more dramatic rise of about 8 cm during the past decade (Hicks, 1972). Evidence for submergence of the coast and subsequent recession is present all along the Outer Banks. Low lying forests shown on maps of Shackleford Banks in 1850 are today salt marshes. Stumps and peat from an outcrop at the low tide mark on Shakleford were dated by the U.S. Geological Survey (Sample W-2307) at less than 200 yr B.P.; such outcrops are frequently found along the ocean side of the banks where they are exposed at low tide. Along with rising sea level, the regular oceanic storms are a major factor in this environment. The Outer Banks are in a particularly hazardous storm path (Mather, et al 1967) and front on a high energy sea; hurricanes frequently recurve up the East Coast and pass over or by the Banks. At least 150 hurricanes have affected the Outer Banks between 1585 and 1967 and there is no reason to expect these storms will suddenly stop (Carney and Hardy, 1967). Yearly, winter "northeasters" moving along this coast are often even more destructive than hurricanes. The storms create surges that invariably wash over the lower parts of the islands, where they meet little resistance on natural sections. However, the surges are prevented from overwashing the developed islands and instead erode the dunes. The natural response of a barrier island to storms and sea level rise is to retreat (Shepard, 1963; Hoyt, 1967), with wind and overwash moving the sand back. Such a pattern of natural retreat can still be seen on Core Banks, but no longer at Cape Hatteras. Nevertheless, the sea still tries to push these islands back and severe erosion results.

THE NATURAL SYSTEM – CORE BANKS

Core Banks (Fig. 4) is an example of an overwash barrier combined with relic inlet deposits similar to the barriers of New Jersey described recently by Kraft (1971). Storms maintain the wide, bare berm as waves dissipate their energy on the back shore and pass across the island. The berm is the "elbow room" for the

Figure 4. Central Core Banks south of Drum Inlet; a typical unstabilized barrier island. Features are a wide berm; low, open dune zone; grasslands of varying density on dunes and overwash terraces; scattered shrubs and thickets; salt marshes that formed on tidal deltas and overwash deposits. Ocracoke and other parts of Cape Hatteras National Seashore were similar before development. Atlantic Ocean on the left, Core Sound on the right.

Figure 5. Core Banks and Cape Lookout Lighthouse, looking south towards Cape Point. In contrast with Cape Hatteras, there is a wide berm and dune zone, such as Cape Hatteras had in the 1930's. The island is covered by water during storms, but no damage results.

storm surge. Whatever dunes or vegetation build up on the active berm during quiet periods are usually destroyed by the surge. Those dunes that survive are back far enough to escape the main wave force. As the overwash occurs, sand is carried from the beach and berm crest onto and across the berm to the interior of the island, where it is deposited as a veneer of sand in the form of a terrace or fan. Winds then blow the fine sand from the deposit into dunes, leaving a coarse shell pavement on the surface. Such layers of shell and coarse sand occur regularly in stratigraphic profiles and represent surfaces of the island in years past. As sand is moved into the interior of the island by overwash, and on dunes by wind, the height of the land increases and tends to keep pace with the rising sea. The buildup of the interior of Core Banks from overwash alone was measured relative to permanent beach marks set up in 1960 by the Corps of Engineers at 8.4 cm in ten years (Godfrey, 1970). The thickness of the overwash veneer and the distance it is carried is a function of the storm surge. Thin layers are added nearly every year, with major buildups during severe storms.

Based on historical aerial photographs and corings of the island, we have been able to construct a cross-sectional diagram of Core Banks showing surfaces approximatley as they were at previous times (Fig. 6). We know that the underlying marsh in the center of the island was exposed during the 1940's and was then buried by overwash. Oceanic shells indicate this material had to have come from the beach. During the late 1950's and early 1960's a series of storms pushed more sand across the island and created the present terraces. Natural vegetation and new dunes have developed on these terraces during the past decade (Fig. 7). Thus, much of the island consists of a series of quite recent overwash terraces. New dunes form on the terraces only to be buried or eroded by overwash in their turn. The sand, however, is carried farther back into the island during an overwash instead of being lost to the littoral currents, and so it is conserved.

Grassland is the typical vegetation of these barrier islands, with shrub savannas and thickets where protection by dunes or where overwash has not occurred recently. With long term protection, woodlands can develop. The presence of this grassland makes overwash a constructive rather than an erosive force. On some islands, such as Portsmouth, where there are no grasslands perhaps as a result of excessive grazing in the past plus severe storms, overwash is lowering the land surface rather than building it up. *Uniola paniculata,* (sea oats) is the primary dune builder and it will grow on the berm wherever its seeds are buried. As it grows, sand accumulates around the plant, and both the plant and dune grow vertically. The development of dunes by beach grasses such as *Uniola* has been well documented by ecologists. Other species have this ability as well; one very common grass, *Spartina patens* (salt meadow cordgrass) builds dunes as well as growing on flats and in salt marshes. It is a very versatile and resistant species that plays a major role in the bio-geomorphology of the barrier island. It grows rather sparsely in the passes between dunes, and on the dunes themselves, but becomes quite dense on the oldest, least recently overwashed terraces. Figure 8

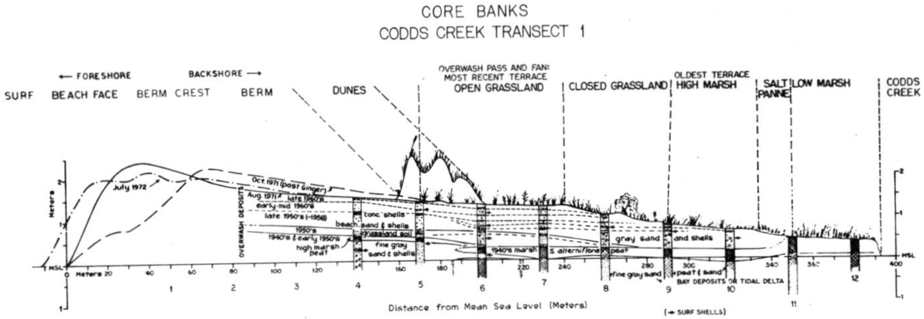

Figure 6. Stratigraphic diagram of Core Banks at Codds Creek showing locations of piston cores from which overwash strata and marsh deposits were identified. Vertical scale greatly exaggerated relative to horizontal scale. Changes of the beach profile due to Hurricane Ginger are shown; note overwash build-up and progradation of beach a year later.

Figure 7. Dune zone on Core Banks at the location of the transect shown in Figures 6 and 8. Note the open nature of the dunes, their rounded shape, and the relatively sparse vegetation of sea oats *(Uniola paniculata)* and cordgrass *(Spartina patens)*.

shows the general distribution of dominant species along a transect across Core Banks. It is quite clear that the vegetation is directly related to the overwash terraces, their elevation, and the frequency of overwashing those areas close to the berm and between dunes being overwashed most frequently.

Spartina patens is remarkably well adapted to an overwash environment. It is resistant to salt spray and salt water flooding. The tough wiry leaves and culms dissipate wave energy and catch sand. It is often completely buried, but will come up through new material by rhizomatous growth and begin a new grassland. Following an overwash burial it seems to grow more vigorously than before. Several artificial burials were made on *Spartina* grassland by covering the grass with clean beach sand in 1 m square boxes to depths of 10 and 20 cm. After two growing seasons, the growth of grasses in the boxes was the same as the controls outside. A major natural burial and recovery were observed following Hurricane Ginger in September 1971. The storm crossed the Outer Banks near Drum Inlet and storm surges deposited up to 50 cm of overwash sand on a *Spartina* grassland. A year later, it was almost impossible at first inspection, to tell that an overwash had occurred; the regrowth of the grassland was nearly complete. The overwash sediment was still there and the grasses grew up through the new layer. It is important to note, however, that for all their important roles in building dunes, dissipating wave energy, and trapping and colonizing overwash

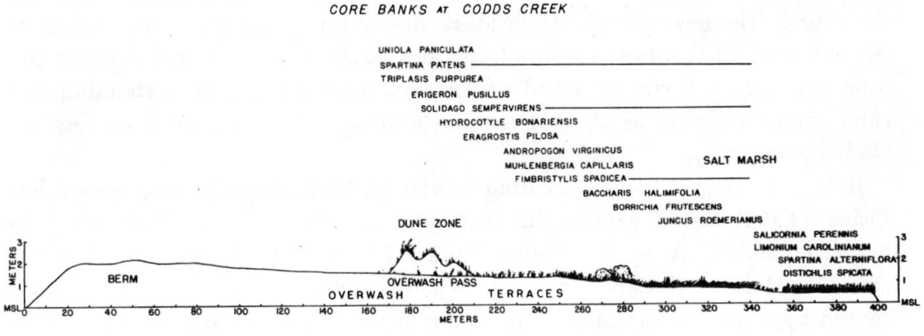

Figure 8. Diagrammatic vegetation sequence and elevation profile along Codds Creek transect. The position of a plant name indicates general zone of occurrence of that species with lines indicating more extensive distribution. Note that vegetation occurs in the overwash pass as well as on dunes and rear terraces.

material, *Spartina* and *Uniola* rarely completely stabilize the sand in the active zone, unless the area is well protected from oceanic forces.

Sand dunes play just as important a role as does overwash in barrier island dynamics. They keep overwash from being so frequent that the vegetation in the interior of the island has no time to recover between events. When a storm big enough to create overwash does occur, the dunes act as a first line of resistance. The waves are forced to dissipate some of their energy passing between the dunes, so that when they reach the inner part of the island they have slowed down and can deposit their sand load rather than cause erosion. Natural dunes can make the difference between overwash being a constructive or a destructive event. Another function of dunes is to serve as a sand reservoir; a really severe storm will flatten the first line of dunes on the berm and spread out their sand over the middle of the island. Low, broken dunes are most desirable in these respects, but even a continuous, artificial dune is better than none in some ways. For example, some parts of northern Core Banks, Portsmouth, and Ocracoke Islands became very low and bare apparently due to past abuse. Overwash was so frequent and erosive that no plants could take hold. Once on northern Core Banks and Ocracoke artificial dunes were built (Woodhouse and Hanes, 1967). However, succession proceeded to the point where a healthy vegetation cover now exists that is normally found on natural overwash terraces along the southern section of Core Banks behind the dune zone (Fig. 8); clearly, this low dune line improved the ecological conditions and the islands should now be able to withstand an overwash event without damage to the vegetation at this stage in succession. Indeed, this is now the case on northern Core Banks where the experimental dune is no longer maintained and is being pushed back in places onto the new grassland. Overwash is an example of a force which is constructive, but only if tempered, and the ability of the dunes and grasslands to temper it is one of the check and balance systems that keep the barrier island stable.

Overwash provides new substrate for salt marsh development. This happens when overwash completely crosses an island and sand spreads out as a fan into the sound. The new material is an ideal surface for colonization by marsh grass. An overwash of this type occurred on Core Banks 10 yr ago, and the new sand now supports a luxuriant stand of *Spartina alterniflora* with a standing crop three times as great as that of the neighboring, older marshes (Godfrey and Godfrey, in press).

It is also natural for a retreating barrier to be broken by temporary inlets. Fisher (1962) found geomorphic evidence, as well as historical accounts, for numerous inlets along the Outer Banks; 14 on Core Banks alone between Ocracoke Inlet and Cape Lookout. When a temporary inlet is open, usually caused by passage of a major storm and remaining open for as long as a century, or only a few decades, sand moves into the opening and spreads out to form tidal deltas. As the inlet closes, and the shoals become well developed, ecological succession begins and procedes at a rapid rate once the inlet is completely closed. The deeper flats are invaded by underwater beds of *Zostera* and *Ruppia* shoals exposed part of the time become *Spartina alterniflora* salt marsh, and the highest areas acquire a *Spartina patens* grassland in which small dunes begin to appear. These dunes may grow high enough to support shrubs and even woodlands in time. Figure 9 shows the changes at Drum Inlet which opened in 1933 and closed in 1971 and illustrates the general pattern of succession on tidal deltas. An inlet a few km south of Drum Inlet, Cedar Inlet, which was open in the last century, now has a typical salt marsh system on the former shoals and barrier island vegetation where the inlet used to be. Thus inlet dynamics combined with vegetation can greatly widen a barrier island in a relatively short time.

THE ALTERED SYSTEM – CAPE HATTERAS

On the islands of Cape Hatteras National Seashore, these natural processes have been greatly altered by the heavy hand of man. Much of this alteration – that is, dune building – in years past was done in good faith with the belief that such action was necessary to prevent the islands from being destroyed and to correct past abuses, with perhaps some justification in places. However, this approach has led to the attempts at total stabilization, the desire to hold everything in place, and to prevent the sea from washing over the islands. Such a program has come about to protect roads and buildings and with this development it has been necessary to try to hold all the sand in place and prevent dune movement, overwash, and the opening of inlets. A continuous artificial dune line has been built the length of the Seashore, planted with American Beach grass (*Ammophila breviligulata*) which must be fertilized and sprayed with biocides to combat insects and disease so that it produces a dense stand stopping all blowing sand. The beach grass is an exotic on the Outer Banks south of Nags Head while the native grass, sea oats, does not require the elaborate programs aimed at culturing the beach grass dune. As the beach grass dies out, sea oats take its place by natural propagation. The man-made dune is effectively a dike which stops all ordinary overwash. The beach has become very narrow and irregular, except at naturally accreting points such as the southwestern end of Ocracoke Island, and the south side of Cape Hatteras. Therefore storm waves impinge directly on most of the dune, since they cannot roll over the island, and rebound from the dune faces, eroding beach and dune sand into the littoral drift and out to deep water (Dolan, 1972). Grass cover cannot be stabilized or stand against this kind of erosion. The grasses are adapted to burial but not to undermining, and a dune that cannot retreat can only erode. Occasionally overwash occurs but it is hardly considered constructive (Fig. 10). There is usually an immediate response to plug the gap in the dunes and rebuild the dike.

The managed dune is strikingly different from natural dune zones. The use of fences, planting machines, and occasionally heavy equipment has produced a single, straight, steeply sloped structure compared to the rounded, irregular and scattered natural dunes. The heavy use of fertilizers produces a much denser

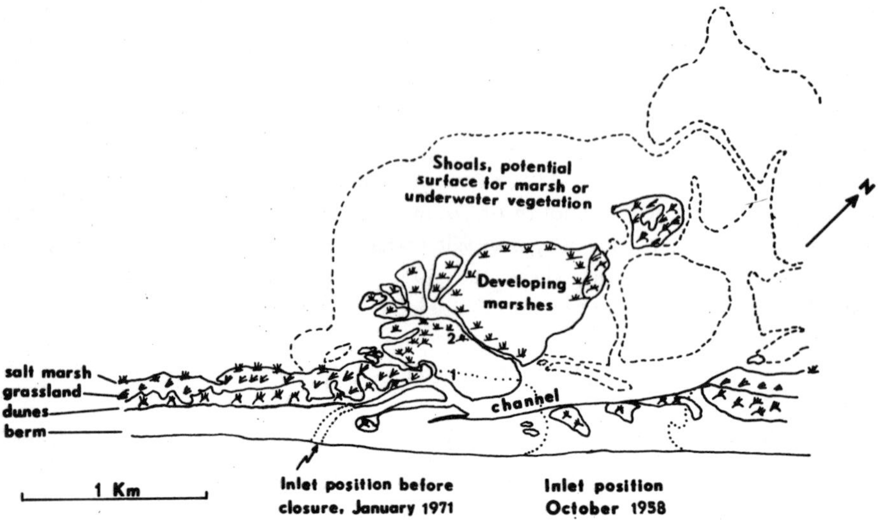

Figure 9. Map of old Drum Inlet showing inlet location in 1958, and in January 1971 just before closure. Shoals generally above low tide are being invaded by salt marsh cordgrass *(Spartina Alterniflora)*. Dunes are building on the spit as well as on some of the old shoal area. (From Godfrey and Godfrey, in press.)

Figure 10. Overwash occasionally occurs along Cape Hatteras National Seashore despite dune construction; Avon during a winter storm on 28 November 1971. Here flooding of property and damage to buildings, not on pilings, is considered most undesirable. (Photo courtesy of Cape Hatteras National Seashore.)

vegetation on the dune which effectively traps all moving sand (Fig.11). This probably accounts for the steep backslope. In most areas this dune is up to 6 m in height, the calculated height that is supposed to prevent most storm surges from going across the dune. The effect of the barrier dune on the vegetation behind it has been to accelerate ecological succession in the direction of species less tolerant of flooding and overwash burial. Shrubs and other plants which would ordinarily be expected in the most sheltered parts of the island are now coming up very close to the sea, along with normal dune plants. A storm severe enough to breach the dune would reverse this trend, eliminating all but the most tolerant species, such as *Spartina patens*.

Figure 12 is a cross-section of Hatteras Island at Sandy Bay near the area shown in Figure 11, illustrating the nature of the high man-made dune, with its dense vegetation and steep backslope, and the development of terrace-type vegetation in what was once the active dune and berm zone. Compare this profile with Figure 8. It is obvious that dune building on the berm has been a success here, but now there is concern over what will happen should the dune break. The road was built on what was once apparently part of the active beach zone; it is now below the level of the upper tide mark. The pre-development profile was probably much like Core Banks since similarities can still be seen in the basic profile, excluding the managed dune. The old overwash terrace corresponds to the major terrace on Core Banks and both probably formed during the same storm period. The low dunes to the right of the road are in apparently the normal position compared to Core Banks and probably formed before the main dune was built. Extrapolating the profile from these low dunes to the berm, suggests this was the active berm zone. The loss of the berm zone and natural overwash, because of the road and dune, has led to major difficulties in this section. Here the island is eroding from both sides – sound and ocean – since no sediments are moving back to widen the island and extend the protecting salt marsh fringe. An artificial widening process is being planned for the narrowest places by using fill and marsh propagation. (Godfrey, 1970b; Woodhouse, Seneca, and Brown, 1972).

Should a storm break through and overtop this dune line, sand from the dune will be pushed back over the road and onto the present vegetation. We do not know how long it would take for this system to recover from such a major burial; we do know that regular burial is easily survived by natural Outer Banks grassland. In the earlier stages of grassland succession, such as could be seen on the southwestern end of Ocracoke Island following the dune building programs there, recolonization should be fairly rapid since overwash tolerant species dominate in that stage. However, as time goes on and overwash is prevented, vegetation will grow which will be less likely to survive a major overwash.

To allow the normal opening and closing of inlets has of course been considered impractical on a developed barrier island. Inlets necessary for navigation are kept open by dredging, such as Oregon (which has been steadily migrating to the south), Hatteras, Ocracoke, Barden, and Beaufort Inlets, while

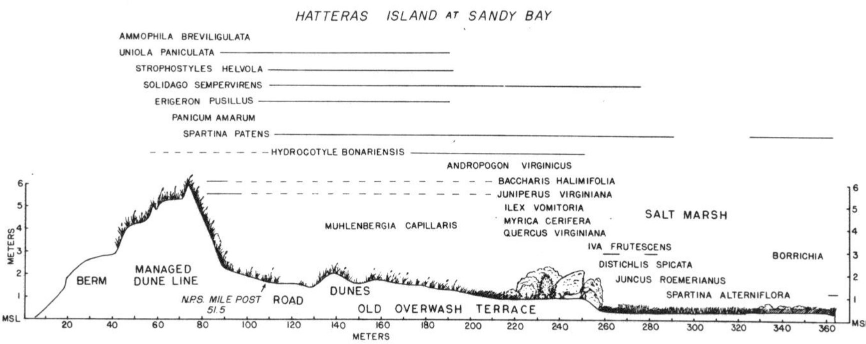

Figure 11. The man-made, managed dune line near the transect shown in Figure 12 on Hatteras Island at Sandy Bay. The dune is high, continuous, unbroken, eroding on the seaward side in numerous places, steeply sloped on the sound side, and covered with dense dune vegetation.

Figure 12. Elevation profile and vegetation sequence across Hatteras Island at Sandy Bay. Note the shape of the dune and the narrow beach compared to Core Banks. Dune vegetation is typical of the region except for the American beach grass *(Ammophila breviligulata)*. While the species distribution is similar to that of Core Banks, zones are much closer to the sea and much denser on the dune, a result of fertilization. While not listed here, typical salt marsh species such as those on Core Banks are present at Sandy Bay. Of particular note is the shrub zone developing along the backside and advancing seaward.

others are closed usually at great expense and effort, for example the inlet that opened just north of Buxton in 1962. An experiment in inlet creation was made on Core Banks by the Corps of Engineers following the closure of Drum Inlet. A new inlet was made opposite Atlantic where Core Banks was very narrow, and the spoils were placed behind the island in a more or less natural pattern. This spoil was then planted with marsh grasses in an attempt to stabilize the sand and create a new marsh thus widening the island at that point. A typical tidal delta has formed within the year that the inlet has existed and only time will tell if it will be navigable, which was the main reason for its creation.

The overall result of dune management at Cape Hatteras National Seashore is that the islands are eroding from the front but not retreating as a unit. Overwash and inlet closure are the two main processes which widen the islands, and also the two most important sources of new substrate for salt marsh development. Since the younger salt marshes are the most productive, the trophic condition of the sounds cannot help but suffer if present practices are continued indefinitely. Another function of overwash is to build up the interior of the islands; without it, the islands become not only narrower and narrower, but ever lower with respect to rising sea level. Management emphasis has been on replacing beach sand, trying to move sand from the back of the island to the front, while natural forces tend to move it from front to back. Recent projects have moved sand back up the beach from a downdrift accreting area so the sand could make another trip in the littoral drift.

The results of interfering with normal island dynamics can be seen in the contrast between the effects of Hurricane Ginger on the Cape Hatteras Islands and Core Banks, mentioned above (Dolan and Godfrey, in press). At Cape Hatteras and all along the developed beaches there was serious erosion, even though the storm passed over Core Banks. In contrast to Cape Hatteras, it was very difficult to see any permanent change on Core Banks and Cape Lookout (Figs. 13a, 13b). These islands were completely under water except for higher dunes, yet there was no damage. Instead, there was overwash buildup along the whole barrier island. The berm crest was pushed back, and the beach flattened; new sand was added to the backshore, and the level of the interior was raised; new fine sands were available for dune growth; and the grasslands recovered as usual from burial. Soon after the storm the beach was building seaward again and in 10 months had built out beyond its pre-Ginger position (Fig. 6). Meanwhile, at Cape Hatteras, the severely damaged dune scarp remained.

COMPARATIVE SEQUENCES ON NATURAL VERSUS ALTERED BARRIER ISLANDS

One of the main goals of our studies on National Park Service managed islands is to provide alternative management models, or options, that can be considered by the Service as it develops or reexamines master plans for such areas. We are

Figure 13. Core Banks at Codds Creek, showing effects of Hurricane Ginger September 1971. Figure 13a before storm and Figure 13b soon after. From the air little change could be seen, while at Cape Hatteras extensive dune erosion was evident. Ground surveys on Core Banks showed changes involving overwash deposition. Although the island was completely flooded except for higher dunes, it was very difficult to tell that a storm had passed over this area, other than by drift lines on the dunes. Dotted line indicates the location of profiles shown in Figures 6 and 8.

now in a position to compare natural coastal systems and those modified by man, and point out certain features that typify both. These comparisons can be used by resource managers to predict the type of situation they will have on a barrier island should certain fundamental decisions be made. Since geomorphic and ecological patterns vary from region to region and even within a single island, each system has to be considered individually.

Figure 14 compares a wholly natural barrier island system, such as Cape Lookout National Seashore, with a developed and stabilized system such as Cape Hatteras National Seashore. The patterns shown here in cartoon form can be readily seen on these seashores respectively, and illustrate the two basic types of change we are now witnessing – one controlled solely by natural forces, the other by man's interference with these forces. Two options are offered to the seashore manager – what happens if the natural system is allowed to operate, and what happens from stabilization efforts. Most of the information on beach processes at Cape Hatteras has been derived from Dolan (1972).

CONCLUSION

When a barrier system is not altered, its vegetation and geomorphic processes are closely interdependent. The grassland is adapted to prevailing conditions; its zonation depends upon elevation, distance from the sea, frequency of overwash, and salt exposure. This ecosystem is well able to withstand continuous natural stress, and can be expected to maintain itself in its present form for the foreseeable future.

An artificial system, such as Cape Hatteras, presents a disturbing contrast. Thanks to the barrier dune, both vegetation and developments are now very close to the ocean. If the dune were to be breached now, most of the plants right befind it belong to species which could survive overwash and flooding. If present successional trends continue, however, shrubs and small trees will take over the land just behind the dunes. These species are not able to handle overwash, and if a break in the dune occurs there will be a major ecological perturbation instead of the gradual change that we see on Core Banks. The ocean system can be expected ultimately to reassert its control of these islands, with inevitable human dislocations, unless prohibitively expensive engineering projects are undertaken. The ecological future of the area is not bright unless natural processes are allowed to restore themselves, at least partially; to deliberately allow this to happen would be politically difficult. Several alternatives are possible in managing the Seashores. Present undeveloped seashores should be preserved in their natural condition, such as Cape Lookout. On the managed seashores where no permanent structures are present, retreating beaches and dunes can be allowed to follow the natural course of events, but in a controlled fashion. Dune growth can be encouraged, perhaps as dune zones instead of lines, but it must be recognized that the dunes will retreat with the migrating island by the means of overwash and wind. It is necessary to get away from ideas that all dunes must stay in place, and, that storm flooding will destroy the island. A

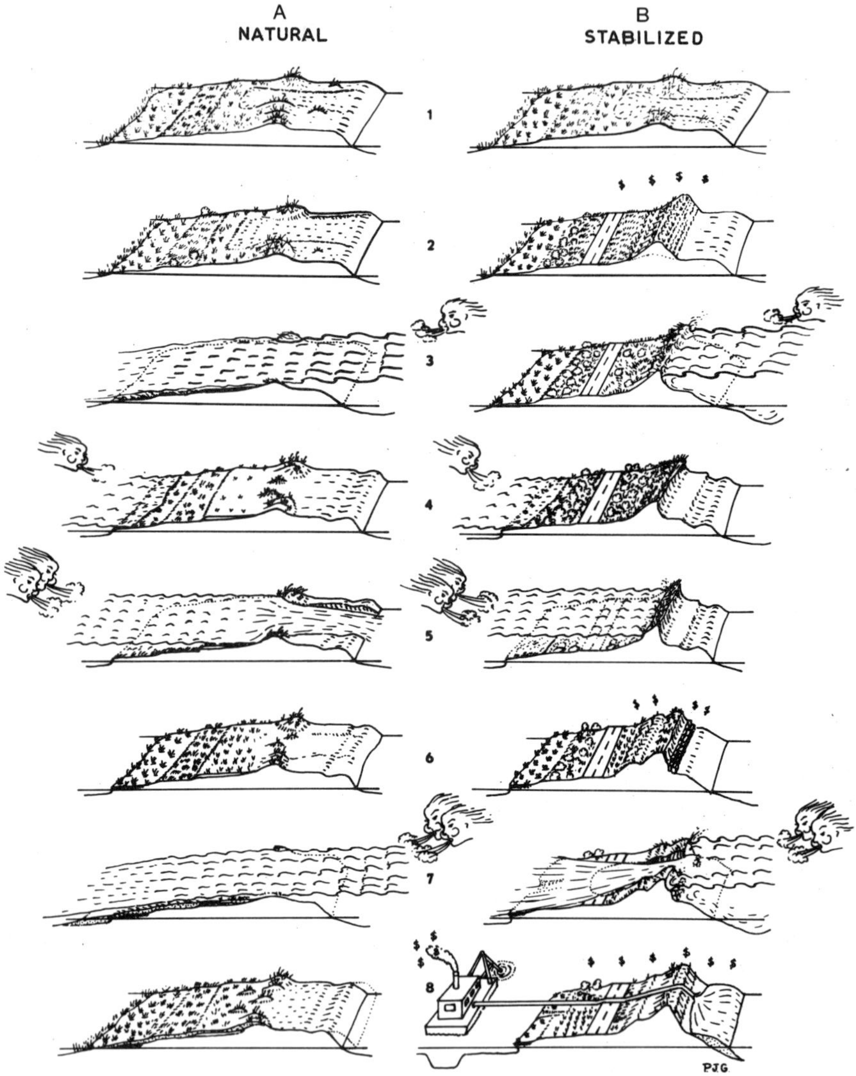
A
NATURAL
B
STABILIZED
1
2
3
4
5
6
7
8
PJG

controlled overwash passing through dune fields will build the island's interior and in the process beaches will widen. Islands can be widened artificially by techniques which mimic the natural patterns of retreat. It may become possible to predict weak areas that might open into inlets and plan ahead to bridge such areas before the event occurs. The important point is that dunes, overwash, wind, grasses, inlets, and storms are all part of a very flexible, changing environment and successful management must likewise become flexible. The lesson which emerges, however, is clear; man, not the sea, is the worst enemy of barrier islands today.

Figure 14. Comparison of altered and unaltered barrier systems. In 14-1A and 1B, the barrier systems are alike. 2A shows no change in the natural system. 2B shows the stabilization of an artificial dune to protect a road. 3A shows the affect of a storm overwashing the natural barrier. 3B shows erosion of the artificial dune and foreshore and nearshore zones by storm waves, enhanced by positive feedback. 4A the overwash terrace is raised and recolonized, while 4B shows badly eroded stabilized dunes and rebuilding berm. 5A shows waves from the sound overwashing the natural barrier. 5B shows the sound storm tide eroding the land side of the dune and damaging property and vegetation, supposedly protected by it. 6A shows the natural barrier undamaged by the storm, while 6B shows further human interference, rebuilding the dune artificially. 7A a very violent storm causes overwashing, but no permanent damage to the natural barrier. 7B the same violent storm breaches the artificial dune, with disastrous results to human installations. 8A shows new salt marsh growing on overwash sediment and stability of the natural barrier. 8B artificial beach nourishment, at great cost, attempts to repair the damage on the narrowed seaward slope of the artificial barrier and erosion of the marsh further narrows the barrier from the land side. Soundside dredging for borrow material has fortunately been discontinued in many areas, because sediments are not as sandy and easily wash away. Accreting spits, inlets, or other sources are now considered more suitable for beach nourishment.

REFERENCES

Carney, C.B. and A.V. Hardy. 1967. North Carolina Hurricanes: Weather Bureau, ESSA, U.S. Dept. of Commerce, Washington, D.C., 40 p.

Dolan, R. 1972. Barrier dune systems along the Outer Banks of North Carolina: A reappraisal: Science, v.176, p.286-288

Dolan, R. and P.J. Godfrey. In press. Effects of Hurricane Ginger on the Outer Banks of N.C.: Geol. Soc. of Am. Bull.

Dunbar, G.S. 1958. Historical geography of the North Carolina Outer Banks: La. State Univ. Studies, Coastal Series No. 3, La. State Univ. Press, Baton Rouge, 234 p.

Fisher, J.J. 1962. Geomorphic expression of former inlets along the Outer Banks of North Carolina: M.A. Thesis, Univ. of N.C., Chapel Hill, 120 p.

Godfrey, P.J. 1970a. Oceanic overwash and its ecological implications on the Outer Banks of North Carolina: National Park Service, U.S. Dept. of Interior, Washington, D.C.

Godfrey, P.J. 1970b. The use of *Spartina alterniflora* to control erosion and build new salt marsh: National Park Service, U.S. Dept. of Interior, Washington, D.C. 63 p.

Godfrey, P.J. and M.M. Godfrey. In press. The role of overwash and inlet dynamics in the formation of slat marshes on North Carolina barrier islands: Physiology of Halophytes Symp. Academic Press, New York.

Hicks, S. 1972. As the oceans rise: Shore and Beach 40(1): p. 20-23.

Hoyt, J.H. 1967. Barrier island formation: Geol. Soc. of Amer. Bull., v. 78, p. 1125-1136.

Kraft, J.C. 1971. Sedimentary facies patterns and geologic history of a Holocene marine transgression: Geol. Soc. of Amer. Bull., v. 82, p. 2132-2158.

Mather, J.R., R.T. Field, and G.A. Yoshioka. 1967. Storm damage hazard along the east coast of the United States: Jour. Applied Meteor, v. 6(1). p. 20-30.

Milliman, J.D. and K.O. Emery. 1968. Sea levels during the past 35,000 years: Science, v. 162, p. 1121-1123.

Shepard, F.P. 1963. *Submarine Geology:* 2nd ed., Harper and Row, N.Y., 557 p.

Woodhouse, W.W. Jr., and R.E. Hanes. 1967. Dune stabilization with vegetation on the Outer Banks of North Carolina: U.S. Army Coastal Engr. Res. Center Tech. Memo. No. 22.

Woodhouse, W.W. Jr., E. Seneca and S. Broome. 1972. Marsh building with dredge spoil in North Carolina: Bull. 445, Agr. Exp. Stat., N.C. State Univ., Raleigh, 28 p.

PART 3

APPLICATIONS OF GEOMORPHOLOGY

Man lives on the surface of the earth. Since geomorphology is the study of landforms and those processes that shape terrain, its application should have wide use in today's world. This section of the book contrasts two different aspects of the utility of geomorphic principles. Two chapters indicate the importance of geomorphology for coastal planning, and design of engineering structures; three chapters illustrate how geomorphology provides basic data for increased levels of understanding in other academic disciplines.

What happens in the coastal zone will be a matter of increasing concern to mankind during the 1970's, a period that is now being called "the environmental decade." The United States has 94,000 mi of shoreline, including the Great Lakes. All but one of the nation's 13 largest cities occur in the coastal zone, and 75 percent of the population live in coastal states. Many factors affect coasts and contribute to the way in which man uses them. Current demands are increasing for their use in industry, commerce, as home sites, and for resource and recreational development. The fragility of the coast environment, such as the beaches and wetlands, requires careful investigation and analysis to place this cherished real estate in modern perspective and harmony.

For thousands of years man has been accidentally or deliberately changing coastal processes and beach configuration. By 2,000 B.C. the Mediterranean shores were studded with harbors that had changed the coastal regime in Greece, Phoenicia, Egypt, and Syria. By 600 B.C. in Eretrea breakwaters were being constructed in 100 ft deep water with lengths exceeding 2,000 ft. During Caesar's rule of Britain, reclamation of coastal marshes was initiated, and by 1885 nearly 50,000 acres had been reclaimed in the Romney Marsh area of southeastern England. The largest coastal engineering project undertaken was the reclamation of new lands in the Low Countries where in the Netherlands about one-third of the 15,000 sq mi area of the country is the hard-won prize of the sea.

Until recently one of the largest reclamation works was the conversion of the Zuider Zee in the Netherlands into the Ijsselmeer, within new polder lands. The main dike was built between 1927 and 1932. By the 1980's when the scheme should be completed, 226,000 ha of new land will have been reclaimed, and the freshwater reservoir of the Ijsselmeer will cover a further 121,000 ha. A still larger project is the Delta scheme, now under construction. This scheme was initiated after the disastrous flood of 1953 and, by linking the outer islands, 25 km of dikes will take the place of 900 km of sea banks around the islands. This is a coast protection scheme and is not intended primarily for reclamation, although freshwater will be made available.

Another coastal project is the proposed water storage scheme in the Wash in east England. The largest of three possible schemes suggested would enclose the southern third of the Wash. The tidal flux of this area is 1½ times that of the Scheldt delta plan, and it would be the largest engineering project of any tidal area. It is likely, however, that smaller bunded reservoirs will be more in keeping with all the varied interests of the area. The feasibility of these schemes is now being investigated by the Hydraulics Research Station by means of mathematical and hardware models, linked with field observations.

Amounts of sediment reaching the ocean from inland areas is being influenced by two major tendencies. On the one hand, accelerated man-induced erosion of watersheds causes abnormally large sediment yields in some coastal regions (Orme). This is not a new problem because man was probably partly responsible for the 180 mi linear growth of the Tigris-Euphrates delta during the last 4500 yr. On the other hand, the 20th century has witnessed construction of many inland dams which trap sediment before it reaches the ocean. Thus, many beaches are becoming sand-starved, which is a growing problem, where in such areas as the delta of the Los Angeles River, new erosion occurred over a 12 mi stretch and beach recession reached rates of 50 ft/yr.

Knowledge of the dynamics and equilibrium of coastal regimes has advanced rapidly in the last two decades. Occurrences such as that at Santa Barbara, Calif. where construction of a breakwater caused coastal erosion of a 10-mi strip downdrift of the structure are becoming more rare. Unfortunately, some of the early efforts of beach stabilization (see Dolan) have set in motion almost irreversible reactions.

Prior to 1930 the federal government only performed coastal works for protection of federal property or improvement for navigation. In 1930 the Beach Erosion Board was authorized by Congress (Public Law 520, 71st) to undertake studies of beach erosion problems when so requested by other governmental levels. By the time of establishment of the Coastal Engineering Research Center (U.S. Army) in 1963, the Corps of Engineers had been given much broader mandates by Congress which included study and protection of public beaches (in 1946), and of private beaches if instrumental to the safety of public shores or for public benefits (in 1956). Between the years of 1946-64, the Corps undertook more than 100 projects for prevention of beach erosion that cost $180 million. In 1968 a study was initiated for evaluation of erosion problems on all U.S. coasts and resulted in a series of publications called the National Shoreline Study (such as Shore Management Guidelines, Regional Inventory Reports, and Shore Protection Guidelines, 1971).

Recently many regional areas have initiated their own studies. For example on Long Island, N.Y., the Nassau-Suffolk Regional Planning Board instituted a Regional Marine Resources Council that is evaluating six major problem areas of coastlines including, wetland management, shore erosion and stabilization, water use and resource development. Additional emphasis for regional planning has been provided by federal agencies through such arrangements as the Sea Grant Program, patterned after the Land Grant model, whereby states contiguous to the Great Lakes or oceans can apply for funds from the U.S. Department of Commerce (NOAA). Awareness of the importance of coastlines was evident in referendum issues on State ballots in the election of November 7, 1972. California voted for protection of the 1,072 mi shoreline from uncontrolled commercial development, and Washington voted for management of shorelines by State officials. However, Delaware voted against a measure designed to save the shoreline for recreation purposes.

Here in Part 3, Chapter 12 is a bridge with Part 2 and forms a companion treatise with the Godfrey's work of man-induced problems occurring on barrier islands. In many coastal areas, artificial construction of dunes is considered beneficial and necessary for stabilizing beaches and protection of buildings. However, in dune construction on these beaches, cited by Dolan, short range protection is apparently leading to long term deterioration. The un-altered parts of the beach are often several times wider than the beach in front of man-created dunes.

Chapter 13 deals with the evaluation of profile data for Long Island beaches and shows the importance of the geomorphology-engineering interface. Knowledge of the areal and temporal changes in beach profiles is a necessary prerequisite for planning of any type of engineering structure or artificial beach nourishment program. There are numerous case histories showing that failure of groins, breakwaters, and nourishment practices was due to inadequate geomorphic input (see Yasso in Coates, 1971).

With renewed government interest in coastlines, laws relating to littoral rights will increasingly come up for review. Much of the common law dates from a decision by Lord Tentreden in England in the 1820's. When a land owner brought suit because a town-constructed groin caused erosion of his property it was ruled that, "Each landowner may erect such defenses for the land under their care as the necessity of the case requires, leaving it to others, in like manner, to protect themselves against the common enemy." This philosophy of man against the elements (see also Dolan) is a recurring theme in much of the coastal protection literature. Instead of the combative role of fighting against nature, in many cases it would be preferable to work with natural processes. Nearly anything that man does in the delicate environment of coastal regions causes changes in the natural system. It should be one of the designated tasks of geomorphologists to provide information which will minimize these distortions and cause the least amount of environmental alteration.

In contrast to engineering applications, the last three chapters are devoted to the role of geomorphology when related to other academic disciplines. This change of theme is illustrated in Chapter 14 which shows the importance of geomorphology to archeology. On the Washington coast application of geomorphic principles concerning processes and landforms is necessary for a complete understanding of the cultural aspects of earlier man. Furthermore, appropriate landform analysis and the use of analog types is an additional exploratory tool in the location of suitable sites for new digs. Such knowledge can save the archeologist time, energy, and money because of higher probability ratios for success in a new site.

Chapters 15 and 16 indicate the value of geomorphology for interpreting coastal sedimentary rock sequences. Kraft et al show in their study of the Delmarva region how, by using a vertical sedimentary rock sequence model, a greater depth of understanding is reached between geomorphology and sedimentology. Not only does this interlocking relationship aid in determination of past events, but use of the model can also provide new insight into the prediction of the type and scope of continuing future coastal changes.

Chapter 16 closes Part 3, and this volume, by discussion of an environment quite different from others in this book; the action occurs on an island, and the materials instead of being largely siliceous sands, are of calcarenite composition. Vacher shows how an appropriate interpretation of horizontal rock sequences can be reached through knowledge of dune origin and morphology, and of accompanying coastal processes during their formation. Thus geomorphic principles provide a significant data base by which sea level history can be reconstructed and the sedimentary rock history deciphered.

Donald R. Coates

Chapter 12

BARRIER ISLANDS: NATURAL AND CONTROLLED

Robert Dolan

ABSTRACT

The Outer Banks of North Carolina extend from the Virginia-North Carolina border south to Cape Lookout, a distance of 240 km. These islands were originally one continuous biophysical system, but the upper section has experienced three decades of dune and beach stabilization, coupled with extensive public and private development. In 1957 the stabilized islands became the Cape Hatteras National Seashore. The southern segment, which includes Core Banks, remains in a natural state and is authorized to become the new Cape Lookout National Seashore. The responses of these two different systems to the inshore processes are of special interest for both preservation and management reasons.

Natural barrier islands are much better adapted to steady-state processes and extreme events than are the man-manipulated islands. Since there is little resistance to the storm surge movement across the natural barriers, wave energy is dissipated across the wide berm, among the low dunes, and finally in the grasslands and marshes behind. These islands actually gain material from the beach as the surge moves across the islands, and such deposits serve as sources of supply for new dune growth.

Within the stabilized segment of the Outer Banks, the massive and unbroken man-made dunes act as an impenetrable obstruction to a storm surge, so the energy dissipation and sediment transfer processes are significantly different. The most important differences are associated with the restrictions of the run-up profile and the elimination of oceanic overwash. Overwash is the major means by which low barrier islands retreat before the rising sea. It is the only way massive quantities of coarse sediment can be moved inland from the beach.

INTRODUCTION

Man's role as an agent of landscape change was a topic of interest of geomorphologists long before the 'environmental crisis' became a national issue in the United States. The lists of geologists who have made reference to the

Department of Environmental Sciences, University of Virginia, Charlottesville, Virginia 22903

effects of man is long and distinguished, but I find R. F. Peel's statments (1952, p. 264), in a rather obscure book for Americans, is one of the most articulate:

"The earth is not an inert and haphazard collection of materials that can be altered piecemeal without danger. Land, water, and air are all active, and all interact on one another through a host of connected processes. The physical environment of any region is somewhat like a highly complex machine, and a machine which is often rather delicately balanced. Unthinking alteration of one part of it will inevitably affect others, and may set in motion a sequence of changes having unforeseen consequences which may spread far beyond the immediate locality. Man has been producing such changes on the face of the earth for thousands of years through his practices of agriculture and animal husbandry, and the effects, until recently little heeded or even suspected, have in many areas been wholly disastrous. In considering the contribution a knowledge of physical geography can make to human welfare we may thus usefully glance at the effects, often quite unintentional, of man's interference with nature in the past."

Problems within the coastal zone offer excellent opportunities to investigate the impact of man's manipulations of the natural environment. This is a case study of an investigation conducted along the Outer Banks of North Carolina.

BACKGROUND

For many years the policy of the Army Engineers, and some of the other land management agencies, has been to control or prevent certain natural events that were considered to be harmful or destructive. In fact, this effort to control the beach-energy system has been described in terms of actual "combat" with nature. For example, in the Coastal Engineering Research Center the U.S. Army Corps of Engineers concludes its technical report (1964), *Land Against the Sea,* with this sentence (p.43): "Our campaign against the encroachment of the sea must be waged with the same care that we would take against any other enemy threatening our boundaries." I believe this strategy led to unbalanced conditions which eventually culminated in undesirable shoreline responses.

Through experience with sea walls, groins, land fills, and much trial and error engineering, we are now beginning to realize that natural change is often essential to the maintenance of geologic and ecologic systems; however, it is also clear that an uncontrolled natural system, undergoing periodic perturbation, creates serious land management problems. This means that extensive areas must somehow be maintained safely for those who wish to use them for recreational purposes, and yet these same systems must be allowed to remain in a dynamic state.

This management contradiction is a serious problem in areas subject to the greatest physical stresses – the Barrier Islands of North Carolina provide an excellent example of the dilemma land managers face today along the mid-Atlantic coast (Fig. 1). The low profile of these islands, their narrowness, sand compostion, exposure to high wave energy, coupled with a gradually rising sealevel, have created a state of almost continual physical and ecological change (see also Godfrey, this volume).

OUTER BANKS

The Outer Banks extend from the Virginia-North Carolina border south to Cape Lookout, a distance of 240 km (Fig. 1). Originally, these islands were one continuous biophysical system, but the upper section has been altered by more than 40 yr of dune and beach stabilization, coupled with extensive shoreline development. In 1957 the stabilized islands became the Cape Hatteras National Seashore. The southern segment of the Outer Banks, which includes Portsmouth Island and Core Banks, remains in essentially a natural state and is authorized to become part of the new Cape Lookout National Seashore. The responses of these two systems to waves and storm surge are of special interest to the National Park Service for both preservation and management reasons. They can also be compared to determine some of the geologic implications of man-made modifications, as well as the economics associated with maintaining altered environments.

THE NATURAL SYSTEM

The earliest known descriptions of the Outer Banks date from the late 18th century; however, by this time the islands had been settled, and cattle, sheep, horses, and hogs had been introduced for at least 100 yr (Dunbar 1958). Therefore, some uncertainty exists concerning the extent to which the islands were originally more extensively vegetated, and that grazing and woodcutting by the early settlers combined to reduce the island cover or grasses and trees, creating a somewhat desolate, unnatural condition.

The presence of man and domesticated animals have surely had some effect on the vegetation of the Outer Banks; however, the observations of Price (1926), Welch (1885), Pinxhor and Ashe (1897), Kearney (1900), and Engels (1942), all indicate that the islands have always been sparsely vegetated. For example, the regular occurrence of storm surges, oceanic overwash, and extreme tides from Pamlico Sound preclude the development of permanent forests except at a few broader areas.

The unaltered barrier island system adjusts to periodic storms since there are no natural obstructions in the path of the waves and surges. Most of the initial storm stress is sustained by the broad beaches. Because no resistance is created by impenetrable landforms, water flows between the dunes and across the islands with the result that energy is rapidly dissipated. On the sound side the fringes of marsh act as a buffer to reduce erosion from waves and surges

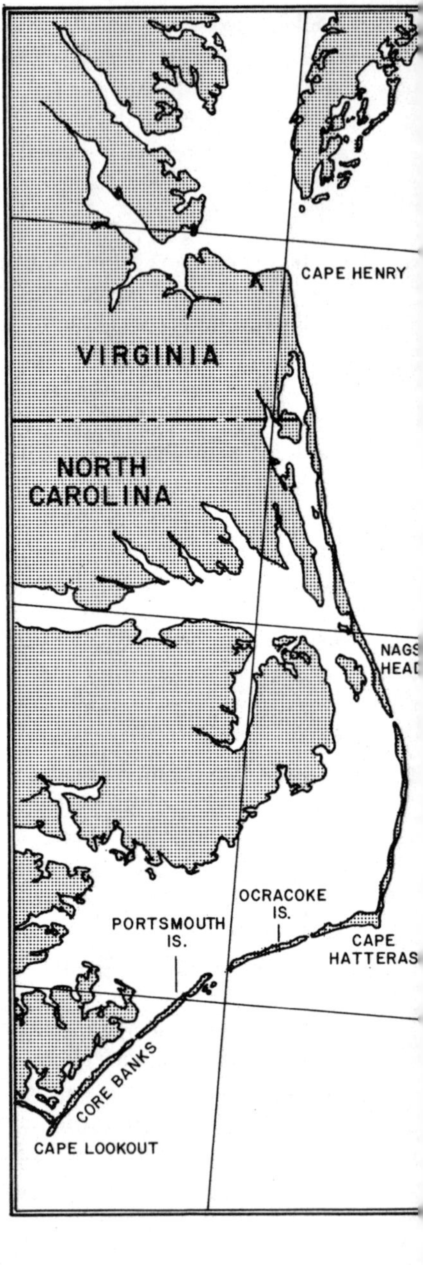

Figure 1. The North Carolina barrier island system. Cape Lookout is the southern segment and Cape Hatteras the northern reach.

Figure 2. Sand fencing with natural materials in the 1930's. Photo by the National Park Service.

generated on Core and Pamlico Sounds.

The combination of extreme high tides and large waves occasionally succeeds in eroding the low-lying beach foredunes, carrying sediment completely across the island and into the marshes. This process of "oceanic overwash" has been well documented by Godfrey (1970). Godfrey has also documented the important role overwash plays in marsh formation by replenishing sediments and creating new land on the sound side of barrier islands. Because of steadily rising sea level (Hicks, 1972), the beaches along the Core Banks have in most places receded, resulting in increased wave attack on the dunes and subsequent overwash and build-up of the marshes. The net effect has been a gradual westward movement of the islands. During severe storms inlets may be closed.

THE STABILIZED SYSTEM

The frequency of destructive storms along coastal North Carolina, with ocean overwash, precluded the establishment of a permanent road network on the Outer Banks until the 1930's. It was determined at that time to construct a protective dune system between the proposed road and the beach. In the period between 1936 and 1940, the CCC and WPA, under the direction of the National Park Service, erected almost 1,000,000 m of sand fencing to create a continuous barrier dune along the Outer Banks (Fig. 2) – including Hatteras, Pea, and Bodie Islands (Stratton, 1940). Most of this construction took place in the zone comprising the original low beach dunes and a strip 30-100 m wide behind the foredune. The sand which collected around the fencing was further stabilized with some 2,500,000 trees and shrubs, and enough grass to protect 3,254 acres of dunes and sand flats. This was augmented in the late 1950's by the National Park Service, so that at present almost a continuous mass of vegetation blankets the barrier island from south Nags Head to the southern tip of Ocracoke Island (Fig. 3).

COMPARISON: CONTROLLED VERSUS NATURAL SYSTEMS

Thirty years of artificial dune stabilization has greatly altered the geological features of the Cape Hatteras sector of the Outer Banks. A comparison of cross sections of Hatteras Island (Fig. 4), indicating the altered condition, with the profile of Core Bank representing what many interpret as the natural condition, demonstrates the extent to which stabilization has brought changes in the beach and dune morphology. Analysis of aerial photographs shows there is a most striking difference between the natural and the controlled barrier islands. In addition the artificial barrier dune system is associated with a change in beach width. The unaltered parts have beaches ranging from 125 to 200 m wide, averaging about 150 m. Along many of the Hatteras and Bodie Islands beaches which were altered 30 yr ago, the shoreline has receded to widths of 30 m or less. Ocracoke Island, which was extensively altered 10 to 15 yr ago, has intermediate width beaches with barrier dunes, ranging from 50 to 100 m, and averaging 75 m. This beach narrowing process, combined with the presence of permanent barrier dunes, has created a situation in which high wave energy is concentrated in an increasingly restricted run-up area, resulting in a steeper beach profile, increased turbulence and accelerated sediment attrition. The net effect is increased erosion and further narrowing of the beach. Ultimately, the beach face becomes so narrow that wave up-rush is applied directly to the stabilized dune (Fig. 3). This is occurring at several places within Cape Hatteras National Seashore, and has resulted in a gradual undercutting of the dune front, with eventual destruction of the dune system and endangering man-made structures behind the dunes (Fig. 5).

Coarser beaches are steeper, and if they become steep enough to initiate wave reflection, a new set of interactions is started which may result in a

Figure 3. Stabilized barrier dunes of the Outer Banks. This photo was taken in the early 1940's—note that already the shoreline is moving in on the barrier dune system and that overwash is all but eliminated.

concentration of the energy dissipation process in narrower zones. The increased stress and turbulence across the narrower beach results in higher attrition and winnowing rates and leads to accelerated losses of fine sand (1.25 mm).

There are few plants that can tolerate the extreme conditions in the vicinity of a beach subject to high wave and surge action. Figure 4, the representative transect across the Core Banks, illustrates two important points: (1) a wide zone of no vegetation between the shoreline and the extreme backshore and (2) low dune and overwash deposits in the central part of the island. In contrast, along the transect across the stabilized and fertilized dunes on Hatteras Island, grasses may be as close as 10 to 20 m of the high tide line, because of increased protection from storm surge. The great height of the stabilized dunes, up to 10 m, provides a region of protection from salt spray and flooding, allowing extensive growth of shrubs within 30 m of the high water mark.

The high stabilized dunes not only divert salt spray from the zone immediately behind the dune, but they also prevent flooding and overwash. Because of this protection the shrub community normally found near the back of the island has spread seaward, and in many places forms an impenetrable thicket 3-4 m high (Fig. 6). The most successful shrubs in this invasion have been wax myrtle, silverling, yaupon holly, and salt marsh elder. Attempts are now underway to check the spread of shrubs by several methods including controlled fires.

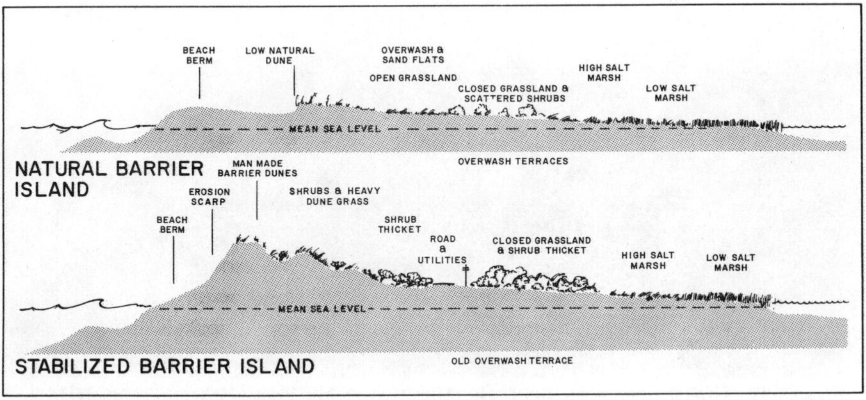

Figure 4. Cross sections of the barrier islands. The upper diagram is typical of the natural systems and the lower typical of the stabilized barriers of North Carolina.

Figure 5. The motel area near Buxton (Cape Hatteras) on Hatteras Island. All that remains of the one-time barrier dune is the sand fencing. The sands have been stripped away by wave action.

Figure 6. A dense scrub growth has completley replaced open sand flats and marsh grasses on Bodie Island. In the 1920's this area was open grasslands and shifting sands.

DISCUSSION

From a geomorphic point of view, the Outer Banks of North Carolina are one of the most dynamic systems in nature. Because oceanic overwash plays an essential role in this process, an unbalanced situation develops wherever artificial barrier dunes have been built. Further compounding the seriousness of the situation has been the false impression of safety and stability offered by the barrier dune. Numerous structures, including motels, restaurants, beach cottages, park facilities, and a U.S. Naval Base at Cape Hatteras have been built immediately behind the barrier dune in the belief that the dunes provide permanent protection from encroachment by the sea. Instead the beach has steadily narrowed. Subsequently, the barrier dune has eroded away leaving these structures with little protection from extreme events.

When Hurricane Ginger moved inland near Cape Lookout, N.C. on September 30, 1971 it provided an excellent opportunity to compare the impact of a severe storm on the two adjacent barrier island systems. Simultaneous field investigations were underway at both Cape Hatteras and Cape Lookout, and NASA (Wallops Island) provided before-and-after Ginger color I.R. aerial photography.

The patterns of change along most of the Core Banks, as determined from the Wallops Island aerial photography and site visits were similar – flattening of beach face, retreat of berm crest, burial and deposition on the backshore, entrapment of sand by grasslands, and general flooding.

Wihin the stabilized segment of the Outer Banks, the massive and unbroken man-made dunes acted as an impenetrable obstruction to the surge, so the energy dissipation and sediment transfer processes were significantly different. Photographs taken three days after the seas of Ginger had subsided (Figs. 7 and 8), and the before-and-after profiles (Fig. 9), illustrate the important implications of these differences. The man-made dunes absorb the full impact of the surge and storm waves and erode rapidly. The run-up profile is restricted to the narrow beach face. Sands are not transferred inland, but rather offshore or alongshore. In the case of Ginger, recession of the seaward face of the barrier dunes near Cape Hatteras averaged between 3 and 5 m.

The storm drove sea water completely across the islands from Cape Lookout north to Portsmouth Island (Fig. 1). Drift lines in the vicinity of the Cape Lookout Lighthouse were up to 0.7 m above that level 43 km to the northeast on Portsmouth Island. The surge was slightly lower near Cape Hatteras.

Profiles surveyed at 2 m intervals across Core Banks and Hatteras Island during the summer of 1971 were resurveyed shortly after the hurricane and again during the summer of 1972. Elevation changes of the berm surface relative to bench marks on the Core Banks profile were also measured. Figure 9 allows comparison of before-and-after profiles along two typical transects, illustrating an increase of land elevation on Core Banks and severe erosion on Hatteras Island. On Core Banks, the storm flattened the beach face by moving sand

Figure 7. Post-Hurricane Ginger on Core Banks three days after the storm. Note the wide beach berm and stringers of overwash sand.

Figure 8. Post-Hurricane Ginger on Hatteras Island. The abrupt scarp of the man-made dune is clearly evident.

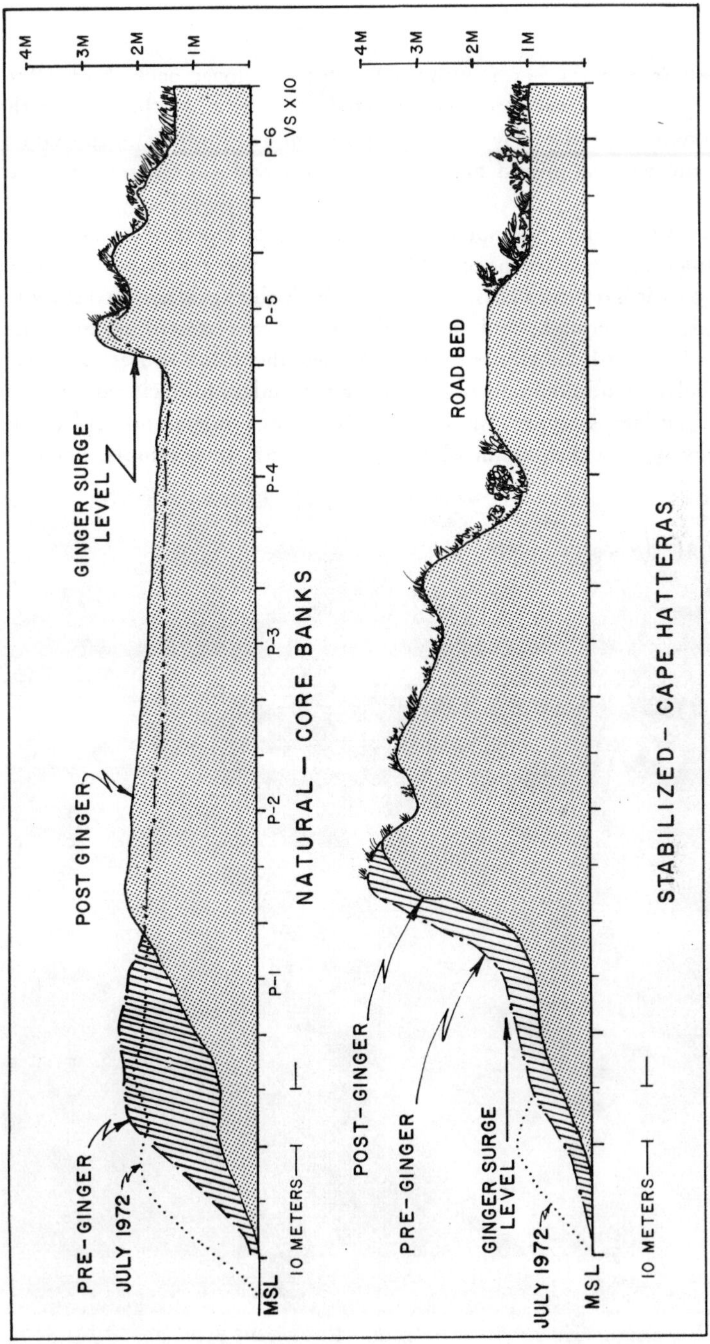

Figure 9. Profiles of Core Banks and Hatteras Island before and after Hurricane Ginger.

offshore as well as over the berm, with the beach crest moved back about 40 m. However, the mean sea level intersect retreated very little. At the time of the first resurvey, nine days after the storm, a new berm was forming. A substantial part of the berm material moved onto the backshore and was transported up to 100 m inland.

Oceanic overwash, and the opening and closing of inlets, creates serious problems in maintaining a permanent highway down the center of the Outer Banks. In the past it has been necessary to clear the highways when covered with sand deposited by overwash, and the roads have been rerouted several times when erosion destroyed the dunes and threatened the fixed routes (Fig. 10). Bridges have been abandoned and roads built where inlets have closed. In other cases it has been necessary to fill in recently formed inlets and replace the destroyed highway. The Ash Wednesday storm of 1962, for example, opened a

Figure 10. The receding shoreline has forced the relocation of several sections of the main road along the barrier islands.

new inlet between Buxton and Avon which required $1.4 million to rebuild the dune system and replace the roadway. This same storm destroyed segments of over 15 mi of the artificial dune system which also had to be rebuilt.

Although the present system is undependable, endangered, and expensive to maintain, alternatives are even more expensive and somewhat questionable in terms of application and economics. One approach has been to attempt to maintain the beaches by constructing groins at right angles to the beach, or by dredging sediments and pumping them into the beach (Fig. 11). The cost of extensive groin fields runs into millions of dollars, and has not been very effective on the Outer Banks at Hatteras Lighthouse. Dredging and beach nourishment may cost $1 million per mi, and in most cases this too is only a temporary measure.

Another suggestion has been to build a reinforced dune system at critical sites

Figure 11. The groin field installed near Cape Hatteras had design problems even before it was completed. Note the structural failure in the foreground.

by forming sea walls of sand bags. A structure such as this has been estimated to cost about $4 million per mi and ignores the basic fact that once the beach is gone, nothing will stop surge and surf action.

Survival of the development on the natural beach environment along coastal North Carolina requires a strategy of submission and rapid rebuilding. This is the key to success. Unfortunately, man has attempted to draw a line and prevent the sea from passing. The results have been unexpected and negative. Because the Cape Hatteras portion of the Outer Banks has already developed to the point where it is impossible to remove the highway, it must be maintained; however, as the system continues to narrow, new instances of overwash, erosion of the artificial barrier dunes, and inlet formation can be predicted. Many of the structures which have been built in the proximity of the shoreline will be lost, and the highway will require relocation within a few years (Fig. 12).

The Cape Lookout section of the Outer Banks, on the other hand, presents an entirely different situation. The islands from Portsmouth Island south to Cape Lookout and then west along Shackleford Bay are undeveloped. There are no highways, utilities, or permanent settlements to protect. Placement of a permanent roadway down the island would require a continuous artificial dune system for protection and stabilization. The National Park Service plan is, however, to leave Portsmouth Island, Core Banks, and Shackleford Banks in a

Figure 12. The final chapter of man's confrontation with nature.

natural state.

CONCLUSIONS

With the rapid deterioration of the barrier dune systems along the Outer Banks of North Carolina in recent years, and the large expenditures of resources necessary to reestablish or maintain them, or both, my research suggests that now is the time to review the basic concept of dune construction in light of the geological and economic implications. If sea level continues to rise, as all evidence seems to suggest (Hicks, 1972), the resources required to maintain extensive areas of barrier dunes may exceed the economic and psychologic value attached to their existence. The barrier islands, in their natural condition, can survive severe perturbations of tropical and extratropical storms by the low resistance they present to storm surges. These natural islands are not being washed away, but rather they are moving back by processes that were fundamental in their origin, processes that continue to be important if they are to be preserved in a natural state.

ACKNOWLEDGEMENTS

My colleagues in this effort are Paul Godfrey and William Odum. This research is supported by the Office of Natural Science, National Park Service, Washington, and the Chesapeake Bay Ecological Studies Program, National Aeronautics and Space Administration, Wallops Island, Virginia.

REFERENCES

Brown, R.H. 1948. *Historical geography of the United States:* Harcourt, Brace and Co., N.Y. 596 p.

Dolan, R. 1972. Barrier dune system along the Outer Banks of North Carolina: a reappraisal: Science, v. 176, n. 4032, p. 286-288.

Dunbar, G.S. 1958. *Historical geography of the North Carolina Outer Banks:* Louisiana State University, Coastal Series No. 3, Baton Rouge, 234 p.

Engels, W.L. 1942. Vertebrate fauna of the North Carolina coastal islands, a study in the dynamics of animal distribution. I:Ocracoke Island: American Midland Naturalist, v. 28, n. 2, p. 273-304.

Engels, W.L. 1952. Vertebrate fauna of the North Carolina coastal islands. II:Shackleford Banks: American Midland Naturalist, v. 47, n. 3, p. 702-742.

Godfrey, P.J. 1970. Oceanic overwash and its ecological implications on the Outer Banks of North Carolina: National Park Service, Office of Natural Science Studies, Washington, 32 p.

Hicks, S.D. 1972. As the Oceans rise: NOAA, v. 2, n. 2, p. 22-24.

Kearney, T.H. 1900. The plant covering Ocracoke Island:a study in the ecology of the North Carolina strand vegetation: Contributions, U.S. National

Kearney, T.H. 1900. The plant covering of Ocracoke Island: a study in the ecology of North Carolina strand vegetation: Contributions, U.S. National Herbarium, v. 5, p. 261-319.
Peel, R.F. 1952. *Physical Geography:* The English University Press, London, 290 p.
Price, J. 1926. A description of Ocracoke Inlet: N.C.H.R., v. 3, n. 4, p. 624-633.
Stratton, A.C. 1943. Reclaiming the North Carolina 'Banks': Shore and Beach, v. 11, n. 1, p. 25-27.

Chapter 13

BEACH PROFILE CHANGES ON WESTERN LONG ISLAND

Craig H. Everts

ABSTRACT

Ten beach profiles across a 9 mi section of straight coast in the vicinity of Jones Beach, New York have been surveyed about 10 times per year since the great storm of March 1962. The study area, on the south shore of Long Island, has a restricted fetch from the west, is generally open to waves from the south and east, and is near the western (downdrift) end of a net longshore transport system.

The nominal beach slope, defined as elevation over distance between the front dune crest and the mean sea level (MSL) intercept, is 0.10 at the eastern boundary of the study area and decreases unevenly to 0.018 at the western boundary. The elevation of the upper limit of scour, and the foreshore slope, are greatest for the steep profiles. The elevation of maximum sand loss during storms has a tendency to be higher for the first storm following a period of accretion and to decrease with succeeding storms.

Two extensively studied storms in February 1972 appear to have moved large quantities of sand in opposite directions. On 4 February 1972, a storm deposited 45,000 yd^3 of beach material above -3 ft MSL on the eastern one-third of the study area while eroding 285,000 yd^3 from the western two-thirds. Two weeks later another storm of similar intensity removed 365,000 yd^3 from five eastern profiles and deposited 48,000 yd^3 on the five western profiles. These data suggest that gross longshore transport rates in this area are much greater than net rates. In each storm, the amount deposited above -3 ft MSL was approximately 15 percent of the total quantity eroded. The indicated direction of net transport during each storm correlates well with the principal wind directions during that storm.

Such areal and temporal changes in beach profiles, if not anticipated, can affect the success of beach nourishment programs or the stability of protective shore structures. Data developed in this beach profile study may assist designers

---

Coastal Engineering Research Center, Department of the Army, 5201 Little Falls Roads N.W., Washington, D.C. 20016

on these problems. For instance, the envelope of beach profiles on Jones Beach had a maximum horizontal extent exceeding 650 ft, and a maximum vertical range of 12 ft. Such data are useful in siting and designing groins. Similarly, the indicated reversing transport direction affects areas of deposition and erosion, and thus must be considered in positioning structures and in the location of feeder beaches.

INTRODUCTION

Since the great March storm of 1962, the Beach Evaluation Program (BEP) of the Coastal Engineering Research Center (CERC) in Washington, D.C. has collected and processed some 35,000 beach profile surveys obtained from the Atlantic Coast. The primary purpose of the program is to observe variations on typical beaches in response to waves and tides of specific intensity and duration and to generate a predictive model as a first step in developing a storm warning system for low-lying coastal communities (Galvin, 1968). Another important application of the data, and the objective of this report, is to provide representative values of basic engineering information for use in planning and design of protective structures or remedial measures for stabilizing and maintaining beaches. This report describes, quantitatively, beach changes on 10 profiles (4, 7-15) at a BEP survey locality on Gilgo, Tobay and Jones Beaches, Long Island, New York between 1962-1972 (Fig. 1) and several relations among the data. The changes include short-term variations in profile shape, location, volume and rates of change, and longer term modifications in the total volume of material on the beach and the location of the shoreline.
Long Island, New York between 1962-1972 (Fig. 1) and several relations among the data. The changes include short-term variations in profile shape, location, volume and rates of change, and longer term modifications in the total volume of material on the beach and the location of the shoreline.

PHYSICAL SETTING

The study area is located on the Atlantic Coast of western Long Island 40 mi east of the Battery, New York City (Fig. 1). It encompasses the straight subaerial beach zone extending from 20,000 ft west of Fire Island Inlet to 15,000 ft east of Jones Inlet; a total shore reach of 45,500 ft. The recreational barrier beach is publicly owned and from east to west includes Gilgo Beach, Tobay Beach and Jones Beach (Fig. 1). As shown, this ENE-trending island, typically less than 3,000 ft wide and 25 ft in elevation, is separated from Long Island by Great South Bay, South Oyster Bay, and East Bay.

Exposed sediments on the barrier island are marine and windblown littoral material, much of which appears to have moved westward from eroding headlands east of Southampton, Long Island (Taney, 1961). The beach

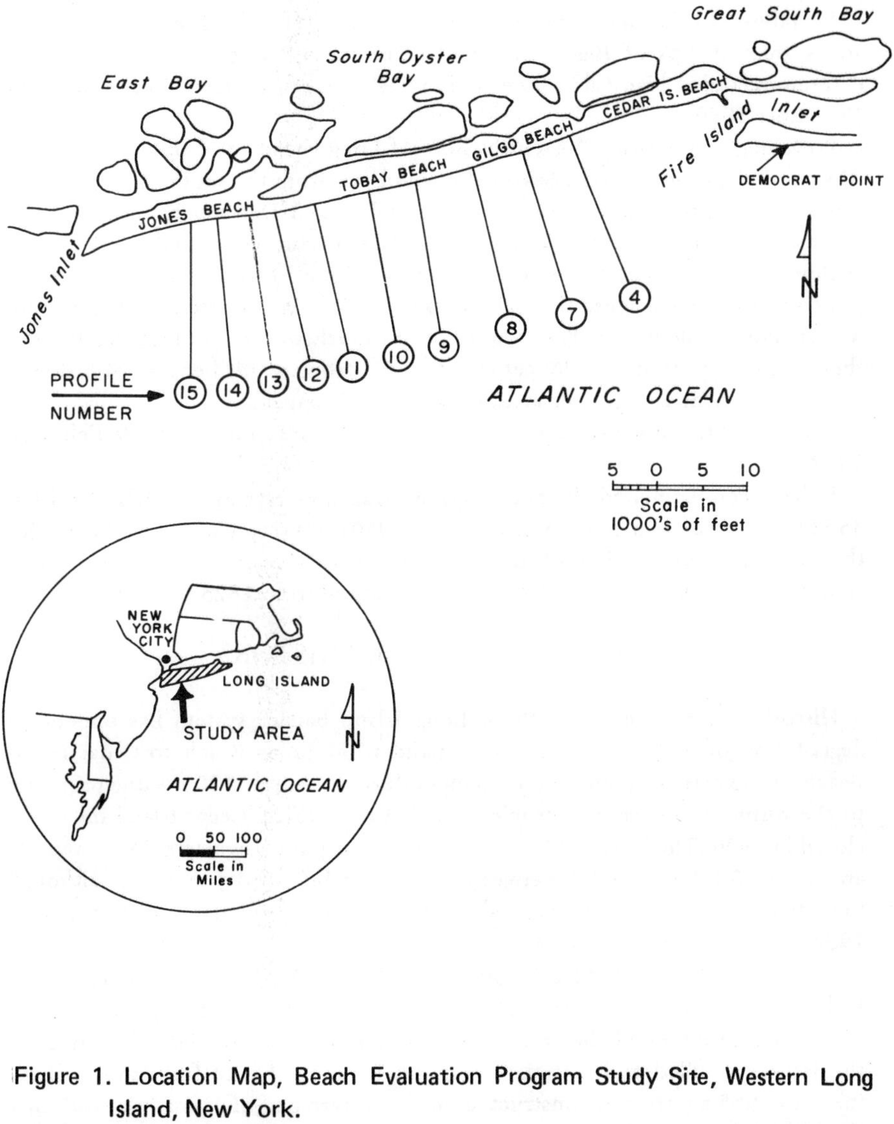

Figure 1. Location Map, Beach Evaluation Program Study Site, Western Long Island, New York.

sediments are composed of well-rounded quartz and a few percent heavy minerals, the most conspicious of which is violet garnet. The mean diameter of material at Jones Beach is near 0.30 mm.

An analysis of 5497 wave observations obtained between January 1955 and December 1958 from a CERC–U.S. Coast Guard cooperative surf observation program near Jones Beach was reported by Galvin and Seelig (1969). They indicate the mean breaker height is 1.7 ft, the wave period is 6.8 secs, and 13 percent of the time breakers are less than 0.5 ft high.

Semidiurnal tides have a mean range of 4.1 ft in Fire Island Inlet and 3.6 ft in Jones Inlet (Corps of Engineers, 1970). The mean spring range is 5.0 ft. An extreme recorded high tide of 9.4 ft above MSL (mean sea level) occurred on 25 November 1950.

Prevailing winds along the south shore of Long Island are from the southwest (Corps of Engineers, 1963). Seasonally, winds are from the southeast from April through October, from the west in November and December, and from the northwest during January, February and March. Hurricanes and extratropical cyclones, or "northeasters", are the most severe storms. Hurricanes occur principally during August through October while "northeasters" are generally winter and spring events. The frequency of "northeasters" is much greater and they commonly produce a longer duration of strong winds because of a slower storm movement and greater areal coverage. This paper includes detailed study of two notable "northeasters" that affected the area on 4 and 19 February 1972.

It has been estimated the net longshore transport rate at Fire Island Inlet is 450,000 yd^3 per year to the west (Saville, 1960). The most striking evidence for this rate is the migration of Democrat Point, the western extremity of Fire Island, which moved westward 4.6 mi in the period from 1825 to 1940.

RECENT CIVIL WORKS HISTORY

Historically, the entire southern Long Island barrier system has undergone degradation of varying severity. The island from Jones Beach to Cedar Island Beach has experienced enormous changes during the past 120 yrs due primarily to the former presence of four inlets, the latest of which, Cedar Island Inlet, was closed in 1930. The National Shoreline Study (Corps of Engineers, 1971) reports an average 3 ft/yr recession between Fire Island Inlet and Jones Inlet although the offshore contours are reported to have remained relatively constant since 1934.

Between 1939 and 1941 a 5,000 ft jetty was constructed at Democrat Point to halt the westward migration of Fire Island Inlet (Corps of Engineers, 1970). The structure inhibited the natural by-passing function of the inlet, but by 1950 the jetty was filled and material began passing around it to form shoals in the inlet. In 1955 a jetty was constructed on the eastern end of Jones Inlet and since that time it has been impounding material moving westward from the study area.

In 1958, the River and Harbor Act provided for Federal participation in the restoration and protection of the shore from Fire Island Inlet to Jones Inlet by dredging the Fire Island Inlet shoals, by providing a deposition area for littoral drift, and by obtaining fill material for a feeder beach (Corps of Engineers, 1970). About 2,000,000 yd^3 of medium sand from the inlet shoal was placed on a feeder beach 5,000 ft east of the location of Profile 4 in 1959 (Fig. 1). However, this proved inadequate in furnishing the required beach nourishment, and in 1961 the State of New York placed another 2,200,000 yd^3 of 0.30 mm median diameter bay-derived material along Gilgo Beach and Tobay Beach west of the feeder beach. In 1964, an additional 1,925,000 yd^3 of inlet material was placed on the feeder beach. From 13 November 1968 to 27 January 1969, 725,000 yd^3 of hydraulically dredged material from Great South Bay was placed on Gilgo Beach between Profiles 4 and 7. After each sand placement erosion of the beach was evident.

ACQUISITION AND REDUCTION OF SURVEY DATA

Figure 1 shows the location of the 10 BEP beach ranges utilized in this study. Additional profiles are surveyed, but are not considered in this report because of their proximity to the inlets. The separation distances for Jones Beach Profiles 12 through 15 in the west-central region of the island are approximately 3,000 ft. Between Profiles 9, 10, and 11 on Tobay Beach near the center of the island, and the east-central Gilgo Beach Profiles (4, 7, 8), the average distance interval is about 6,000 ft. At the present time surveys are made eight times yearly on a regular basis, and additionally as soon after storms as possible.

Beach surveys are made by the New York District, U.S. Army Corps of Engineers using a standard transit and stadia method. Each profile begins at a semi-permanent base station which in turn is tied to the state coordinate system. Profile data are collected normal to the shore at each change in beach slope from the vicinity of the frontal dune crest to about -2.0 to -3.0 ft MSL. The stations taken in the water are occupied using an amphibious vehicle (LARC–V). Surveys are made if at all possible during times of low water. Horizontal and vertical data are recorded to 1.0 and .1 ft, respectively.

Since 1968 survey data recorded in the field have been transferred to scanning forms which are sent to CERC. Upon receipt, the forms are automatically scanned and the resultant data is edited and put on magnetic tape. Five computer programs are used to process the field profile data. These include programs to prepare and edit the data and store it on magnetic tape, compute elevation-distance tables, plot profiles for two sequential surveys, compute area changes, and compute first and second derivatives of profile segment combinations. The first derivative is the slope of a line segment between two survey points, and the second derivative is defined as the rate of change between two of these slopes.

BEACH SURVEY RESULTS

There are two frequency ranges for beach survey information of general interest in the planning and design of coastal engineering works. The first includes short-term changes of a year or less such as profile slope and shape variations, changes in profile cross-sectional area and alongshore volume, and rates of change of these quantities. The second, a longer term frequency range, is on the order of years and tens of years. Important information in this category relates to the advance or recession of the shoreline and net changes in the volume of beach material.

Profile Characteristics

Profile Slope

One of the general attributes of a subaerial beach profile is its nominal slope or degree of inclination to the horizontal. The intercept of the mean sea surface and the beach, the MSL intercept, serves as a useful lower and seaward boundary. For beaches with a frontal dune line, such as those occurring in the study area in 1972, the dune crest serves as the landward and upper bound. Frontal dune crests are identified as that position of the profile exhibiting the largest positive second derivative above zero elevation (MSL). This derivative, which varies between .0002/ft and .05/ft in the study area, commonly exceeds by five times or more other second derivative values on the profile.

Figure 2 illustrates elevation-distance plots of three profiles surveyed during the first quarter of 1972. Successive profiles are superimposed in sets of two to show the changes that took place between the survey dates shown at the left. The hatched regions indicate erosion. Figure 3 is a graph of the profile slope averaged in this time interval for each profile in the study area. Slopes decrease unevenly in gradient from east to west at an average of -0.014 per 10,000 ft.

For the period December 1968 to March 1972, the eastern and central profile slopes (4, 7 to 12) remained nearly constant. On prograding western beaches the slopes decreased uniformly by approximately 10 percent as expected for beaches with a fixed landward limit. Between yearly quarters, the mean slope of individual profiles varied less than 11 percent. The uneven alongshore slope variation between profiles, however, is much larger. The Profile 4 slope, for instance, is 10 times that of Profile 14.

Profile Shape

Profiles shown in Figure 2 exhibit characteristic shapes that represent dominant pre-survey erosional or accretional processes on the upper subaerial profile. As noted by Hayes (1969) the erosional profile is characterized by a steep concave-up landward sector and a gentle, slightly concave or flat lower profile above the MSL line. Table 1 lists representative values of some diagnostic elements observed on selected erosional profiles following two February 1972 storms. For the steep profiles the elevation of the upper limit of scour and the

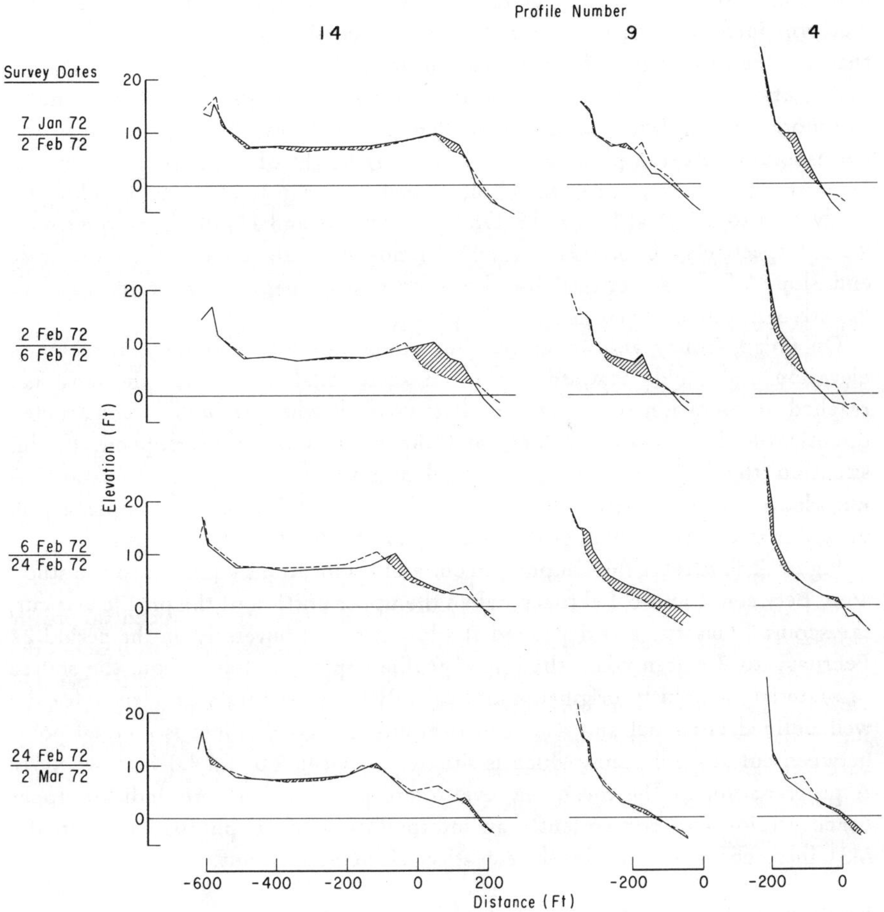

Figure 2. Western Long Island Beach Profiles—First Quarter 1972.

slope near the upper limit of scour are greater.

In the case of Profile 4, for instance, a scarp was observed at the upper scour limit while scarps were absent on eroding portions of the more gentle western profiles. The maximum second derivative, an indication of concavity, is greater for steeper profiles below the upper scour limit. The width of the erosion platform, defined as the concave-up horizontal distance segment from the upper scour limit to the seaward limit of negative second derivative values, is not dependent on the nominal slope of the profile. The elevation range of the platform, however, correlates well with nominal slope. For example, for the 4 February storm the elevation range decreases east to west from 16.0 ft at Profile 4 to 5.6 ft at Profile 15. In general the width and upper limit of the erosion platform increase with repetitive storms. The erosion platform is important in that it is the basal depth of cut for a given storm.

Accretionary profiles are characterized by a definite berm crest and a foreshore several times as steep as that of the seaward 50 percent of the corresponding erosion platform (Table 1). The height of the berm crest formed between 24 February and 2 March decreased from 5.6 ft at the east end of the study area to 4.3 ft at Profile 15. On Profiles 4, 10, and 15, the foreshore slopes were, respectively: .078, .052 and .041. Trends in the elevation of the berm crest and slope of the accretional foreshore for a given period appear to correlate directly with the nominal slope of the profile.

On Gilgo, Tobay and Jones Beaches the berm crest commonly increases in elevation and builds seaward during an accretional cycle once the crest has reached an elevation of about 4 ft. Below this height, the berm is constructed directly on the erosion platform and the crest is often overtopped. In this situation the landward trough is filled and the crest migrates landward. A knowledge of the location and slope of a bermed foreshore is important in designing structures such as groins and seawalls to minimize overtopping.

Figure 2 illustrates the coupling of cut and fill locations common to the study area. Between 2 and 6 February, when the upper portion of the profile was cut, the scoured material was deposited at a lower level. Conversely, in the period 24 February to 2 March when the upper profile experienced accretion, the source of material apparently originated immediately below the fill area. Hence, for the well defined erosional and accretional events discussed, there is a nodal point between cut and fill zones which is situated between 2.0 and 4.0 ft above MSL. A progradation of the mean sea level intercept may, therefore, indicate upper beach erosion and consequently an interpretation of air photos based on the MSL intercept must consider the elevation of the nodal point.

Profile Bounds

Profile bounds are identified as the upper and lower limits of change experienced by a beach profile for a finite number of surveys during a specified time interval. Such data, when plotted, present an easy means for determining the lower and upper profile envelope, and the landward bound for elevation and

TABLE I. Profile Characteristics Indicating Erosion After Storms For Three Beaches on Western Long Island, New York

	1		2	3	4	5	6
	Upper limit of erosion (x,z) x=distance from MSL intercept (ft) z=elevation above MSL line (ft)		Slope at the upper limit of erosion	Maximum second derivative below the upper scour limit (ft^{-1})	Width of erosion platform (ft)	Elevation of erosion platform (ft above MSL)	Slope of lower 50% of erosion platform
Survey Date 6 Feb 72 Profile	x	z					
4	200	16.0	1.90	-.269	200	0 to 16.0	.034
10	225	11.1	0.184	-.0033	175	1.9 to 11.9	.013
15	280	8.1	0.049	-.00037	240	2.5 to 8.1	.030
Survey Date 24 Feb. 72 Profile							
4	215	18.0	1.60	-.061	215	0.3 to 18.0	.030
10	247	12.6	0.260	-.022	169	2.7 to 12.6	.019
15	310	7.5	0.046	-.00041	265	2.9 to 7.5	.015

contour intercept changes. The plots may also indicate accretion or erosion trends.

Figure 4 illustrates the bounds obtained for Profiles 4, 9, and 14 during the years 1969, 1970, and 1971. The bounds for 1971 are superimposed on the 3-yr plots. Zero distance on the horizontal axis references the MSL intercept for the initial profile surveyed in 1962. Maximum horizontal variations occurred during the 3-yr interval at the -3 ft contour, the lowest surveyed elevation available, on Profile 4; at +3 ft on Profile 9; and at +5 ft on Profile 14 which suggests the trend for maximum horizontal excursions is toward an increased elevation as the nominal profile slope decreases. Yearly net changes are somewhat smaller, but follow these trends.

The maximum to minimum elevation experienced by the profiles at specified distances from the initial MSL intercept, and above the –3 ft contour, increases as the nominal slope decreases. For example, in the 1969 through 1971 survey period, the maximum elevation excursion for Profile 4 was 9 ft at a distance location of -10 ft from the 1962 MSL intercept. On Profile 9 the range was 9.5 ft at a distance of -120 ft, and on Profile 14 the maximum elevation change was 12 ft at a +70 ft distance location. In the first quarter of 1968, Urban and Galvin (1969) found the maximum one-week change at Profile 7 to be 5.7 ft of erosion and 4.7 ft of accretion. The maximum erosion, which appeared to be correlated to sand elevation at the start of the one-week interval, commenced at the +8 ft contour.

Profile bounds in Figure 4 for the three-year period compared to those of 1971 indicate Profiles 4 and 9 are retreating while Profile 14 is building seaward and up. An analysis of maximum change by quarter indicates October through December is the period of greatest variation when the change is approximately 70 percent of that observed for the entire year.

Profile Area and Volume Changes

As the shape of a profile changes in response to different dynamic factors, it characteristically loses or gains cross-sectional area in the subaerial zone. The magnitude and location of this displacement is important in many aspects of coastal works design. Area data in this paper are given in square feet (ft^2). For conversion to square yards (yd^2), divide by 9. For approximate conversion to square meters (m^2), divide by 11.

Area Change (elevation)

Between January 1969 and March 1972, the elevation of maximum cut or maximum fill for various events, when averaged for all profiles, is shown on Table 2. The maximum cross-sectional scour occurred at the highest elevation during the first storm of a sequence. In subsequent storms this elevation decreased as the initial berm volume was lost and scour of the erosion platform increased. The elevation of the maximum scour area among the 10 profiles that were analysed increased 1.5 ft between Profile 15 and Profile 4 for the average

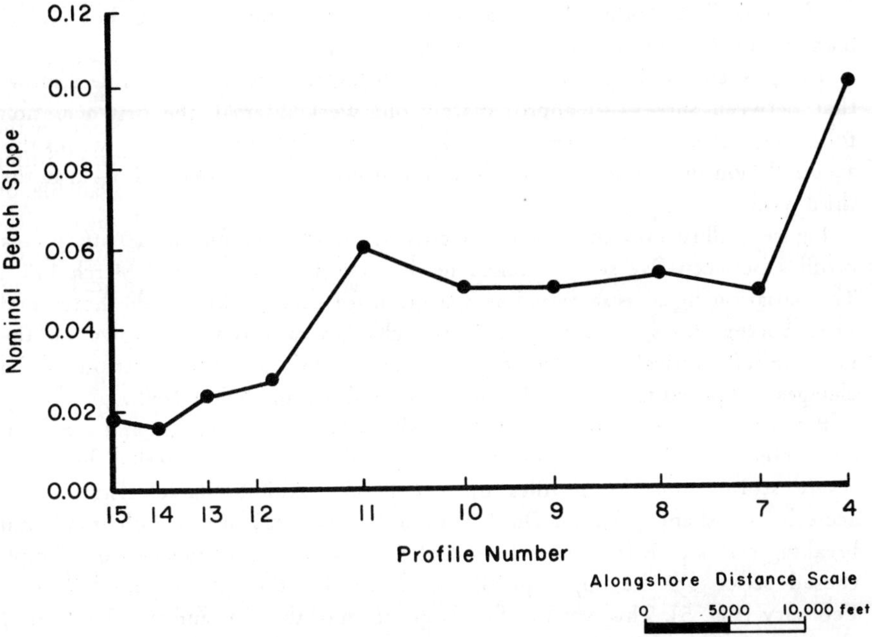

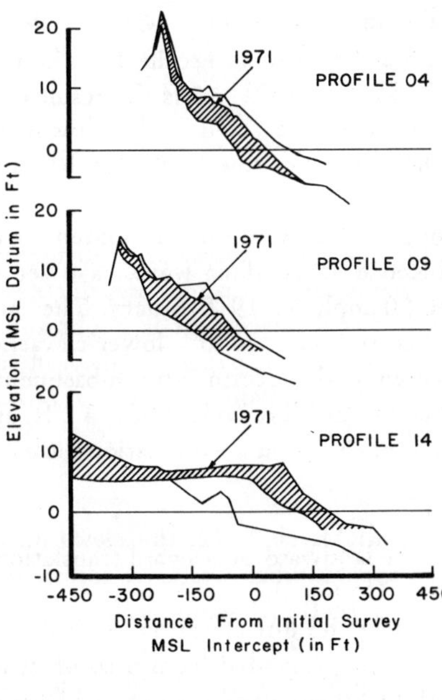

Figure 3. Nominal Beach Slope for Profiles on Western Long Island.

Figure 4. Profile Bounds for 1969-1971 for Beach Profiles on Western Long Island.

of the five first storms. For the second storm in the sequence, however, a maximum cut occurred 2.0 ft higher in the west (Profile 15).

The post-storm elevation of maximum fill increases with time. Table 2 shows that between surveys of approximately one week interval, the first accretion following a storm was a maximum at 4.0 ft above MSL. In succeeding weeks this accumulation increased, and the elevation of maximum fill reached 6.4 ft in the third week.

Figure 5 illustrates the change in cross-sectional area for the 10 study area profiles between five surveys taken in the period 7 January to 2 March 1972. The variation in cross-sectional area is given for each profile on the horizontal axis. Survey dates between which the changes occurred are shown on the extreme left vertical axis. The adjacent vertical axis lists the elevation of the changes computed for 1 ft vertical increments above and below MSL.

Between 7 January and 2 February small fluctuations in erosion and accretion are evident without correlation between profiles. This was a period devoid of severe storms and the profiles on 2 February exhibit characteristics of an accretionary shape (Fig. 2). On 4 February, a two-day storm with maximum breaking waves of 8 ft, a wave period of 5 sec, and a top wind speed of 55 mph caused scour on the upper profile as shown for the survey period 2 to 6 February (Fig. 5). This was the first large storm of the year and the elevation of maximum scour coincides with the berm crest elevation of the 2 February profiles. The elevation range over which erosion occurred was higher on the steeper eastern profiles. For instance, scour is evident between 3.5 and 17 ft for Profile 4, and between 2.5 and 9 feet on Profile 14. The multiple upper scour and accretion sequence shown for Profile 15 is the result of wind activity. The cut-fill node above which scour occurred and below which fill is evident, decreased from the peripheral profiles toward Profile 10 where no fill occurred above -3 ft MSL.

The second large storm of the season, similar in intensity and duration to the 2 February event, produced 8 ft breaking waves, a 6 sec wave period and a maximum wind speed of 50 mph on 19 February. The maximum area losses resulting from the second storm occurred at a lower elevation, especially in the eastern profile area. Following this storm, and subsequent to 2 March, the profiles were aggraded. Maximum fill is evident near 4.0 ft, an elevation, slightly below that of maximum cut at the berm crest location during the first storm.

Area Change (normal to shore)

Figure 5 also illustrates the landward or seaward translation of beach contours between successive surveys. During the 4 February storm a maximum contour retreat of 110 ft occurred near the 6-ft elevation on Profiles 12-15. In the same time interval the MSL intercept prograded from 5 to 45 ft for those profiles. A maximum progradation of 110 ft occurred between 24 February and 2 March at a 2-ft elevation on Profile 14.

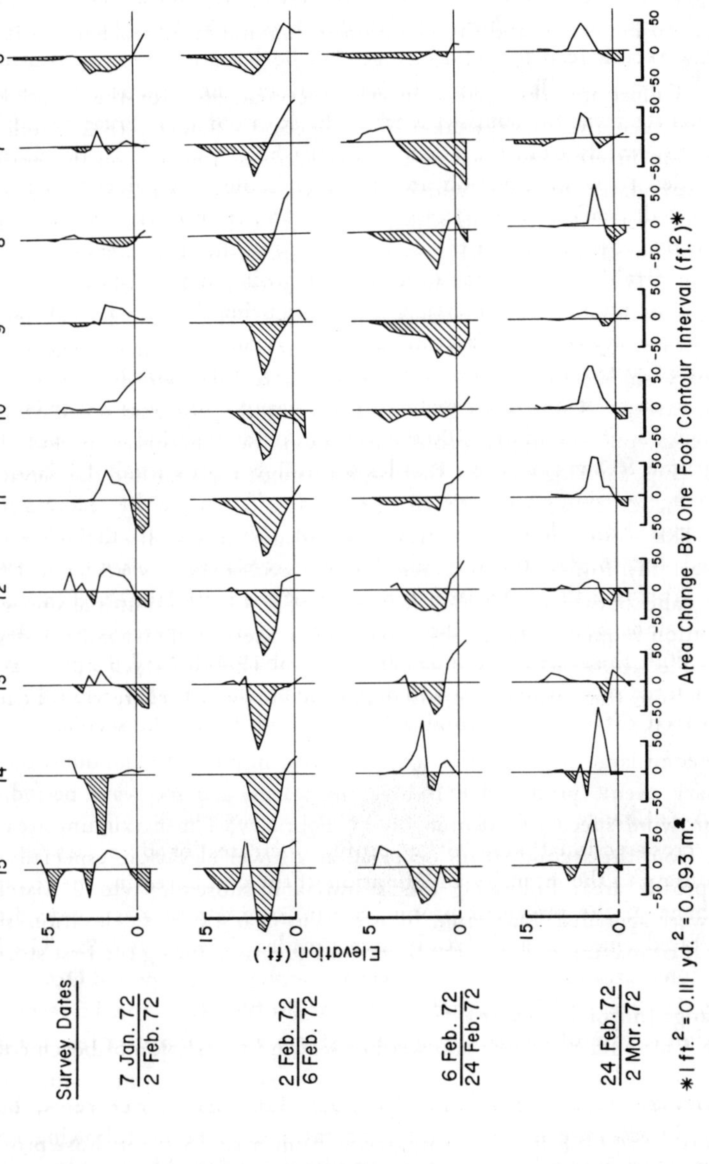

Figure 5. Area Change Between Successive Surveys on Western Long Island Beaches for the First Quarter of 1972.

Area Change (alongshore)

The total cross-sectional area change parallel to the shore in the study region is illustrated in Figure 6 for the two significant winter storms of 1972. Change in total profile area above -3 ft MSL is shown on the vertical axis in square feet. The horizontal axis references profile location. The striking feature shown is that there is net scour on the subaerial beach at one end of the study area while net accretion occurs at the other end. Profiles 4 and 15 at the boundaries of the study area experienced the least change. For both storms the crossover region is located between Profiles 9 and 10. Winds during the 4 February storm were predominantly from the southwest while the storm of 19 February produced its most intense winds from the northeast. Longshore transport of beach material corresponded to wind direction and although scour was evident on the upper portions of all profiles (Fig. 5) a large amount of fill was concentrated at lower elevations on those profiles at the downdrift end of the study area.

Of considerable interest in the design of groins and in beach nourishment projects is the amount of material moving and being deposited on the subaerial profile in the alongshore direction in comparison to that moving and being deposited away from the subaerial zone. Figure 6 shows that for the 4 February storm an average of 355 ft^3 of material per longshore foot was lost from Profiles 10 through 15, or a total of 7,700,000 ft^3. In the same period an average of 51 ft^3 per longshore ft was gained on Profiles 4 through 9, or a total of 1,200,000 ft^3. If we average erosion on the western beaches, and assume that accretion in the east resulted from alongshore movement of this material, the quantity not accounted for, 6,500,000 ft^3, must have been moved elsewhere, probably offshore. Applying these figures, 16 percent of the material moved east and was deposited alongshore during the first storm and 84 percent was displaced offshore. With the same analysis for the storm of 19 February, 13 percent of the eroded material was moved west and deposited above -3 ft MSL while 87 percent was displaced offshore. The total volume eroded during the second storm was 9,900,000 ft^3.

Rate of Area Change

Total cross-sectional area of a profile increases or decreases at a rate commensurate to the intensity and duration of the event creating the change and to the shape of the pre-event profile. Storms of a day or a few days duration produce rapid changes in the study area profiles above -3 ft MSL. The largest cross-sectional area losses between successive surveys in the period October 1962 to March 1972 are 557,609, and 926 ft^2 for Profiles 4, 9, and 15 respectively. The maximum rate of change for a single storm was 550 square feet per day (ft /d).

Accretionary rates are commonly much less than scour rates, but are important in assessing natural restoration rates for a beach following a storm. Accretion, which is rapid when a storm profile is considerably out of equilibrium with the post-storm waves, decreases with time after the storm. Because the rates vary with time, typical rates must be based on survey interval. The largest rate of fill recorded for a seven-day period on Profile 15, was 104 ft^2/d or approximately twice that of the average fill rate following a storm at Profile 15; three times the average rate of 30 ft^2/d on Profile 9; and four times the average rate of 25 ft^2/d for a 7 d interval on Profile 04. These rates, which are only approximate, suggest that profiles with the gentler slopes accrete at higher rates, all other factors being equal. The average accretional rate is difficult to ascertain for periods longer than a week because of the unsteadiness in wave conditions and consequent variation in the rate of fill over the interval. A comparison of survey data indicates, however, that the average rate of accretion for 14 d in the study area is one-half or less than that of the first 7 d following a storm.

Long Term Shoreline Changes

Long term "permanent" shoreline changes may be approximated by a comparison of profiles taken at the beginning and end of the study, by net MSL intercept variations, or by changes in some upper reference location such as a dune crest. The first method is best if a representative profile is chosen at the beginning and end dates. Values based on variations between individual surveys and compared to seasonal and yearly averages are useful to indicate the departure one might expect from the net change. The changes discussed in this

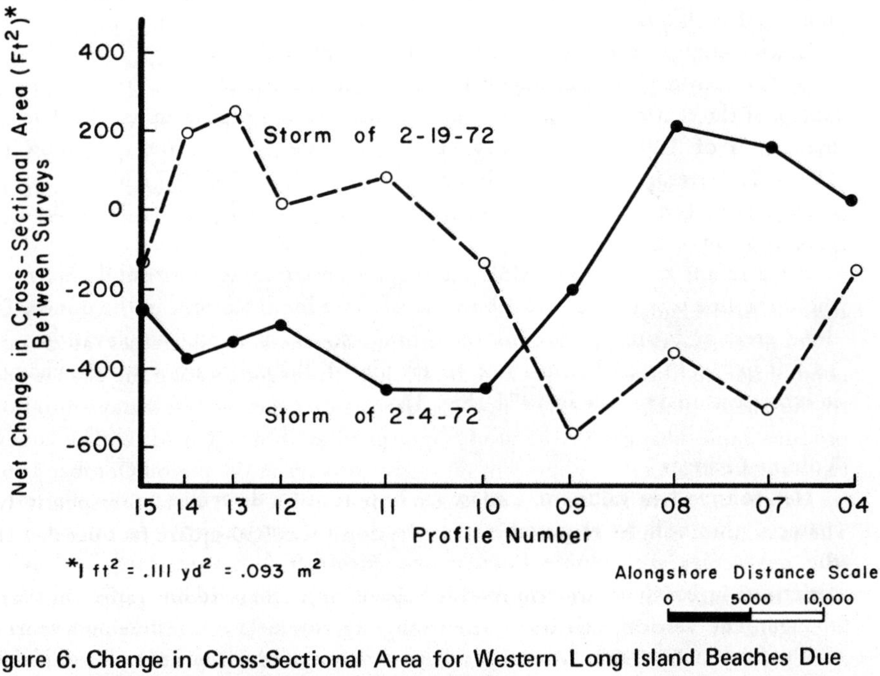

Figure 6. Change in Cross-Sectional Area for Western Long Island Beaches Due to the Storms of 4 and 19 February 1972.

section include those caused by artificial beach restoration.

Intercept Changes

Figure 7 illustrates changes of MSL intercept position for 95 surveys of non-constant frequency during the period October 1962 to March 1972. The horizontal axis defines time and the vertical axis lists the translation distance in ft for each profile about a zero datum assigned for the first survey. Information on quarterly, yearly, and the 1962-1972 average changes with respect to fluctuations of individual surveys is given in Table 3 which was derived from quarters in 1962, 1968, 1969, 1970, and 1971 which included two or more survey dates. The left-hand column lists the quarterly, yearly and 1962-1972 averages from which the MSL intercept difference for a single survey was obtained. Averages of all 10 profiles for a survey date are typically between 50 and 75 percent of the maximum deviation for a single profile when compared to the quarterly, yearly or 1962-1972 average values. The range listed in Table 3 is the sum of the seaward and landward excursions for individual surveys during the period of comparison. In most instances consistent trends were not observed in MSL intercept change between profiles.

Variations in survey frequency render comparisons difficult, but the data from relatively complete years indicates that:

1. The greatest progradation of the MSL intercept occurs in the January-March quarter
2. The greatest retreat is a fourth-quarter event, and
3. The season of least change is July through September.

Seasonal fluctuations about the yearly trends average plus or minus 45 ft with a maximum of 130 ft. The yearly values fluctuate 35 ft above and below the 1962-1972 average with a maximum of 120 ft. The 1962-1972 MSL intercept change, averaged by survey, is shown in Figure 8 for all profiles. Note the large progradation of the shore along the western end of the study area.

An attempt to correlate MSL intercept translation to horizontal changes in the dune line was unsuccessful because of rapid modifications in the dunes. The dune crest at Profile 4, for instance, progressively increased in elevation from 18.7 ft in 1963 to 23.3 ft in 1972. At Profile 14, the entire dune system changed in cross-sectional profile in 1971 (Fig. 3).

Volume Changes

Net volume loss values on a coast are important in determining the amount of material that will be required for restoration or nourishment. As important as the net values are volume fluctuations. Figure 9 shows quarterly and yearly variations in beach volume (in ft^3/ft of shoreline). The solid line indicates yearly change. The vertical axis scales the change in volume for quarters and years in the period 1962-1972. Variations between quarters often exceed 300 ft^3/ft. Yearly variations average less than 70 ft^3/ft. This suggests that volume data obtained in all quarters of a year, when averaged will provide a reasonably

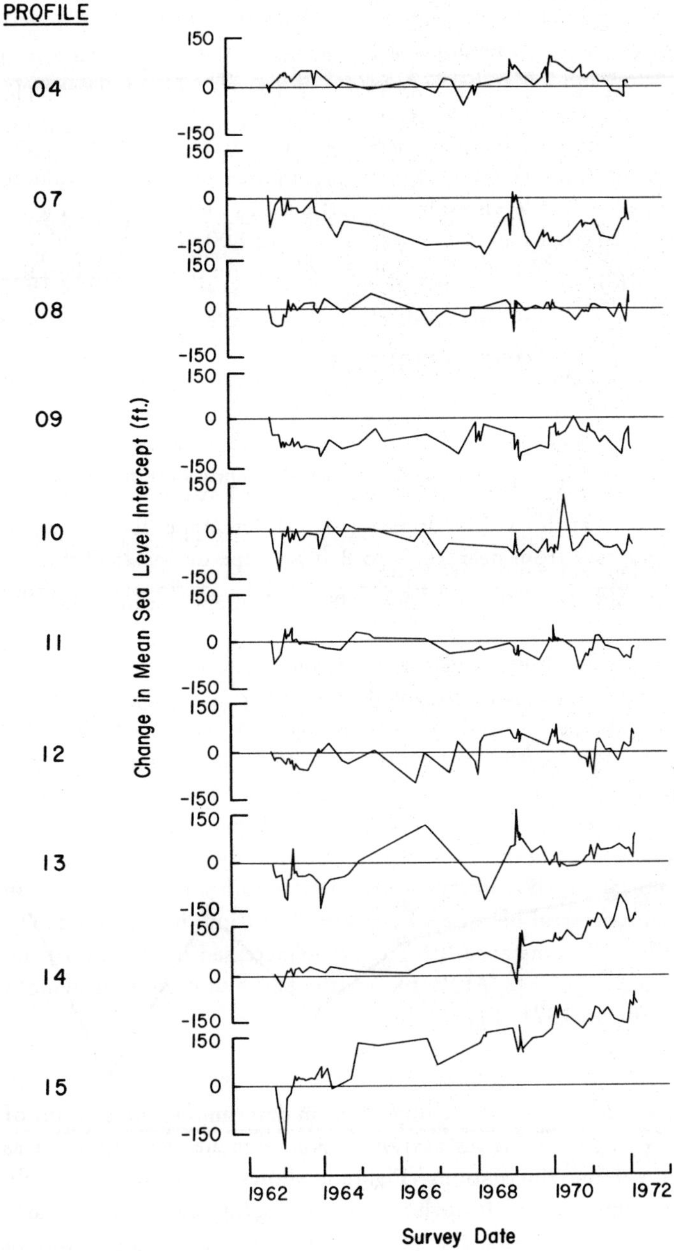

Figure 7. Mean Sealevel Intercept Variations for Western Long Island Beaches Between 1962-1972: Positive Values Indicate Seaward Accretion.

accurate figure for net change.

Figure 10 illustrates the net volume change for the 1963-1972 period. The left vertical axis identifies the cross-sectional area change in two surveyed profiles between 5 March 1963 and 2 March 1972, a total of 9 yr. The March dates were chosen because profiles were each of the accretional type, therefore similar in shape. Volumes were obtained by planimetering the profile bounds from the MSL line to the dune crest intersection of the superimposed profiles. Net volume change for the 9-yr period is referenced by the vertical axis on the right in Figure 10 and the dashed line. Total loss in this period was 15,800,000 ft^3 or 590,000 yd^3. During this interval 6,850,000 yd^3 of fill was placed on the eastern beaches in and near the study area.

CONCLUSIONS

Beach Changes

1. Nominal beach slope values decrease unevenly from 0.1 at the east boundary of the study area to 0.02 at the west boundary (Fig, 3). The upper limit of storm erosion typically decreases from near 16 ft to 8 ft as slope decreases (Table 1). Scarps are more common on steeper profiles (Fig. 2), and accretionary foreshore

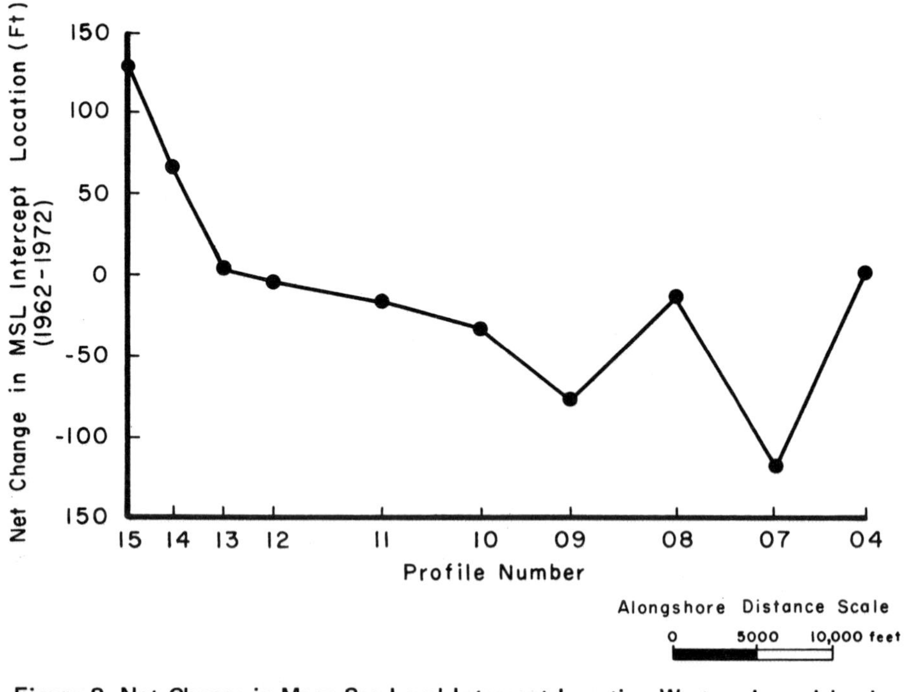

Figure 8. Net Change in Mean Sea Level Intercept Location Western Long Island Beaches: 1962-1972.

TABLE 2. Maximum Area Changes for Selected Erosional and Accretional Events Between 1969 and 1972 for Western Long Island, New York

A. Elevation of maximum cut (-A)			B. Elevation of maximum fill (+A)		
1st storm	5.5 ft	average of 5 storms	1st fill	4.0 ft	average of 4
2nd storm (repetitive)	4.7 ft	average of 2 storms	2nd fill (repetitive)	5.0 ft	average of 2
3rd storm (repetitive)	3.7 ft	average of 1 storm	3rd fill (repetitive)	6.4 ft	average of 1

TABLE 3. Horizontal Fluctuations of MSL Intercept for Individual Surveys*

	MSL excursion seaward of average (ft)		MSL excursion landward of average (ft)		Range (ft)	
	average value 10 profiles	maximum value	average of 10 profiles	maximum value	average of 10 profiles	maximum value
Quarterly	28	50	54	80	82	100
Yearly	84	150	85	150	169	270
1962-72	102	180	101	200	203	380

* From quarterly, yearly and 1962-72 averages for beaches on Western Long Island, New York

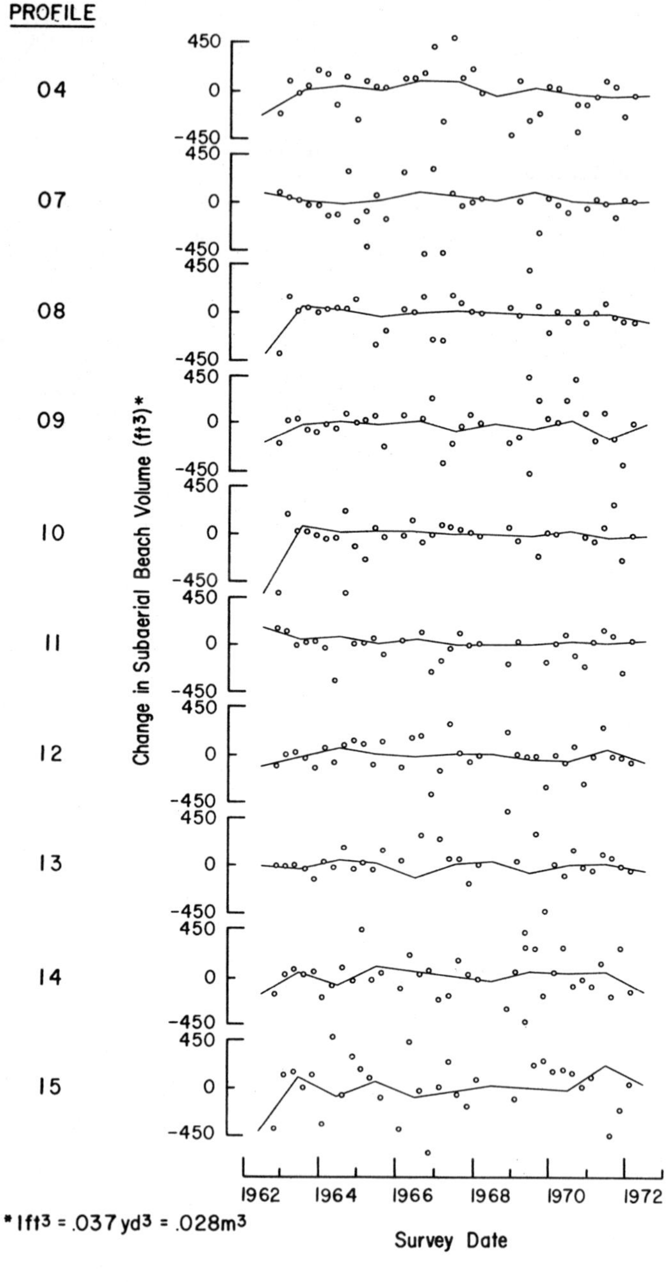

Figure 9. Beach Volume Change By Quarter and Year for Western Long Island Beaches: 1962-1972.

slopes and the height of the berm crest increases with norminal slope from east to west.

2. Between 1969 and 1971, a maximum horizontal change in excess of 650 ft and a maximum vertical change of 12 ft occurred at Profile 14 near the western boundary of the study area (Fig. 4). The maximum horizontal excursion of a beach contour between surveys in the three year period occurred at -3 ft MSL in the east and at +5 ft in the west (Fig. 4).

3. Net movement of the MSL intercept on the profiles over a 10-yr period indicates accretion up to 15 ft/yr on the west and erosion of 2.5 to 11 ft/yr in the central and eastern parts of the study area (Fig. 8).

4. The net loss in total beach volume in the 45,500 ft coastal reach (Fig. 1) between 1962 and 1972 was about 600,000 yd^3 (Fig. 10). In that interval 6,800,000 yd^3 of fill was added to the beaches.

5. Two extensively studied storms in February 1972 appear to have moved large quantities of sand in opposite directions. On 4 February, a storm deposited 45,000 yd^3 of beach material above -3 ft on the eastern one-third of the study area while eroding 285,000 yd^3 from the western two-thirds (Fig. 6). Two weeks later another storm of similar intensity removed 365,000 yd^3 from five eastern profiles and deposited 48,000 yd^3 on the five western profiles. For each storm, the amount deposited above -3 ft was approximately 15 percent of the total quantity eroded. The indicated direction of net transport during each storm correlates well with the principal

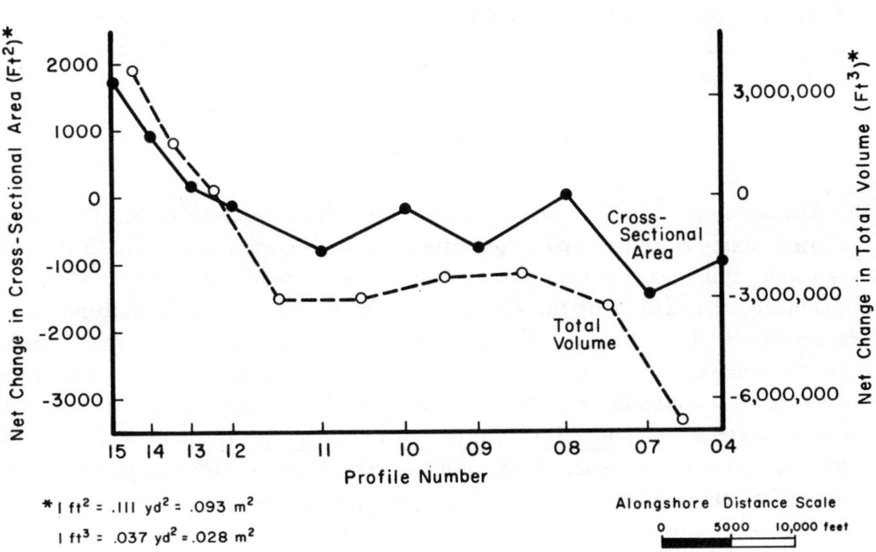

Figure 10. Net Change in Cross-Sectional Area and Total Volume for Western Long Island Beaches: 1962-1972.

wind directions during that storm. Maximum cut occurred in the center of the study area while at either end changes were small.
cut occurred in the center of the study area while at either end changes were small.

6. During certain times of rapid erosion on the subaerial beach, such as occurred during the storms on 4 and 19 February 1972, a node of zero change between cut and fill zones on the beach profile often occurs above the MSL intercept (Fig. 2 and 5). The eroded material is frequently deposited directly below the scour zone causing the MSL intercept to prograde seaward as much as 80 ft. Conversely, during post-storm accretion of the upper profile in the study area, the MSL intercept frequently retreats landward.

Coastal Engineering Applications

1. Protective beach structures may lose their effectiveness by being undermined, overtopped, or flanked. The quantitative vertical, horizontal and alongshore changes in sediment volume, such as described in this paper, provide representative design information.
2. Artificial beach nourishment includes either the establishment and periodic replenishment of a stockpile at an appropriate updrift section of an eroding coast, or restorative filling at various elevations along the entire beach after severe erosion. Indicated gross erosion rates in the study area are much greater than the net longshore transport rate. For several storms in 1972 the central portion of the area appeared to experience the largest changes in profile volume, suggesting this would be the optimum location for a stockpile.
3. Progradation of the MSL intercept, which frequently occurs when the upper foreshore erodes, suggests caution in interpreting air photos from changes in the MSL intercept.

ACKNOWLEDGEMENTS

Data presented in this paper were obtained from research conducted by the United States Army Coastal Engineering Research Center under the Civil Works research and development program of the United States Army Corps of Engineers. Permission of the Chief of Engineers to publish this information is appreciated. The findings of this paper are not to be construed as official Department of the Army position unless so designated by other authorized documents. I am grateful to Cyril J. Galvin, Jr., Coastal Processes Branch, CERC, who supervised the collection of much of the beach profile data, and especially for his many helpful suggestions and his critical review of this paper. I also appreciate the assistance of Linda Mintz, Sally Bruce, William Seelig, Cary Jones, Everett Woo, and Ralf Kohler who aided in reducing much of the data.

REFERENCES

Corps of Engineers. 1963. Atlantic Coast of Long Island, N.Y. Fire Island Inlet and Shore Westerly to Jones Inlet; Review Report on Beach Erosion Control Cooperative Study: U.S. Army Engineer District, New York, 31 p. and appendices.

Corps of Engineers. 1970. Atlantic Coast of Long Island, Fire Island Inlet and Shore Westerly to Jones Inlet, New York Beach Erosion Control and Navigation Project: General Design Memorandum, Dept. of Army, N.Y. District, Corps of Engineers. 28 p.

Corps of Engineers. 1971. National Shoreline Study, Regional Inventory Report, North Atlantic Region: U.S. Army Engineer District, Corps of Engineers, New York, 120 p.

Galvin, C.J. Jr. 1968. CERC Beach Evaluation Program: 1962-1968 Background: Unpublished paper of the U.S. Army Coastal Engrg. Research Center, 17 July 1968, 30 p.

Galvin, C.J. Jr. and Seelig, W.N. 1969. Surf on U.S. Coastline: A Summary of Visual Observations from CERC-U.S. Coast Guard Cooperative Surf Observation Program through 31 December 1965: Coastal Engrg. Research Center, Inhouse Report.

Hayes, M.O. 1969. Coastal Environments of Northeastern Massachusetts and New Hampshire: Tech. Report No. 1–CRG, 462 p.

Saville, T. 1960. Sand Transfer, Beach Control, and Inlet Improvements, Fire Island Inlet to Jones Beach, New York: Proceedings 7th Conf. Coastal Engrg., p. 785-807.

Taney, N.E. 1961. Littoral Materials of the South Shore of Long Island, New York: Coastal Engrg. Research Center, Tech. Memo. 129.

Urban, H.D., and Galvin, C.J. Jr, 1969. Pipe Profile Data and Wave Observations from the CERC Beach Evaluation Program, January-March 1968: U.S. Army Coastal Engrg. Research Center, Miscellaneous Paper 3-69, 74 p.

HOMERTON
COLLEGE
LIBRARY

Chapter 14

COASTAL PROCESSES AND PREHISTORIC MARITIME CULTURES

Maurice L. Schwartz and Garland F. Grabert

ABSTRACT

The importance of an interdisciplinary approach in scientific inquiry is hardly better exemplified than in the relationship between coastal processes and prehistoric maritime cultures. The archeologist may describe the culture of a maritime people and the geomorphologist may interpret the processes that have taken place, but only the two, working in concert, can describe what life was like at the time and how the environment shaped the quality of that life. Five case studies from the Puget Sound-Strait of Georgia region illustrate this thesis. A dig at the landward terminus of Semiahmoo Spit has uncovered gravel overlain by clay, both containing evidence of early habitation. The sequence suggests occupation of an exposed headland beach followed by activity carried out at a later stage in the sheltered lee of a developing spit. Proceeding down the coast the next embayment, Birch Bay, is a classic example of a log-spiral headland bay beach. Raised terraces inland from the present shoreline have a similar geometry as a result of the same geomorphic processes. Extrapolation from middens near today's sea level, has led to the discovery of more primitive middens on the older elevated terraces. The stratigraphy in pits dug on a raised terrace at Cherry Point illustrate a succession from cobble beach to well developed soil horizon. The terrace was crossed by early fishermen when it was a beach, much like that which now lies at the foot of the bluff, then reoccupied as a raised terrace in more recent times.

Farther to the south lie the Conway and La Conner sites. Highway interchange excavation on the Skagit delta near Conway recently uncovered a meandering distributary complete with fish-weir and willow-reed mats. La Conner's Fishtown is a row of summer cabins built on a baymouth bar bordering the Skagit estuary. The cross-section exposed in a dig behind the present bar suggest an earlier inner mid-bay bar, occupied repeatedly between stages of inundation by previous maritime cultures.

Department of Geology/Department of Anthropology, Western Washington State College, Bellingham, Washington 98225

INTRODUCTION

We maintain there is a unique opportunity for interdisciplinary cooperation on investigations which combine the study of coastal processes and prehistoric maritime cultures. The interplay between natural processes and man's adaptation to these is of the utmost significance in unravelling the natural history of these people (Butzer, 1971); and neither specialty alone provides a sufficient interpretation of the environmental regime of an area at the time of habitation. In the context of the approach referred to here as interdisciplinary, geomorphology is not employed solely to provide archeology with facts upon which to construct a theoretical framework. Surely, such techniques as geochronological dating, stratigraphic measurements, and and sedimentological analyses are employed to furnish quantitative data; but there is more to it than that. In our use of the term, interdisciplinary means that the maritime archeologist and the coastal geomorphologist describe *how* and *why* an earlier people in this area lived as they did. Thus, an understanding of coastal processes becomes more than a tool; it is a means of gaining insight into human history.

To illustrate our thesis, five case studies from the Puget Sound region will be described. Digs at Semiahmoo Spit, Birch Bay, Cherry Point, Conway, and La Conner (Fig. 1) have all incorporated the interdisciplinary approach advocated here.

Geographically, the study area lies within the Puget Sound–Strait of Georgia lowland. Bedrock within the region is usually mantled by Pleistocene stratigraply, the last of these being Fraser Glaciation drift and outwash. Holocene coastal features are largely developed upon this late Pleistocene stratigraphy, and it is glacial debris, along with Holocene sediments, that has provided much of the material for the development of the deltas, spits, bars, and beaches found in the region today (Easterbrook and Rahm, 1970). There is little structural control at work in the vicinity of the digs that are discussed here. Configuration of individual segments of the coast is affected essentially by local wind velocity and direction, fetch, and river and tidal currents, at various stands of sea level. Bellingham, Washington wind records indicate predominant winter winds from the south, southeast, east and northeast; while during the summer these give way to west, southwest, and southerly winds (Schwartz and others, 1972). Fetch and shoreline orientation, with this wind regime, combine to produce different conditions at each small section of the coast. Within the location map area (Fig. 1), for example, shore drift can be found moving toward any point of the compass. Each of our archeological sites, therefore, has had a somewhat different coastal process history.

From an archeological viewpoint, the study area lies toward the southern end of the Fraser River delta locality occupied, in recent times, by the Coast Salish Indians (Provincial Archives, 1971). These Indians trace their past to the interior plateau, and have since acquired much of the neighboring northern-coastal culture. Fanning out along the lowland coast, the life-style evolved by these

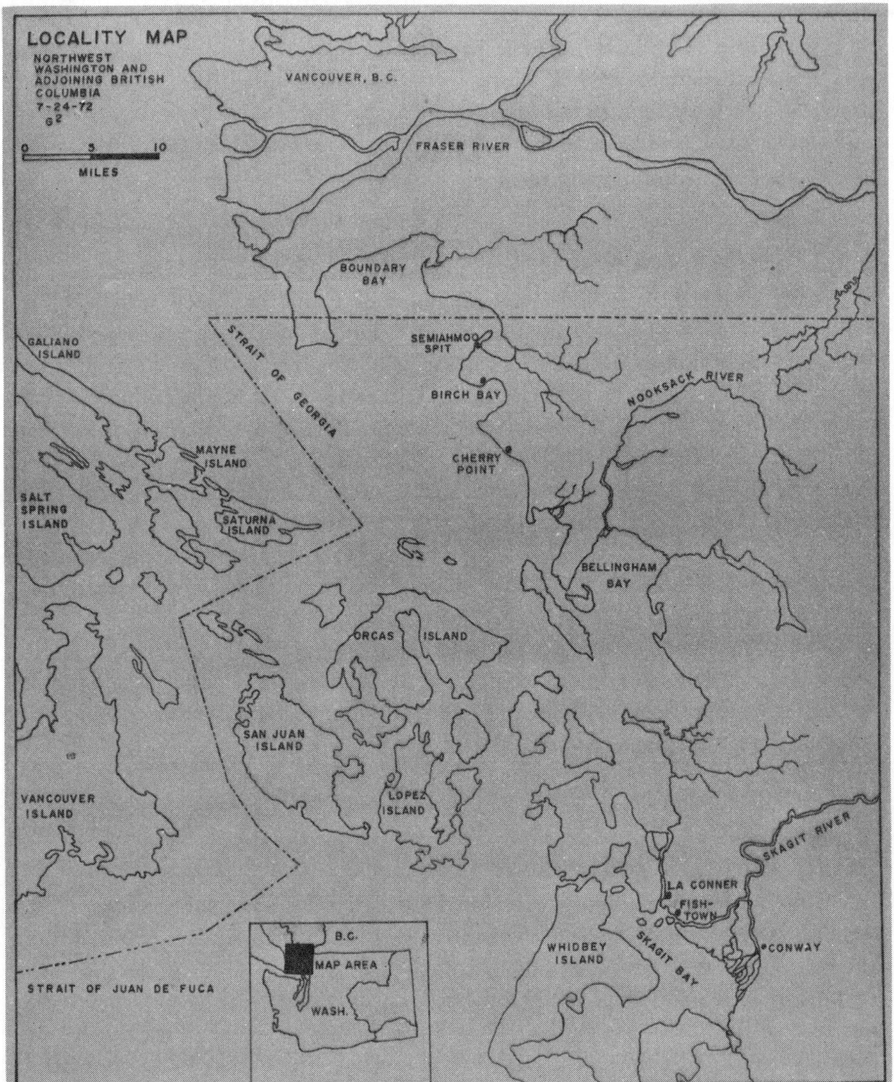

Figure 1. Locality map, showing archaeological sites discovered in this article.

people was basically maritime; dependent upon the sea and river mouths as routes of travel and sources of food. Social organization centered upon family and tribal groups, gathered in a loosely-organized series of villages located up and down the coast.

Until recent investigations, little has been known of pre-Salish occupation in this region. Presumably, their culture was marine-oriented, in many ways, of practical necessity, similar to the later Coast Salish. Most certainly, a better knowledge of these prehistoric cultures will become available as a result of the work now being conducted in the area.

SEMIAHMOO SPIT

A site designated 45-WH-17 at Semiahmoo Spit has been tested by an amateur group of high school students at intervals over the past several years under the direction of Milton Clothier, science department chairman at Sehome High School in Bellingham. While their artifact yield has not been large, it has been significant when combined with the cultural and geologic stratigraphy. The shell midden being examined is one of the larger remaining sites between Portland, Oregon and Vancouver, British Columbia. The deepest remaining shell midden is more than 6m in depth, with extensive thinner areas extending northward along the spit for several 100m. The latter seem to mark the remnants of intensive habitation areas. The maximum habitation area is still uncertainly defined, but is certainly larger than 125 by 250 m.

Artifact assemblages were composed mainly of bone and antler, although a small number of chipped stone projectile points have also been found. Lanceolate bipoints and shouldered forms predominated, with some larger and smaller triangular stemmed and side-notched points. They were typically of Marpole, Whalen, and later forms; although single specimens comparable to ground slate forms of Locarno Beach apsect were also found. Composite antler harpoons were present, and remnants of barbed fish spears were among the specimens.

Semiahmoo Spit, unlike other southward-growing spits in the region, has developed to the northeast since the only open fetch to which it is exposed is on the west and southwest. The midden is at the landward terminus of the spit in the shadow of a 25 to 30 m vegetated bluff constituting the edge of the of the glacial headland from which the spit extends. On the outer mainland beach large erratics weathered from the glacial drift demarcate the extent of the headland, and are conspicuously absent on the outer spit beach. The enclosed bay behind Semiahmoo Spit is shallow with broad tidal flats.

Work carried out in the summer of 1972 at the base of the east (lagoon) side of the larger shell mound revealed a sequence of shell-less and shell bearing cultural strata. These lay upon beach gravel in which there also was some cultural material. The main cultural deposits were slightly more than 1.8 m thick. Since the datum for the current test digs was set at 3.1 m above the mean high tide line, the height of the spit at that point before human use was only about 1.3 m.

Most puzzling, for a while, was the basal beach gravel and the overlying stratum, a greenish clay, containing evidence of human presence (Plate 1). Clay over gravel did not appear to fit any acceptable pattern; and with cultural material in both, the problem was compounded. Obviously, geomorphic considerations took precedence here over human occupation.

Assuming that the large beach erratics, and overgrown glacial bluff at the base of the spit, mark the position of the headland prior to the development of Semiahmoo Spit, the earliest inhabitants of the site must have crossed an openly exposed headland beach to reach the water. As the spit grew, the same site would face upon a gradually developing lagoon in the protected lee of the spit. People crossing the site at this stage moved about, ankle deep, in tide-flat muds. The stratigraphic sequence proposed here is much like that outlined by Selley (1970) for similar geomorphological events. Viewed in this context, cultural material in the gravel and overlying clay can be seen to be an expected series when a beach site becomes protected by a developing spit.

The potential of the Semiahmoo site toward posing and solving cultural problems of the local and regional prehistory is considerable. At present we may make only provisional statements concerning the site. In terms of the man-environment relationships of the past this site supplements findings from several others. It is especially relevant in providing a link between the Fraser Delta sites and those to the south. Like the case studies which follow, it also shows man's ready utilization of maritime environments as these were in the process of formation or expansion.

BIRCH BAY

The Birch Bay shoreline is a fine example of Yasso's (1965) log-spiral headland bay beach. Predominant winter winds approach from the north and northwest across the Strait of Georgia, with the concavity of the bay open to the west. As would be expected, grain size and slope decrease from coarse and steep at the south to fine and low in the lee of the headland at the north. In the latter sheltered area compound recurved spits have grown bayward, in nested sequence, across the mouth of a local stream.

Larsen (1971), a former graduate student in anthropology at Western Washington State College, has investigated some middens on these spits and on a series of raised terraces behind the present shoreline (Fig. 2). It was during this work that Larsen noticed, in aerial photographs (Plate 2), that all of these terraces had the same log-spiral configuration as the beach bordering the bay. Startling at first, it becomes apparent on second thought that as sea level rises and falls relative to a given headland the same geomorphic processes that would develop a log-spiral bay at one stand would generate the same form at all levels. Warren Yasso (personal communication) has acknowledged this previously

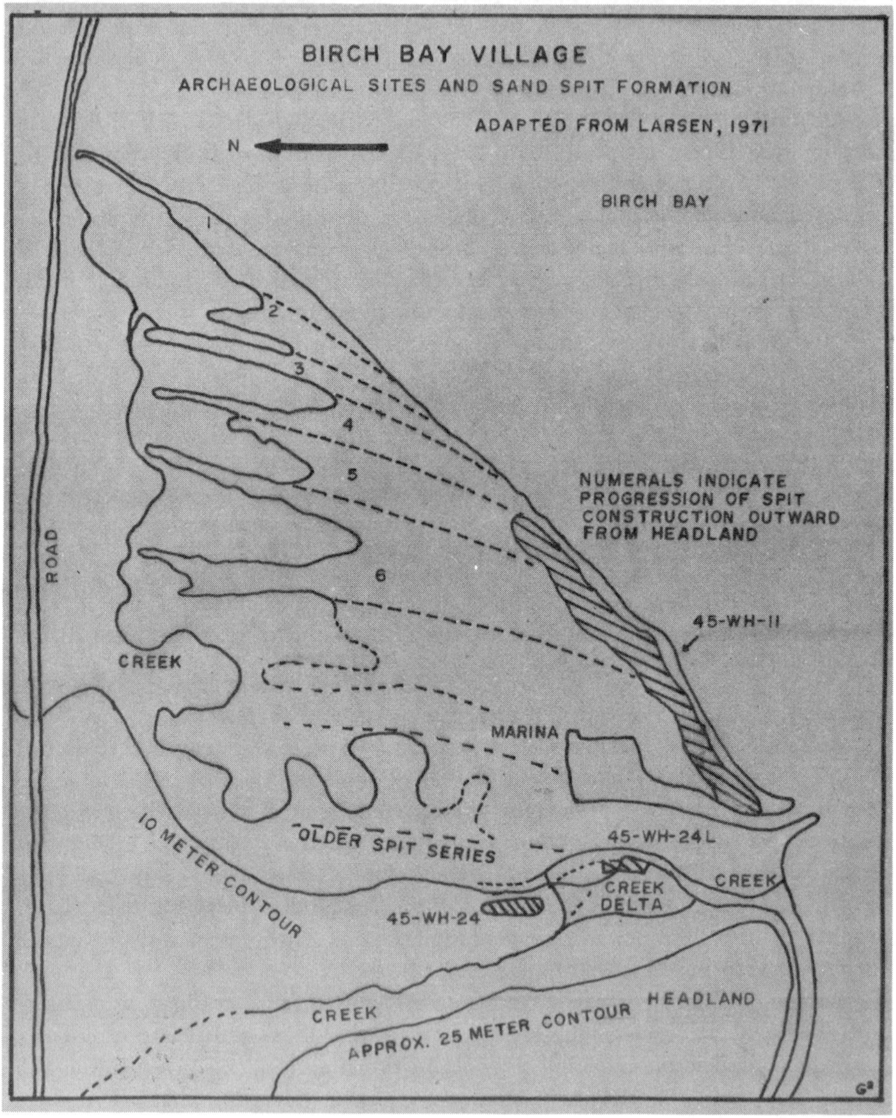

Figure 2. Base map of nested fossil spits and raised terraces along northwest shore of Birch Bay. (Adapted from Larsen, 1971).

Plate 1. Pit wall at Semiahmoo Spit Site. Basal beach gravel can be seen constituting floor of pit and extending up to vicinity of 290 cm mark on measuring rod. Light stratum near 285 is thin clay deposit. Measuring rod is not accurately positioned in this view. (Photograph by Milton Clothier).

Plate 2. Aerial view, looking toward southwest, of dig sites at Birch Bay. Nested fossil spits can be seen in upper center, and raised terraces appear faintly crossing lower center. (Photograph by Curt Larsen).

unstated concept. Not too surprisingly, echo-sounder recordings in Birch Bay, made aboard the R/V Majada in connection with an environmental impact study of local tidal currents, (Schwartz and others, 1972) revealed two submerged terraces, at approximately 7 and 9 m, that appear to have a similar plan shape. The Yasso (1965) form would seem then to extend up and down in space and time.

It was reasoned that if middens were found at the present beach level evidence of similar maritime cultures of earlier times would be found on the raised log-spiral terraces following the method of Giddings (1964) at Cape Denbigh. The approach proved to be a viable one, and Larsen discovered a number of important sites in this manner. Extending the principle to the horizontal, he also dug a series of productive pits on the nested compound recurved spits.

Deeply buried components exposed during marina construction at Birch Bay Village yielded a few artifacts and cooking hearths, located on extinct beaches, buried by progradation at the present sea level. These habitation components were given the same designation (45-WH-11) as a large shell midden occupying the contemporary beach. Traces of the midden can be found for nearly 350 m along the beach line. The major excavated specimens from this area consisted of pebble tools, hammer stones, and some worked bone scraps. However, two antler harpoons, one complete and the other fragmentary, are excellent examples of Borden's (1962) Marpole forms.

Evidence of beach construction and persistent human orientation to shoreline habitats in the region is found in a succession of spits extending from the base of the lowest, 10 m, terrace to the present beach. Three, and possibly more, old beaches were traced in the marine excavation. Each beach has one or more hearths exposed. These hearths show fire-broken rocks, and a few pebble and flake implements. One hearth yielded a broken leaf-shaped projectile point. This hearth, now exposed in the new marina, is presently submerged at high tide.

Site 45-WH-24 lay on the 10 m terrace at the northwest side of Birch Bay. Artifacts collected here were numerous cobble tools and a few cores, predominantly of quartzite, overlying thin lenses of weathered shell and a single hearth. Because of the lack of ground stone, bone and finely chipped stone artifacts from this site, it was assumed to be earlier than the shell midden on the active beach.

A smaller component, designated 45-WH-25L, lay some 120 m to the southeast below the same 10 m terrace, at a former creek delta. Prehistoric occupation was seen in scattered fire-broken rocks, a number of large basalt waste flakes, a flake burin, and several pebble choppers. Another 10 m terrace shell midden has been tested near the midpoint of the bay, east of Birch Bay Village. This site (45-WH-10) yielded only a few artifacts, none of which could be typologically dated. Moving upward in elevation, it was found that a small shell midden and remnants of a shell free camp site (45-WH-29) lay on the 30 m terrace at Birch Point. The best we can say at the present is that the 30 m terrace may have been first inhabited by people who left flake and pebble tools at a time when the site was on either an island or a narrow headland.

The Birch Bay archeological sites show a series of occupations which correlate roughly with successive terrace and beach formation. As glacial rebound raised one terrace up and away from a newly formed shoreline, the maritime people moved down to the lower level. Then as the latest period of present sea level caused spits to develop in rapid, nested succession, the natives then packed up and moved horizontally to the new beaches. The theme is repeated at the Cherry Point dig.

CHERRY POINT

The Cherry Point site (45-WH-1) is located on a wave cut terrace overlooking the Strait of Georgia (Plate 3). The terrace elevation, behind a wave-cut cliff, varies from 6 to 9 m above mean high tide. A cobble beach, typical of the region, lies at the base of the cliff. Landward of the terrace the terraine slopes upward another 10 to 15 m where the gradient lessens. Faint traces of old strand lines may be seen on the slope face, where minor slumps bare the surface of vegetation.

The most obvious feature on the terrace is a shell midden some 200 m long, roughly paralleling the wave-cut cliff, and presently about 20 m wide. It is estimated that the maximum midden width was probably not more than 30 m, even before the cliff eroded into cultural deposits. Sketches of the site made in 1954 and extrapolation of the slope of exposed strata indicates that as much as 1-2 m of shell material has been removed in places by grading and removal for driveway ballast. Some areas of the site still possess up to a meter's depth of shell, although mean thickness is closer to 25 cm.

Pits dug on the terrace surface reached to Bellingham Glaciomarine Drift at depths of 1-2 m (Plate 4). There was noticeable weathering of the upper drift zone. A coarse sand layer 10-15 cm thick occurred above the weathered drift. Next, moving upward in the stratigraphic section, there was a cultural-depositional layer; more than 1 m in depth in some shoreward areas, and wedging into the modern soil zone at the northward edge of the habitation area. This stratum consisted of dark brown to locally black earth with numerous firepits, stone "pavements", and occasional remains of stakes and wooden fragments. This unit yielded over 150 pebble "choppers," many with well battered edges, chipped slate knives or blanks for ground-slate knives used in fish preparation, approximately 80 drilled net and line sinkers, 10 large anchor stones, and elongate slim pebble hammers used in "drilling" the sinkers and anchors. A small number of other artifacts from this zone consisted of 2 ground slate points of Locarno Beach phase affinities (Borden, 1962), a few flake tools, a handful of ground slate and steatite objects resembling the carved pendant Borden assigns to the Locarno Beach phase, and some labret-like fragments. Animal bone was almost absent, as were bone artifacts, and shell material was

Plate 3. Site of dig on Cherry Point raised terrace overlooking the Strait of Georgia. (Photograph by Garland Grabert).

Plate 4. Cross-section exposed in dig at Cherry Point showing stratigraphic section described in text. (Photograph by Garland Grabert).

lacking entirely. Sparce patches of gravel and cobble were found at the base of this layer in some pits.

The top layer in the sequence was the aforementioned shell midden which presently averages 25 cm in depth with local thickness to 1 m. Locally there were fire hearths, intrusive stake molds, and clusters of fire-broken rocks. The artifacts contained in this shell midden layer fit neatly into Borden's (1962, 1970) Fraser Delta sequence. These included ground serpentine adzes and their antler hafts, unilaterally barbed harpoons and fish leisters, a few parts of composite toggling harpoons, bone herring rake teeth in abundance, harpoon arming points, and a few chipped sone projectile points. There also was a continuation, in much reduced numbers, of the assemblage from the deeper stratum. A few miscellaneous objects help to place the border layers of the shell midden in chronological context. Three quartz microblades indicate that the base of the shell midden may have been laid down as late as about 400 A.D. (Borden, 1970); composite harpoons of antler, a variety of fishing and fowling leisters, bone fish gorges, and hook barbs support the bone-antler implement assemblage. Chipped stone artifacts are rare and confined to a half dozen projectile points, simple pebble choppers in limited numbers, and a few flakes of basalt.

A total of more than 1600 artifacts have been cataloged at the Cherry Point site. Despite the propensity of the Northwest Coast peoples for use of bone, wood, and antler there was a majority of lithic materials in the repertoire, C^{14} dating of four charcoal samples taken from the bottom to the top of the shell midden stratum yielded ages of 2,340, 1,640, 1,300, and 960 yr B.P. (all + 200 years) respectively.

Occupation of the Cherry Point site followed a series of geomorphic events. As the Fraser Glaciation waned rebound in the vicinity of Cherry Point raised the Bellingham Glaciomarine Drift to positions where the previously mentioned faint strand lines, landward of today's terrace, were formed. Erosion at surf base (Dietz, 1963) led to development of the terrace, as the nearshore bottom, and a layer of sand was deposited over the wave-cut drift surface.

Continued rebound raised the terrace furthur, exposing it as a part of the coastal zone. The patches of gravel and cobble at the base of the lower cultural zone may represent remnants of the receding shoreline. Human traffic across the new coastline is indicated by the numerous firepits and abundant artifacts. Presumably the people lived back of the beach and discarded their shells in the backshore. Their maritime culture is attested to by the net and line sinkers and anchor stones. Gradually a soil zone developed and thickened over the backshore region, incorporating furthur cultural material.

Over a period of time, glacial rebound, moving the terrace upward at a rate faster than that of post-glacial sea level rise, raised the terrace to near its present elevation. While the various occupants of the terrace rode upward, elevator-like, a new beach was formed at the base of the receding cliff face. Still marine oriented, as shown by their then more sophisticated artifacts, these cultures no longer left their sinkers and anchors on the terrace midden. These tools of their trade are probably to be found under the present beach and on the nearshore bottom , where they lost them while crossing the foreshore just as their ancestors once did at the earlier shore. Shell middens with artifacts developed where people lived, artifacts alone are found where they plied their trade.

As glacial rebound raised Cherry Point, and the shoreline was translated seaward, the maritime cultures moved in the same direction in their relentless pursuit of their favored niche. History repeats itself today as an archeology student tosses a lunch-break chicken bone on the ground at the dig; then wanders down to the beach to watch the nearby salmon fishing boats, while losing some manufactured possession among the cobbles.

LA CONNER

Farther south, at La Conner, there is an archeological site at Fishtown (Plate 5). Fishtown is a local name for a cluster of dilapidated summer cabins, on stilt-like foundations, built along a former baymouth bar facing a distributary of the North Fork of the Skagit River. Here too, human habitation has come full circle.

Nestled between two headlands that bracket the old baymouth bar (Plate 6), in the marshy area behind Fishtown, is a small elongate stretch of dry ground that has yielded an amazing amount of cultural material. Chipped stone is not abundant, but bone and antler artifacts are. Several ornately carved bone specimens, not typical of the Fraser Delta region, have been recovered by Astrida Onat, of Seattle Community College; who has supervised extensive digging here for the past few summers. In addition to these, and the usual midden debris, a few human bones have also been recovered (Plate 7).

Excavations along and across this inner rise show five alternating zones of sand and soil, with the soil layers being rich in their upper portions, in habitational material. Within each sand and soil zone, the soil layers are 20 to 30 cm thick and the sand about 5 to 6 cm. Transverse profiles across the rise reveal that the bedding of these zones dips gently seaward and more steeply landward. C^{14} dating of charcoal found in the uppermost soil zone places occupation at about 1000 yr B.P.; the chronology of the lower zones is not presently known. Reconstruction of events at this site, based on geomorphological and archeological considerations, would proceed somewhat along the lines of the following.

"Fishtown" really started several thousand years ago when some maritime people encamped at the dig site when it was a small mid-bay bar across a cove facing the open water of a Skagit River distributary. Raised a few feet above mean high tide, and covered with sparse vegetation, it was a hospitable site near salmon, molluscs and waterfowl. The concentric hilly structure on the landward side provided protection from hostile marauders, while the depression or pan behind the bar offered a source of brackish drinking water. Meltwater floods, storm waves, or sea level rising faster for a while than the rate of glacial rebound

Plate 5. Archaeological students digging into former mid-bay bar at Fishtown site near La Conner. (Photograph by Philip Schwartz).

Plate 6. Aerial view of Fishtown region. Note bowl-shaped topography behind cabins in center. View is toward the northeast. (Photograph by Maurice Schwartz).

(or some combination of these) eventually flooded the bar and laid down a deposit of sand over the structure. Under these dismal conditions the bar was abandoned and lay fallow and deserted. With the passing of time, however, littoral zone vegetation became established and a soil profile gradually developed. The site was once again ready for some long-lost descendant of the former occupants to wander in and reestablish a community on the bar.

The foregoing postulated sequence appears to have been repeated at least five times on the first or inner bar, and is now going on again at the present outer barrier. The significance of the Fishtown site lies in its use of what was originally an insular environment, and supports the proposal of early pioneering use of tideflat and semi-marshland localities, as Mattson (1970) suggests for his Skagit Delta I Phase.

CONWAY

The most southerly and inland dig discussed here occurred at Conway, where Interstate 5 crosses the Skagit River delta. An excavation there for the foundation of a cloverleaf ramp uncovered interesting organic material covered by deltaic deposits (Plate 8). In a crash-program investigation David Munsell, Washington State Highway Department Archeologist, worked the site thoroughly before ballast rock and cement sealed it from view.

Plate 7. Cairn built over burial in cedar box. Found during dig at Fishtown. (Photograph by Astrida Onat).

The base level of this site (Plate 9) lay below the water table, and slightly below present sea level, though now about 3 km from the saline water of Skagit Bay. Preserved in the silt were antler tools, some bone implements, and large quantities of basketry and matting made of birch and willow. Thin poles and stakes were found in and adjacent to a sinuous 2 m wide pavement of pebbles that extended along the length of the excavation. A trench dug across the elongation of the pavement showed a concave upward pebble lens approximately 20 cm thick at the middle. The site, as described here, is interpreted as a fish-weir set up on a former distributary of the Skagit River, when this location was closer to the edge of the delta and the bay. Still, some questions remained to be answered. What could account for the complete, *in situ,* burial of a fish-weir together with the mats, baskets, and tools accompanying such an endeavor? Slow sedimentation would allow decay of exposed portions of the wooden products, assuming abandonment of the enterprise in the first place; while flood conditions, bringing rapid sedimentation, would wreck and redistribute these objects. Then too, what have been the sedimentation rates following burial; and since the newly-exposed channel section was not elongated in a direct line from the Cascade Mountains on the east to Puget Sound on the west, in which direction did the distributary flow?

Noting that the small weir crossed the channel at the center of a meandering bend, the most plausible hypothesis for quick burial appears to be the sudden collapse of an undercut bank at this point. Thus we now find a well-preserved moment in past life on the delta. C^{14} dating of the site places it at about 700 yr B.P.. With the pebble channel now lying 2.14 m below the surface, which now stands at 1.53 m above sea level, the *average* rate of deposition over the years would be on the order of 30.5 cm/100 yr. This undoubtedly represents a subtle blend of glacial rebound and concurrent sea level rise; how much of each, here and elsewhere in the region, is a question that only furthur investigation can reveal. The direction of flow was more easily determined from a perusal of the imbrication fabric of the channel.

Taken all together, the archeologic and geomorphic insights at the Conway dig have led to the restoration of this veritable museum piece of Indian industry.

SUMMARY

The foregoing case studies are an illustration of an application of method. They demonstrate the efficiency of interdisciplinary investigations of coastal archeological sites. Where human communities are so closely bound to the primary products of their environments, to limit investigations of prehistoric communities to studies of artifacts and structures is to neglect the larger portion of natural-cultural relationships. While this study is by no means new in its approach, the application of coastal process studies to the selection and interpretation of archeological sites is especially appropriate to the region. The conjunction of terminal and post-Pleistocene isostasy and eustasy, and human

Plate 8. Present-day appearance of Conway archaeological site. Excavation shown in Plate 9 was located under car in center portion of highway ramp. Snowcapped summit of Mt. Baker can be seen in background. (Photograph by Philip Schwartz).

Plate 9. **Highway** construction excavation at Conway. Imbricated gravel pavement, tracing course of former deltaic distributary, is being exposed in shallow cut made by workman between the two wheelbarrows at left. Highway department archaeologist, David Munsell, wearing white safety helmet, can be seen standing at center right, talking to senior author. (Photograph by Garland Grabert).

adaptation to insular and maritime environments, climaxed in the historic Northwest Coast cultures. What more powerful tool can we presently bring to bear on the problems of the formation of these historic Indian people? We are convinced that interdisciplinary studies such as these have been productive of understanding in the past; that they will become ever more useful methods in the future is even more certain.

While this paper cannot do more than summarize the coastal process and archeological investigations involved in the accumulation of data from this limited region, we believe these studies have been useful in building upon speculation of the past, have added to a better picture of the culture history of the region, and for those interested in paleoecological studies, contribute to the aggregate of knowledge about coastal processes in one of the world's geologically unusual areas.

ACKNOWLEDGEMENTS

We are deeply indebted to Milton Clothier, Curt Larsen, Astrida Onat, and David Munsell for sharing with us the fruits of their labors at Semiahmoo, Birch Bay, Fishtown, and Conway respectively. Nick Kolby determined the Cherry Point radiocarbon dates, utilizing the Western Washington State College C^{14} laboratory. The Bureau of Faculty Research at Western provided research funds for work at Birch Bay and Cherry Point, students in field classes carried out excavations at seven archeological sites (not all reported here), and several other persons provided volunteer labor at Birch Bay and Cherry Point. To all these kind people, we offer our heartfelt thanks.

REFERENCES

Borden, C.E. 1962. West Coast cross-ties with Alaska: in Campbell, J.E., ed., Prehistoric Cultural Relations Between the Arctic and Temperate zones of North America: Arctic Institute of North America, Technical Paper No. 11, Montreal, p. 9-19.

Borden, C.E. 1970. Culture History of the Fraser Delta region: an outline: B.C. studies, Vancouver, B.C., No. 6-7, p. 95-112.

Butzer, K.W. 1971. *Environment and Archaeology:* Aldine-Atheron, Chicago, 703 p.

Dietz, R. S. 1963. Wave-base, marine profile of equilibrium and wave built terraces: a critical appraisal: Geol. Soc. of Amer. Bull., v. 74, p. 971-990.

Easterbrook. D.J. 1966. Glaciomarine Environments and the Fraser Glaciation in Northwest Washington: Guidebook First Annual Field Conference, Pacific Coast Section, Friends of the Pleistocene, Western Washington State College, Bellingham, 52 p.

Easterbrook, D.J. and Rahm, D.A. 1970. *Landforms of Washington:* Bellingham, Union Printing, 156 p.

Giddings, J.L. 1964. *The Archaeology of Cape Denbigh:* Brown University Press, Providence, 321 p.

Larsen, C.E. 1971. An investigation into the relationships of change in relative sea level to social change in the prehistory of Birch Bay, Washington: M.A. Thesis, Western Washington State College, Bellingham, Wash., 89 p.

Mattson, J.L. 1971. *A Contribution to Skagit Prehistory:* Washington State University, Pullman, 205 p.

Provincial Archives 1971. Coast Salish:British Columbia Heritage Series, Victoria, Provincial Museum, ser. 1, v. 2, 71 p.

Schwartz, M.L.,Fackler, R.C., Hoerauf, E.A., Larsen, C.E., Lingbloom, K.L, and Short, M.A. 1972. Nearshore currents in southeastern Strait of Georgia: Syesis, v. 5, p. 117-130.

Selley, R.C. 1970. *Ancient Sedimentary Environments:* Cornell University Press, Ithaca, 237 p.

Yasso, W.E. 1965. Plane Geometry of headland-bay beaches: Jour. Geol., v. 73, p. 702-714.

Chapter 15

MORPHOLOGY AND VERTICAL SEDIMENTARY SEQUENCE MODELS IN HOLOCENE TRANSGRESSIVE BARRIER SYSTEMS

John C. Kraft, Robert B. Biggs, and Susan D. Halsey

ABSTRACT

Studies of the transgressive barrier elements of the mid-Atlantic Delmarva Peninsula show four major types of morphologic and vertical sedimentary sequence models. Each model shows a response to present wave and current conditions as well as to the paleotopography inundated by the ongoing transgression. The geomorphic features of the shoreline areas of a transgressive barrier coast, as studied in the Delmarva Peninsula area, include numerous discreet geomorphic features that are in fact single sedimentary environment lithosomes. These sedimentary environment lithosomes are part of a dynamic condition in which geomorphic environment-sedimentary lithosomes are moving landwards and upwards in space and time with the presently continuing coastal transgression. As the present coastal transgression is in part a product of late Holocene relative sea level rise and in part the product of coastal erosion, vertical sequences and the sedimentary environment lithosomes of the late Holocene stratigraphic record at the edge of the transgression are in part truncated by erosion and, accordingly, are incomplete. However, detailed geomorphic and subsurface studies of the various elements of the estuarine and Atlantic Ocean barrier coast have shown that the linking of studies of coastal processes, surface geomorphic features, and subsurface stratigraphic sections make possible a relatively complete understanding of the sequence of geomorphic events that have occurred in the coastal zone during late Holocene time.

Four major geomorphic-sedimentary environment sequence models established are as follows:

1. Barrier Systems: *barrier islands, baymouth barriers,* *and, barrier island chains* vary in amount of separation from the mainland by lagoons, distance from other islands, and number of inlets. All are characterized by a vertical transgressive sequence of sedimentary lithosomes identical to the horizontal sequence in the direction of the transgression.

Department of Geology, University of Delaware, Newark, Delaware 19711

2. *Beach against highland:* beach system impinges on low lying (20-30 ft) pre-Holocene highlands formed of Pleistocene sediments which erode and provide a partial source of sands, gravels, and mud to the system.

3. *Estuarine barriers:* long, arcuate barriers, small in width (150 ft) and thickness (10 ft) but of extreme length (up to 50-75 mi), occur along shorelines of large and small estuaries. Internal structures of these sand and gravel barriers are variable and complex; mainly composed of washover features. Four variants are identified.

4. *Spit complex:* extends into open marine and bay areas. All expected coastal transgressive environments of the barrier systems are seen in addition to regressive spit accretion sands and gravels. Consequently, vertical sequences may be disrupted and out of order.

Model studies of the variants of coastal morphologic units of the Delmarva Peninsula may be used to summarize the geologic history of the edge of the ongoing Holocene transgression. In addition, these models may be used to predict continuing changes in coastal morphology, assuming present and past processes will continue into the future. These predictions are useful to man in understanding and developing a more effective use of the coastal zone as man's intensity of involvement with coastal geomorphic environments continues to increase.

ACKNOWLEDGEMENTS

The majority of the work presented herin was supported by the Office of Naval Research, Contract Number N00014-69-A0407—Research in the Coastal and Oceanic Environment—The Coastal Sedimentary Environment. Work in the Assateague Island-Chincoteague Bay area by Biggs was supported by the National Park Service, Contract Number 14-10-5-590-36, Research at Assateague Island National Seashore. The NPS research was conducted at the University of Maryland.

INTRODUCTION

The present day mid-Atlantic coast of the northeastern United States includes a series of transgressive coastal environments that are migrating across the drowned continental shelf-coastal plain. This coastal plain and continental shelf is the upper surface of a geosyncline comprising up to 30,000 ft of Miocene, Paleocene-Eocene Cretaceous and possibly Jurassic-Triassic and older sediments covered by a thin veneer of Pleistocene sediments. From a Wisconsin glacial low sea stand of 11,000-14,000 yr B.P., the ocean level has risen and will continue to rise until the world's ice caps melt or a new glacial stage begins. Local relative sea level rise curves, constructed from over 50 radiocarbon dates show that the late Holocene marine transgression continues at a rate of approximately 0.5ft per century (Fig 1). Actual rates as measured by tide gages fluctuate widely. For instance, Hicks reported a relative sea level rise of 3 in over an 8 yr period in the decade of the 1960's. The long term averaging effect appears to be apporximately 0.5 ft per century over the past 3800 yr, with progressively greater rates previous to that time. Therefore, it is a happenstance of coastal plain tectonics, erosional patterns and world climate that the Delmarva shoreline is in its present position at this stage of the Holocene epoch.

Pleistocene geomorphic history has had an enormous effect on the Delmarva Peninsula (the coastal plain of Delaware, and the eastern shores of Maryland and Virginia between the Delaware and Chesapeake Bays). The eastern part of the Delmarva Coastal Plain clearly shows evidence of a higher Pleistocene sea stand. Clearly defined barrier shoreline features may be identified within the Pleistocene sedimentary sequences exposed on the coastal plain. These coastal Pleistocene geomorphic features have been correlated both with a mid-Wisconsin

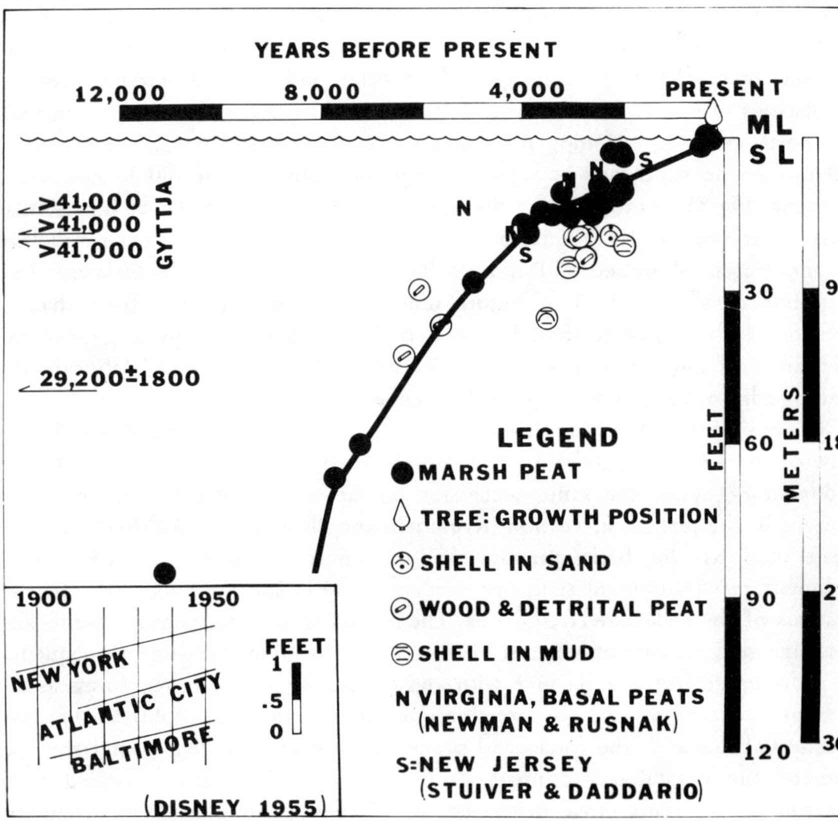

Figure 1. A relative ses level rise curve for the Delaware coastal area, based on marsh peats encountered in a drilling program in the study of the Holocene sedimentary lithosomes at the leading edge of the Holocene transgression. Radiocarbon dates are in years before present, 5580 yr. ½ life (Kraft, 1971a).

high sea stand and with the high sea stand that occurred in pre-Wisconsin-Sangamon interglacial time. Best evidences suggest that the Pleistocene coastal barrier features are a product of the gradually lowering late Sangamon-early Wisconsin ocean at the onset of Wisconsin glaciation (Kraft, 1971 b).

The morphology of the present transgressive coastal environments of the shoreline area is partially controlled by the nature of the pre-Holocene morphologic surface underlying the transgression. For instance, the large estuary, Delaware Bay, is a product of drowning of the ancestral Delaware River Valley and the location of the nearby Atlantic Ocean shore appears to be strongly controlled by the location of the axis of the ancestral Delaware River. Figure 2, an index to the area of study, shows the relationship of the coastal morphology of the Delmarva area to the geomorphic setting of the mid-Atlantic Bight. The submarine morphology as delineated in Figure 3 clearly indicates the axis of the ancestral Delaware River across the north-western flank of the submerged Atlantic Coastal Plain. In addition, on Figures 3 and 4, one can project present morphologic coastal features into the submarine area by similarities of shapes of submarine contours in the offshore area. For instance, the position of Cape Henlopen at the mouth of Delaware Bay can be projected 10 mi to the southeast as evidenced by the shape of the -60 ft subsurface contour (Fig 3), approximating the cape location of about 7,000 yr B.P. Similar examination of the -10 ft contour in Delaware Bay shows a very close parallelism to the present shoreline of Delaware Bay and approximates the Delaware Bay standline of about 2,000 yr before present. Evidences derived from shallow offshore drilling indicate that the area is one of a rapidly transgressing shoreline. The areas of salt water marsh, as indicated in Figures 3 and 4, delineate the leading edge of the present ongoing Holocene transgression.

The basic concept used in this study of the Holocene transgression is Walther's Law or the Law of Correlation of Facies. Walther postulated that within a given sedimentary cycle, the same succession of facies that occurs laterally is also present in the vertical succession (Krumbein and Sloss, 1963). Walther's Law has been used as the basic premise for the construction of detailed vertical sedimentary environment sequence models in the Holocene transgressive barrier systems of the Delaware coastal area. The coastal sedimentary environments and resulting sedimentary lithosomes in the study area include a dynamic sequence of geomorphic forms in balance with and constantly altering their forms as the present Holocene marine transgression continues. Each geomorphic feature is in precise balance with the forces and processes impinging on the coastal area. As part of the coastal environment sequences are eroded, newly formed areas develop as sediments move in the coastal system. Figure 5 shows a schematic analysis of Walther's Law as applied to the coastal geomorphic forms studied. A vertical sequence of sediment lithosomes is formed in the same order as the geographic-geomorphic forms of the present coastal sedimentary environments.

An interesting coastal geomorphic model has been presented by Fisher

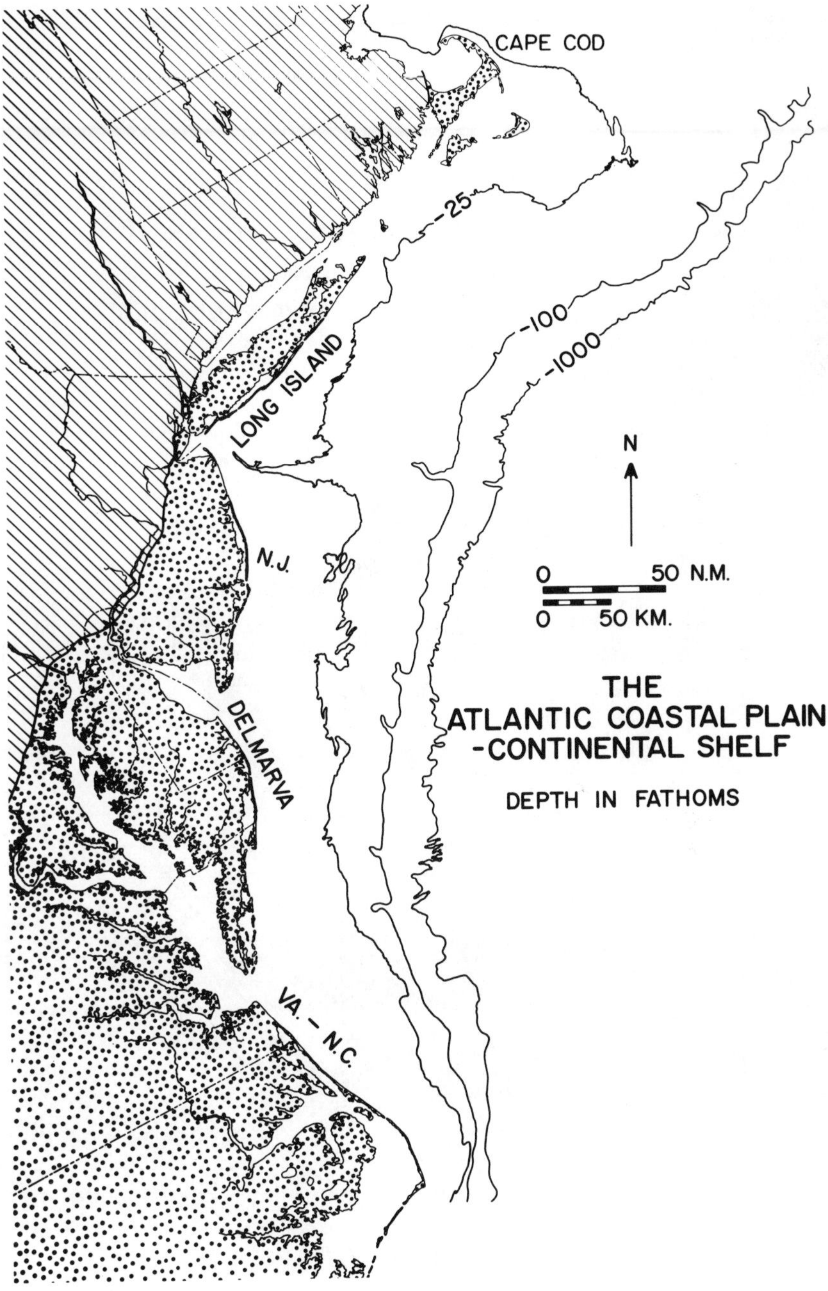

Figure 2. An index to the coastal barrier elements of the mid-Atlantic Bight showing the relationship of the coastal elements of the Delmarva Peninsula to similarly developed coastal morphic forms in New Jersey, New England, and mainland Virginia. Striped area=pre-Cretaceous; stippled area=emerged Atlantic continental shelf.

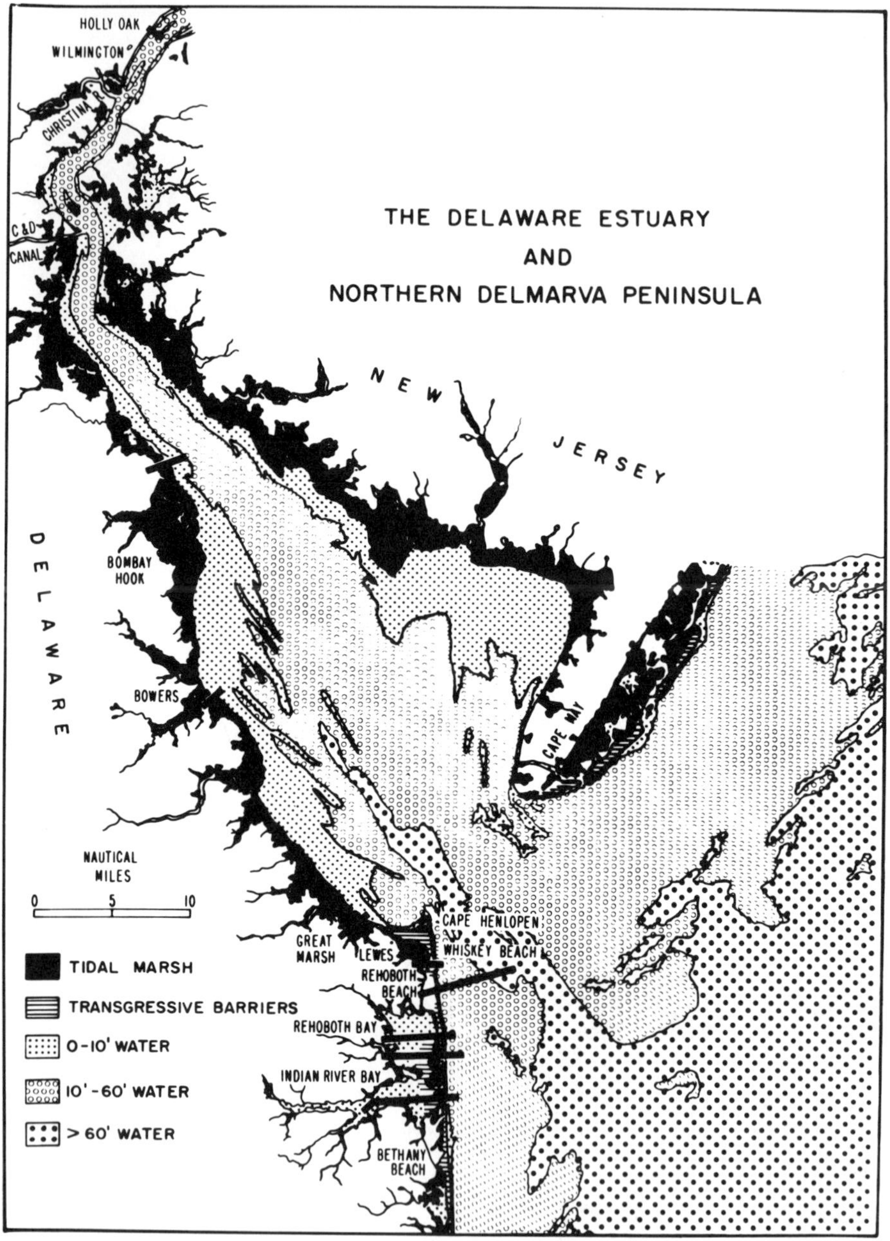

Figure 3. Transgressive coastal geomorphic elements in the Delaware Bay-Atlantic coastal area. Submarine morphology of the Delaware Bay and nearshore Atlantic Ocean is shown for comparison with the coastal environments forming at the edge of the ongoing transgression of the Atlantic Ocean.

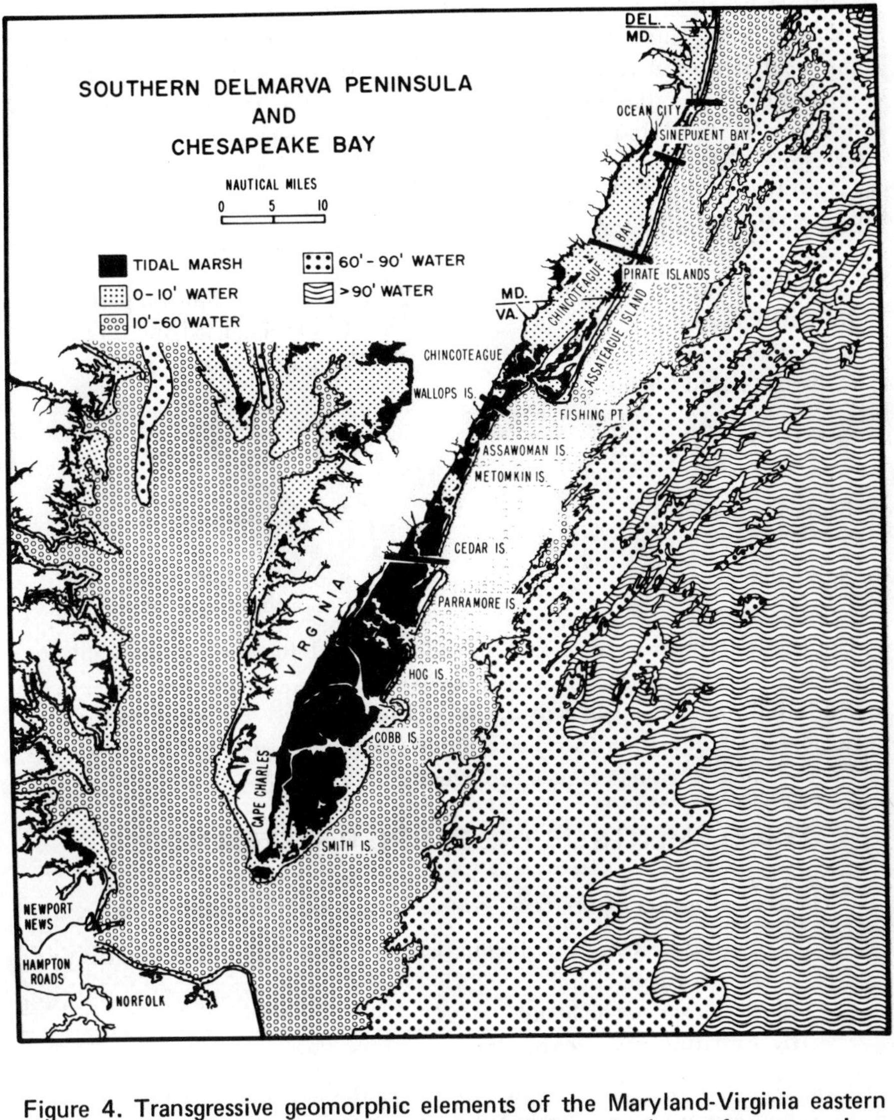

Figure 4. Transgressive geomorphic elements of the Maryland-Virginia eastern shore coastal area. Submarine morphology is shown for comparison with the coastal environments forming at the edge of the ongoing Late Holocene transgression of the Atlantic Ocean.

(Fisher, 1967, in Swift, 1969). He called attention to the fact that the large coastal segments of the Middle Atlantic Bight have strikingly similar surficial geomorphology (Fig 6). The coastal oceanic segments of Cape Cod, Long Island, New Jersey, Delmarva and North Carolina each have four compartments: (1) a northern of cuspate spit (2) an eroding headland (3) a southern spit, and (4) a barrier island chain. He believes these barrier features grew from headlands by distally prograding spits due to the prevailing easterly wind and wave direction in the Bight. However, analysis of the vertical component is needed to determine more fully the formation of these coastal compartments and the movement and types of the resultant sedimentary lithosomes in time and space. Therefore, the following table is presented contrasting Fisher's classification with the author's model that will now be developed:

DELMARVA PENINSULA	FISHER'S COASTAL COMPARTMENTS, 1967
1. Barrier System A. Barrier Islands B. Baymouth Barriers C. Barrier Island Chains	III. Southern Spit (in part) II. Eroding headland (in part) IV. Barrier Island Chain
2. Beach against highland	II Eroding headland
3 Estuarine barriers	———
4. Spit complexes	I. Northern or cuspate spit III. Southern spit (in part)

The models constructed should provide a useful framework applicable to many of the world's coastal areas. The barrier models presented here appear to be useful in both recent and ancient coastal studies. However, the ultimate test of a geologic model is that the sequence models developed are applicable in both recent and ancient sequences on a broad basis.

THE ELEMENTS OF THE BARRIER SYSTEM MODEL

The four elements of the transgressive barrier model presented herein are:

I. Barrier Systems, that include barrier islands, baymouth barriers, and barrier island chains.
II. Beach against highland.
III. Estuarine washover barriers.
IV. Spit complexes

Each element will be considered in detail taking into account its geomorphology as well as its three dimensional aspects.

I. BARRIER SYSTEMS.

By far, the largest part of the oceanic coast of Delmarva is composed of barrier systems that include barrier islands, baymouth barriers, and barrier island chains. These three barrier types vary in surface geomorphology by the amount of separation from the mainland by lagoons and distance from other islands and the number of inlets, but all are characterized by a vertical transgressive sequence of sedimentary environments identical to the horizontal sequence in the direction of the transgression.

An interpretive cross section of the barrier system shown in Figure 7 includes the typical vertical sequence as found in extensive borings in the Delmarva area. This sequence from bottom to top includes: (1) the Pleistocene sediment surface or land surface undergoing transgression (2) a marsh fringe composed of sandy mud with abundant rootlets or organic debris (3) organic muds of a coastal salt marsh with abundant salt marsh peats (4) gray, soft lagoonal mud with abundant molluscan fauna such as *Crassostrea virginica, Mercenaria mercenaria,* and *Nassarius obsoletus.* and (5) sands and gravels of the baymouth barrier.

The sand barrier part of the barrier system may be exceedingly complex. The bulk of the barrier sequence is comprised of beach and washover sands and gravel, with most sediments being similar to those of the present beach area. Fine to medium well sorted eolian cross-bedded dune sands cap the sequence

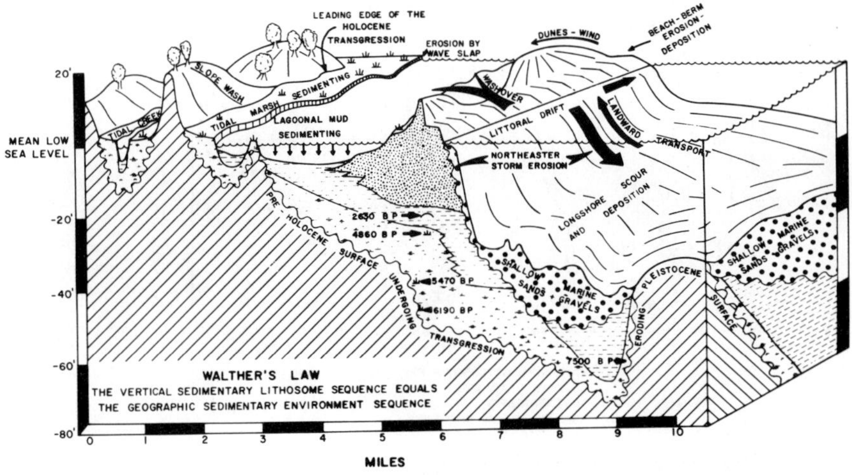

Figure 5. A schematic illustration of Walther's Law as applied to the Delmarva Coastal area. Sedimentary erosional-depositional processes are indicated to show the dynamic nature of this model of constant change but uniformity of process and resultant sequence.

and are typical of the sands of all the coastal dunes in the area of study. Dune sand may in fact be only an ephemeral feature in the barrier system sand bodies and may not be preserved in the subsurface. As transgression continues, sands from the beach are washed through the dune fields and the dunes tend to be destroyed by storms of highly variable frequency and intensity that occur along the coast. Sand transported by wind across the barrier probably forms only a minor fraction volumetrically of the sands transported into the coastal lagoons, but large parts of the beach areas can be stripped by storm waves and washed across the barrier into the coastal lagoon, thus forming washover fans. The tidal delta unit (Fig 7) is an important unit of the barrier sands. The tidal delta sands are medium to pebbly, rather clean, but sometimes include poorly sorted sands and gravel, with abundant foreset crossbedding in sets of 0.5 to 1.5 ft, dominantly in the landward direction. This dominance of landward oriented bedforms is due to the segregation of ebb and flood channels in a tidal delta and can be seen in present day tidal deltas. The flood tide, deeper as the wave advances, has its highest velocity currents near the crest of flood as the water spreads over the whole tidal delta. In contrast, the ebb, lowering as the wave withdraws, has its highest velocity currents near lowest water, when the water is confined to channels and other low spots in the lagoon or estuary (Coastal Research Group, University of Massachusetts, 1969).

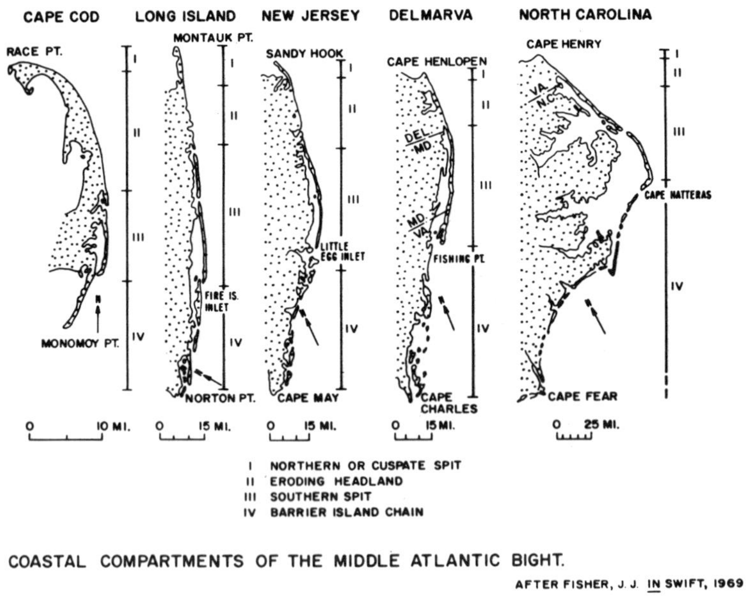

Figure 6. Coastal compartments of the Middle Atlantic Bight.

The superficial *Spartina patens* and *Spartina alterniflora* marsh on the back barrier area expands in width to accommodate the sand spread across the exposed sands of the tidal delta into the lagoonal area. As the inlet currents may cut deep tidal channels, sediments from previously deposited coastal environments can be eroded. Detrital masses of marsh peat and other organic matter, accordingly, may occur at lower physical positions in relation to relative sea level as determined from radiocarbon data. Many of these masses of eroded organic material can be recognized as being detrital. They are rolled in ball-like shapes and frequently occur against the unconformity.

Tidal inlets with their associated tidal deltas are features that migrate rapidly along the barrier systems. It is possible that the recognition of multiple tidal delta units in cores could be used to determine inlet migration along the barriers realizing that inlet tidal intrusions cause the normal sequence to be disturbed and out of order. Historically recorded positions of the present Indian River Inlet, Del., can be shown to have varied north and south over a range of 4 mi. The channels of the tidal delta from previous inlets are still open and full of water, indicating a sediment starved condition. They merge in a fan to a point at which the barrier has migrated across and sealed the old inlet with sand and gravel moved into place by littoral drift (Fig 8).

The sand barriers are formed from sublagoonal sand, washover sand and tidal

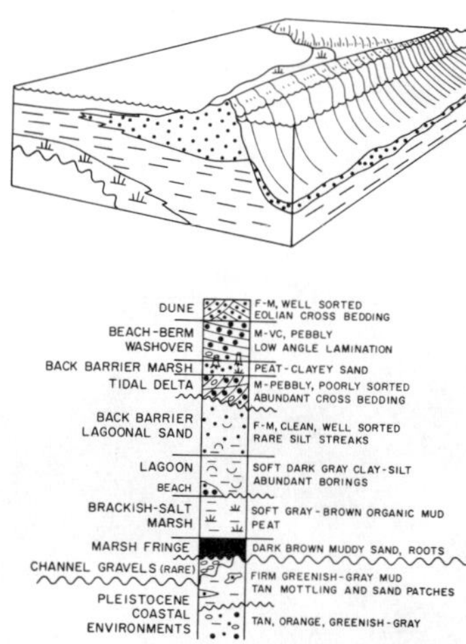

Figure 7. A block diagram of the baymouth barrier south of Rehoboth Beach, Del., between Rehoboth Lagoon on the left and the Atlantic Ocean on the right. Note the lenticular form of the barrier which is migrating landward and upward in space and time, extensive erosion occurs in the nearshore area in the form of wave generated littoral drift processes and washover and wind erosion in storm periods. A typical vertical sequence through a baymouth barrier as encountered by the drill through the Holocene coastal sedimentary environment sequences is shown.

delta sediments which are progressively overridden by a back barrier marsh area, then the beach-washover system, and finally capped by eolian dune sands. Should the transgression proceed, shallow marine sands and gravels, shell debris would override the barrier sequence. As shown in Figure 7, the barrier is lenticular in cross section and elongate and parallel to the shoreline of transgression. This lenticular sand body rests on coastal environment sequences such as lagoonal muds and fringing salt water marsh sediments that in some places may be seen to outcrop in the surf zone. The beach face of the lenticular body is an area of active erosion, transportation, and deposition, with a net loss of sediment as the foreshore area migrates in a landward direction.

Longshore drift or littoral drift transports large amounts of sand along the beach face, both emerged and submerged. Source of sand in motion across barrier systems and in littoral drift remains in question although generally it can be said that the shelf provides one of the sources for the sand in the littoral drift system (Shepard, 1963). A far more important source, however, is the zone of submerged foreshore of the barriers themselves. Projection of the present land surface across the barrier area and into the submarine area clearly shows a nick point with a relatively steep foreshore going from the beach area seaward to -30 ft in the Atlantic coastal area (Fig. 9 and 10). At that point, the shallow nearshore marine area flattens out and remains in a relatively planar -30 to -60 ft

Figure 8. An air photo oblique of the baymouth barrier south of Rehoboth Beach, Del., with superficial marsh covering flood tidal deltas of former inlets behind the transgressing barrier. Note abandoned anastomosing channels of old tidal delta. They are now silting up. The baymouth barrier is transgressing westward (landward) at a rate of 3 ft/yr averaged over 150 yr of relatively accurate map data.

submarine surface. This suggests that major erosion of sand occurs in the nearshore marine area at the relatively sharp rise towards the beach along the submerged foreshore. It appears that sand and other sediments are being removed from the beach face and both transported in the littoral drift system and spread as a lag sediment in the shallow marine area.

Another logical source of the sediment in transport along the Delaware and northern Maryland coast is the eroding Pleistocene highlands undergoing transgression. The Pleistocene sediment surface has relief of approximately 25 ft and is undergoing transgression and disappearing by coastal erosion. Accordingly, the bulk of the sediment in motion in the shoreline area was, and probably continues to be derived from the land. Deposition is in part in the Holocene shoreline environments and in part in the relict sand, gravel sheets strewn across the shallow continental shelf area as the transgression continues.

Much discussion in the literature is related to the origin of barrier features of

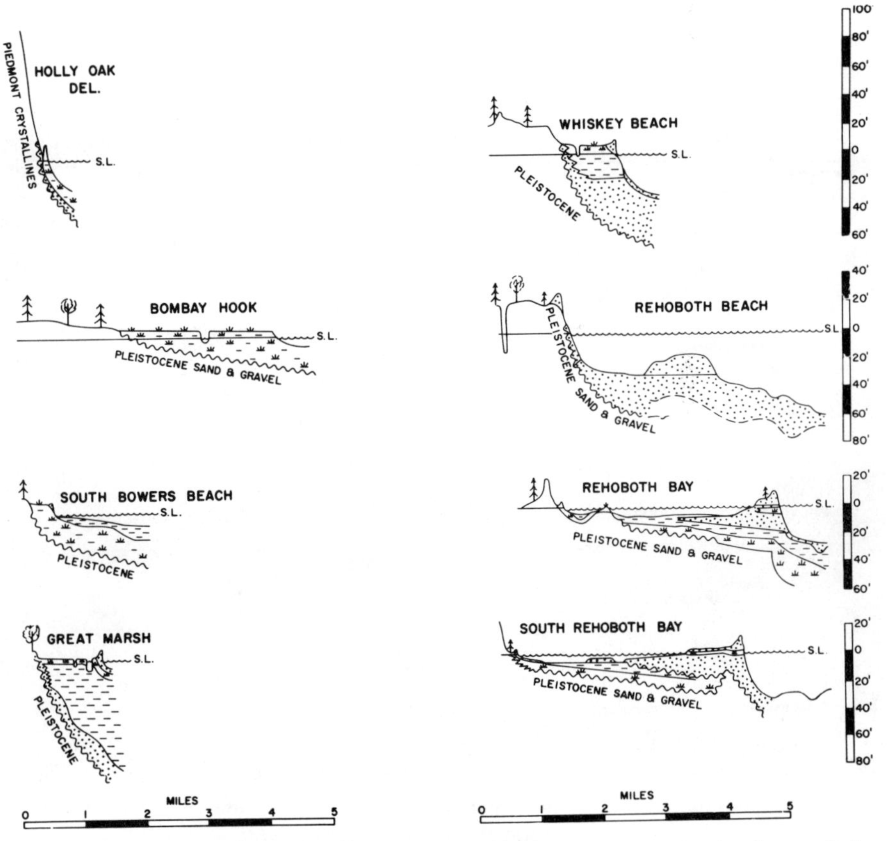

Figure 9. A series of schematic cross sections from the northerly shore of the Delaware estuary southward to the mouth of Delaware Bay and thence south along the Atlantic coast. Specific locations of section are shown on index map Figure 3. An interpretation of the geology of the Holocene sedimentary sequences is shown.

this type. Evidence derived from the area of this study shows the barriers that exist along this coast did not form in place. Rather, they have migrated to their present position from elsewhere on the Atlantic continental shelf. Projection by means of submarine morphology suggests that present barriers have migrated from a previous position at least 5 mi seaward on the shallow continental shelf. Possibly, these barriers were formed by the projection of spits between highland ridges and then began to migrate landward as sea level rose and coastal erosion proceeded. Evidence must be sought from the offshore submarine area to prove origin. Extensive studies are in progress in the nearshore submarine zone using drilling, seismic and geomorphic techniques on this problem.

Shepard (1960) has enumerated several characteristics of barrier islands. These include: (1) an outlying belt of sand separated from the mainland by a shallow body of water (2) much greater length than width, and (3) straight or gently

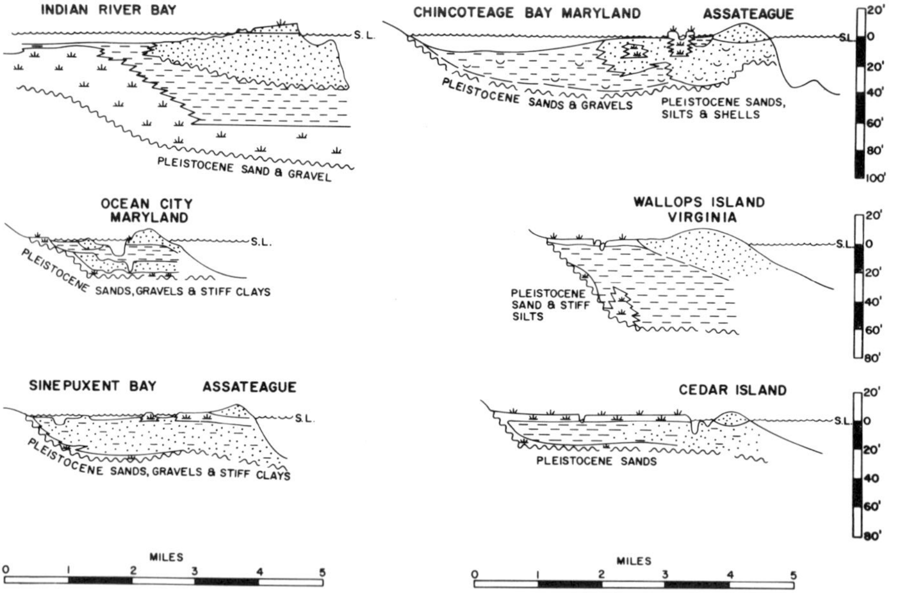

Figure 10. A series of schematic cross sections of Holocene transgressive sequences from the Indian River Bay area on the Delaware Atlantic Coast southward along the Delaware, Maryland, and Virginia eastern shore with the Atlantic Ocean. A geologic interpretation of the Holocene transgressive sedimentary units is shown. See index figures 3 and 4 for location of schematic sections. The section entitled "Cedar Island" is after Newman and Munsart (1968).

curving seaward margin in contrast to a crenulate lagoonal shoreline. The area between the island and the mainland may vary in width from less than 1 mi to more than 20 mi, and consists of open bays, mud or sand flats, and marshes.

The barrier islands bordering the Delmarva Peninsula are of three distinct geomorphic types. Assateague Island is a long, straight, or gently curving barrier whereas Wallops, Assawoman, Metomkin, Cedar, Parramore, Hog, Cobb, and Smith Islands, form a segmented chain with numerous inlets (see Fig 4). The third type is the baymouth barrier that forms a line across the mouths of coastal lagoons such as found at Rehoboth Bay and Indian River Bay between Rehoboth Beach and Bethany Beach, Del., and Assawoman Bay north of Ocean City, Md.

Data presented in Table 1 demonstrate the differences in geometric character among the islands. The coastal lagoons are also quite distinctive. The lagoon behind Assateague Island is open water with fringes of marsh and mud flats, while all of the lagoons to the south of Assateague Island include mud flats and marshes with open water restricted to deep, narrow, channels and a few open bays. Inlets are spaced at about 10 km intervals from Cobb Island to Wallops Island (Table 1); northward lies the unbroken Assateague Island, 60 km from Chincoteague to Ocean City. Shepard (1960) has suggested that lagoonal sediments may be derived from the ocean by way of inlets, and perhaps that is why former lagoons south of Assateague Island are infilled with mud.

There are four significant marshy areas behind Assateague Island. Most of the remaining marshes may be considered as fringing Chincoteague Bay without extending significantly into open water. Most of the large marshes are being eroded, except for the Chincoteague marshes which are stable or growing due to the proximity of the inlets to Chincoteague Bay (Fig. 11). These marshes of Chincoteague probably are developing on the sand flats of flood tidal deltas associated with the tidal influx through the inlets. Tidal range is higher near inlets (2.5 ft at Chincoteague versus 0.5 ft in mid-Chincoteague Bay) and the currents associated with inlets are capable of depositing sand on flats that eventually build high enough to be populated by marsh grasses.

Two of the extensive marshes in Chincoteague Bay are now isolated from existing inlets but historical accounts and maps indicate persistent natural inlets at the location of these extensive marshes (Truitt, 9169). These inlets were located in the proper position to have had flood tidal deltas, thereby stinulating marsh development. Now, however, these marshes are retrograding due to bay currents and wave slap and because the inlets that were associated with their formation have closed, thus decreasing the tidal range and cutting off the source of sediment to create new flats for marsh encroachment.

A prominent feature of central Chincoteague Bay is a large shoal extending 5 km westward from Assateague, and topped by the Pirate Islands (Fig 4 and 10–Chincoteague section). Sub-bottom data indicate that Pleistocene sediments lie at shallower depths beneath the Pirate Islands. Thus, the major shoal area aligned east-west across Chincoteague Bay seems to be the surface expression of

Table 1

GEOMETRY OF DELMARVA BARRIERS AND LAGOONS

	Length Barrier Km (mi.)	Maximum Width Barrier Km (mi.)	Maximum Width Lagoon Km (mi.)	Total Area Lagoon Km^2(mi^2.)	Area Open Water Km^2(mi^2.)	Area Flats & Marshes Km^2(mi^2.)	Area Open Water / Area Lagoon Km^2(mi^2.)
Assateague	57.96(31.30)	4.52(2.44)	11.59(6.25)	428.90(231.60)	320.12(178.86)	108.18(58.47)	0.75(0.40)
Wallops	9.04(4.88)	2.51(1.36)	4.02(2.17)	21.93(11.84)	3.54(1.91)	18.38(9.89)	0.16(0.09)
Assawoman	5.04(2.72)	2.01(1.09)	2.51(1.36)	13.31(7.19)	3.05(1.65)	10.25(5.54)	0.23(0.12)
Metomkin	10.54(5.69)	1.38(0.75)	3.52(1.90)	30.35(16.39)	15.77(7.02)	14.52(7.84)	0.52(0.28)
Cedar	9.43(6.09)	3.76(2.03)	5.54(2.99)	47.94(25.89)	9.09(4.91)	38.85(20.98)	0.19(0.10)
Parramore	13.07(7.06)	3.52(1.90)	7.73(4.17)	95.18(51.30)	18.07(9.76)	77.10(41.63)	0.19(0.10)
Hog	10.15(5.48)	3.52(1.90)	12.56(6.78)	113.18(61.12)	10.74(5.80)	102.43(55.31)	0.09(0.05)
Cobb	9.05(4.89)	7.05(3.80)	13.07(7.06)	113.44(61.26)	16.31(8.81)	97.31(52.45)	0.14(0.08)

a buried Pleistocene topographic high (Biggs, 1970). All of the barrier islands are oriented northeast. Most of the major offshore shoals are aligned significantly more to the northeast than is the present coastline, with the exception of Wallops and Chincoteague Islands.

Beginning at Cedar Island, all of the barriers to the north form a smooth curve bending eastward and ending at Chincoteague. Chincoteague Island exhibits numerous northeast trending dunes or beach ridges which are nipped off at the northern end of the island. Apparently, Chincoteague was once a barrier island (possibly even a cape feature–Shepard and Wanless, 1971), and Assateague Island has extended southward to "capture" the coastline. Assateague Island has been extended southward by the formation of the hooked spit, Fishing Point, since 1850 (Gawne, 1966), (Fig. 11 and 12) and there is little doubt this southerly extension had been occurring before that time. It is tempting to extend to the north, the arc formed by Cedar, Metomkin, Assawoman, Wallops, and Chincoteague Islands because such an extension would predict an inlet at the northern tip of Chincoteague (perhaps that is why the dunes are nipped off at the northern end), and another barrier island extending northward.

It is hypothesized that Assateague Island has coalesced from a series of smaller islands similar to the chain of barriers to the south or to Fire Island, Long Island, N.Y. (Hayes and Kaczorowski, personal communication). The lack of numerous inlets on Assateague Island may be related to the nearshore profile of the island, which is much steeper than the nearshore profiles further south. If the steep profile accentuates longshore drift, then inlets might become clogged with sand, whicle the more gently sloping bottom further south would decrease the competence of longshore currents and allow inlets to remain open (Biggs, 1970). More detailed drilling on Assateague Island will be necessary to determine the morphology of the subsurface topography in order to ascertain whether Assateague Island did coalesce from smaller islands or developed from a pro-grading spit.

From Wallops Island southward to Cape Charles the barrier island chain is composed of segmented barrier islands separated by numerous inlets with broad tidal marshes and flats infilling former coastal lagoons (Fig. 13 and 14). The Holocene vertical sequence of a barrier island chain (Fig. 13) is in many ways simpler than the Holocene vertical sequences for barrier islands and baymouth barriers (Fig. 7 and 10).

As the narrow southernmost mainland of Pleistocene sediments on the Delmarva Peninsula is sharply truncated in a straight, linear scarp, a very small drainage area is exposed (Fig. 4). Accordingly, it is probable that the muds that have infilled the lagoons and marshes of Wallops Island and southwards were transported into the area by waters coming through inlets located further east of the presently transgressing sand-starved washover barriers. Limited evidence, based in part on radiocarbon dating (Harrison and others, 1965), suggests that the Delmarva Peninsula is subsiding basinwards towards the north-east. Possibly this area has an extremely complex tectonic and sedimentary history. If so,

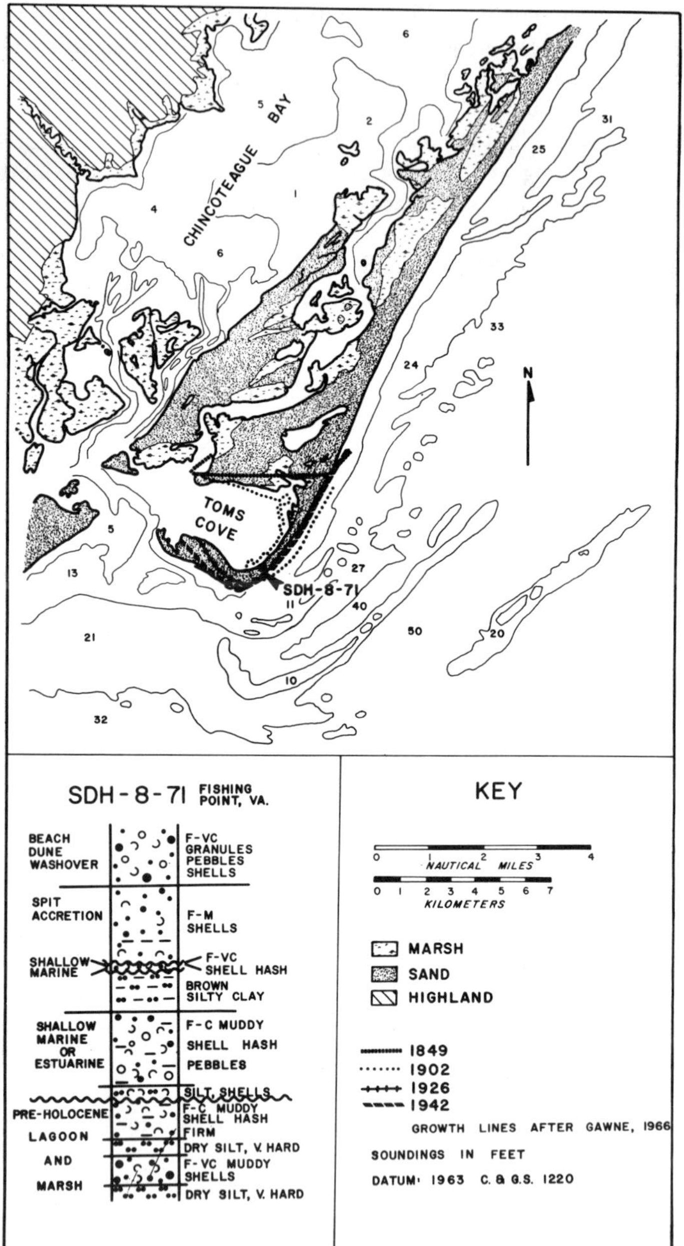

Figure 11. Holocene coastal environments in the Fishing Point area at the southern end of Assateague Island-Chincoteague Bay. A log of a drill hole on the tip of Fishing Point is shown with an interpretive sequence of Holocene transgressive sedimentary environments.

much more work remains before a full understanding of the origins of the broad tidal marsh flats south of the Wallops Island area is attained (Fig. 14).

II. BEACH AGAINST HIGHLAND

A small but significant part of the barrier coast of the Delmarva Peninsula consists of beaches built against low lying (20-30 ft) pre-Holocene highlands of Pleistocene or pre-Pleistocene sediments. These highlands are areas of coastal erosion with a thin sequence of sands of the dune, beach and offshore bar encroaching on the highland (Fig. 9, 15, and 16).

With a significantly high intensity coastal storm event, the entire beach-berm and offshore bar are lifted off the base of the highland and dumped back into the submarine area or removed into the littoral drift system. At that point, significant erosion of the Pleistocene sediments occurs and a major contribution to the littoral drift system is made. During storms, sand will be moved offshore, alongshore and inland. With the change back to normal coastal conditions, the offshore bar migrates shoreward and the beach begins to develop as the berm rebuilds upward at a position further inland.

After a particularly destructive coastal storm, the beach system rarely takes more than one-half year to build back to equilibrium. In the interim, however, the shoreline area itself may have migrated landward over an increment of tens

Figure 12. An air photo oblique of the spit at Fishing Point.

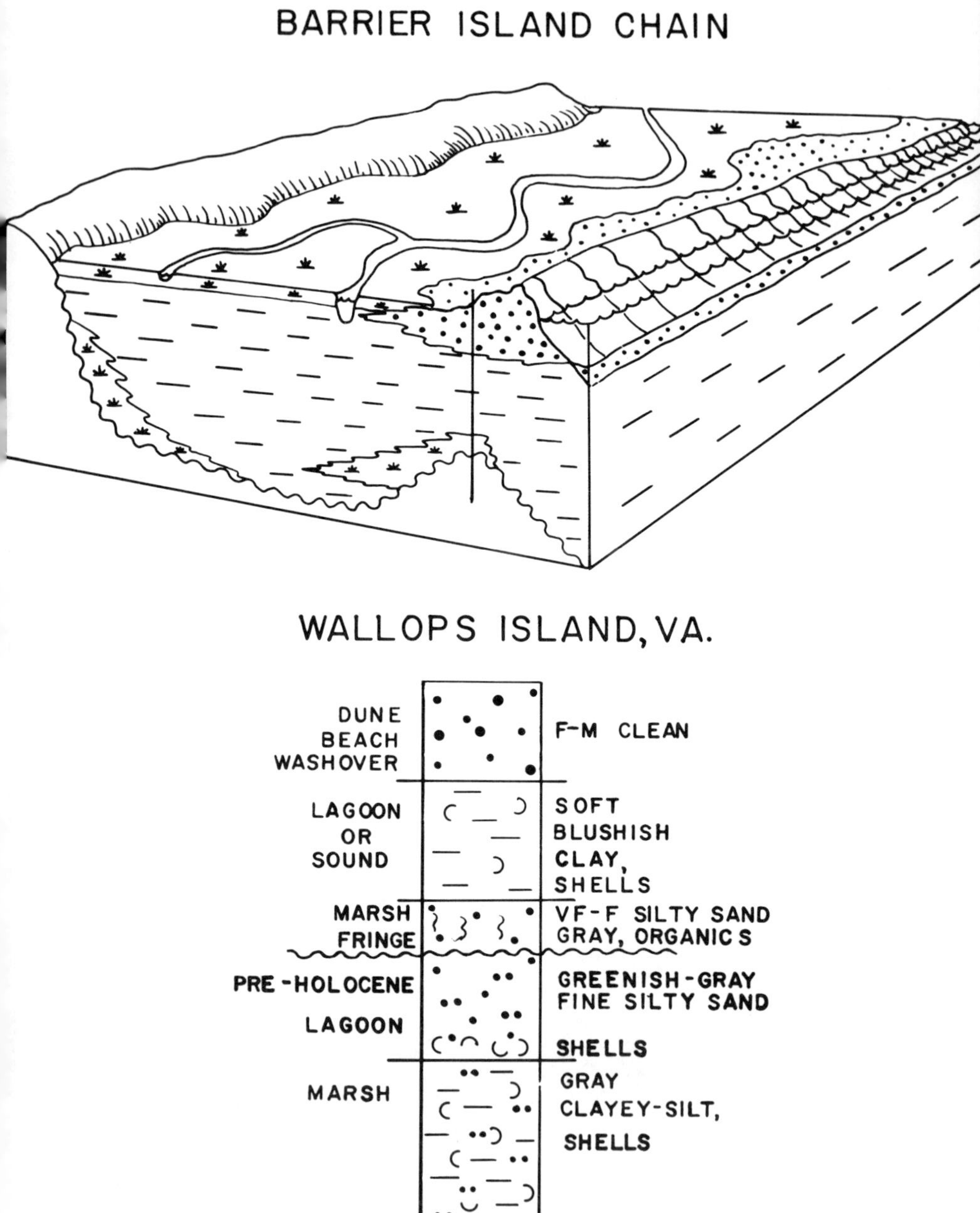

Figure 13. A perspective interpretation of the Holocene transgressive environmental sequences and subsurface geology of the Wallops Island Area, Va.

of feet. As shown in Figure 3, the Pleistocene or pre-Holocene surface can be traced into the submarine area. This has been confirmed by means of seismic reflections by R. E. Sheridan (oral communication). One can make direct correlations between the pre-Holocene hills of Pleistocene sediment that are undergoing erosion at the present coastline area and the valley of the ancestral Delaware River and other stream erosion topography underlying the Holocene sequence.

The beach area is composed of medium to very coarse sands with low angle seaward dipping lamination parallel to the beach face while the back beach shows low angle landward dipping laminations parallel to the berm surface. Only during post-storm recovery periods do these beaches show higher angle landward dipping foresets of the constructing ridge and runnel system so common to the New England beaches studied by the Coastal Research Group of the University of Massachusetts. Often a dune forms just inland of the beach area and includes fine to medium well sorted sands with eolian cross bedding.

A question remains as to whether the sedimentary units formed against the highland are preserved within stratigraphic sequences; however, some evidence found in drilling to the north, under the Cape Henlopen spit, dune, and marsh tract suggests a beach sand and gravel section occurs against the pre-Holocene unconformity in an area of coastal transgression. It is, therefore, extremely

Figure 14. An air photo oblique of the Cedar Island marsh and washover barrier of the southern portion of the Atlantic shore of the Delmarva Peninsula. Broad marshes have infilled former coastal lagoons between Pleistocene highlands to the west and offshore "barrier" islands which formerly extended to the east.

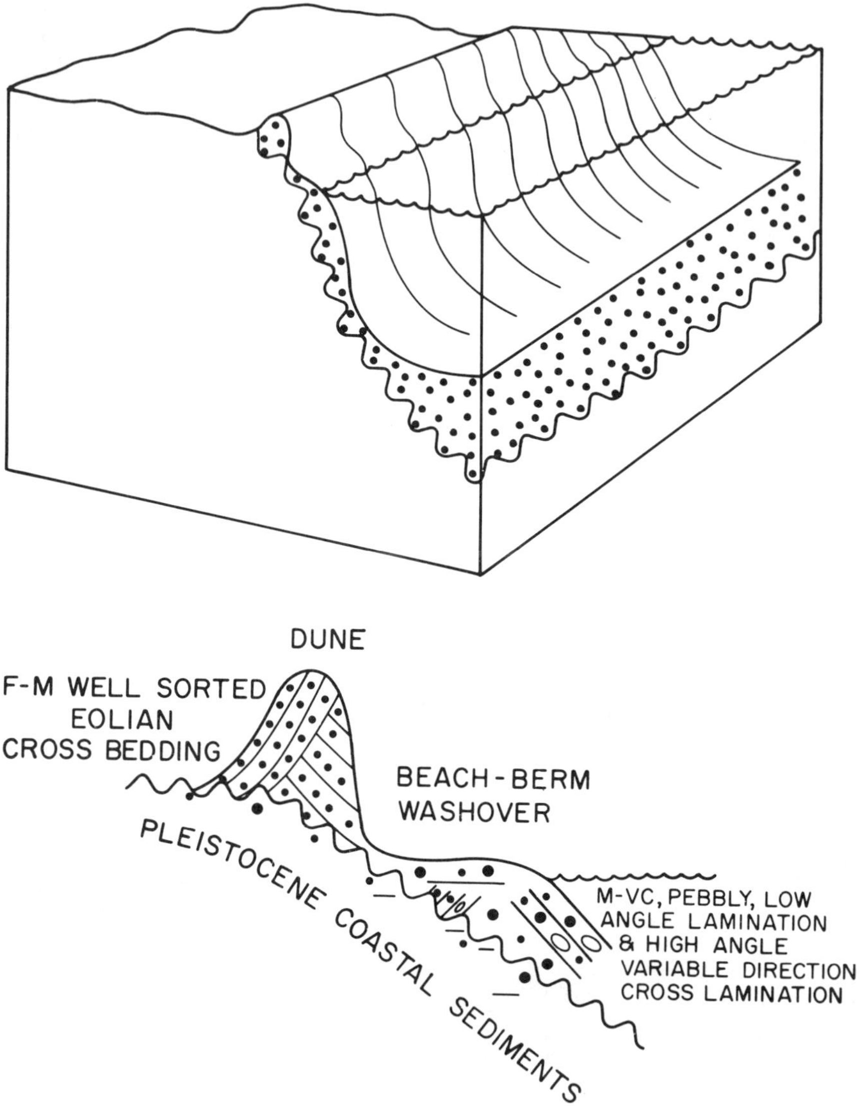

Figure 15. A block diagram interpretation of the eroding Atlantic shore against the pre-Holocene highland at Rehoboth Beach, Del. A true vertical sequence of Holocene sediments does not develop in this situation. Rather, a dune field forms in advance of the beach berm and washover area. During high intensity storms, the entire beach berm system is removed and the Pleistocene coastal sediments of the highland area are truncated by erosive wave action. It is believed that the truncation and erosion of highlands during high intensity storms provides a major source of sediments into the littoral drift stream.

important that workers in coastal transgressive units pay strict attention to their core log's vertical sequences in order to determine the proper paleogeomorphic interpretations.

III. ESTUARINE WASHOVER BARRIERS

Estuarine barrier beaches form on the shorelines of the large and small estuaries, bays, and lagoons in the area of study. They are predominantly found on the westerly and southwesterly sides. This is probably because the dominant energy source in the formation of estuarine barriers is the storm wind-generated waves from the "northeasters" and other storms that frequently occur along the coastal area. Muds of the turbulent Delaware Bay are transported by the tidal streams across the broad coastal marshes where the mud is entrapped and deposited. As a result, the marsh surface steadily grows landward and upward in space and time.

Figure 17 is a diagram showing a cross section of an estuarine washover barrier. One estuarine washover barrier studied in detail is located near South Bowers on the west central shoreline of Delaware Bay (Fig. 3, 17, and 18). The barrier at South Bowers is approximately 100 ft wide and 5-10 ft thick. The sedimentary units within the estuarine washover form thin 0.5-2 ft thick sand-gravel beds, frequently with high angle cross lamination facing landward,

Figure 16. An areal photo oblique showing the 10-20 ft above sea level pre-Holocene highland of Pleistocene coastal sediments at Rehoboth Beach, Del.

indicating formation by strong washover currents towards the land. These units are truncated at the eroding beach face from which the majority of the washover sands and gravels are derived. In other words, the estuarine washover barrier is feeding itself from one side to the other as it moves inland. Additional sources of sediment are derived from the longshore drift system and from further winnowing of the sands mixed within the poorly sorted *Spartina* marsh muds that are eroding in the intertidal nearshore Bay area.

There is a tendency for coarser sand and gravel to occur seaward and upward in the washover barrier. The finer sands and silts tend to be more abundant landward in the barrier sequence. Large amounts of sand and silt are washed over the barrier in storms and flood across the muds of the *Spartina patens and Spartina alterniflora* marsh possibly occurring as thin streaks of sand and silt within the mud sequence. The sands and gravels of the estuarine washover barrier are surprisingly "clean" and lacking in silt and clay. This is probably due to the high energy applied in the littoral drift system and by storm waves. The fine sediment derived from erosion of the dead marsh surface in the nearshore marine area is deposited dominantly in the offshore bay area and transported by waves across the barrier system and by tidal waters which spread across the *Spartina* coastal marshes. Accordingly, a highly efficient selective mechanism of deposition and transport occurs in the formation of estuarine washover barrier features.

The vertical sequence expected in an estuarine washover barrier is shown in Figure 17. Overlying the unconformity on Pleistocene coastal sediments is a marsh fringe comprised of dark brown muddy sand and roots overlain by soft gray-brown organic mud and peat composed of abundant *Spartina alterniflora* and *Spartina patens* fragments as well as other lesser elements of the marsh flora. The surface of truncation at low tide cuts the salt marsh section. The thin washover barrier forms by a winnowing process. At some points estuarine washover areas are capped by thin dunes of fine to medium well sorted sand. As in the barrier system and beach against highland areas, the thin dune sands may not be preserved in the stratigraphic record. Thin washover barrier sands can be shown to be continuous over extremely long distances, up to 50-75 mi. in the area of study. The only breaks in the estuarine washover barriers occur at the inlets into the tidal creeks. The sands formed are essentially "shoestring" sands occurring in long arc-like sections along the western Delaware Bay coast.

Another area in which estuarine washover barriers were intensively studied is the southwestern corner of Rehoboth Bay, the coastal lagoon behind the baymouth barrier south of Rehoboth Beach shown in Figures 3 and 9. Here, as in the Delaware Bay model previously discussed, a miniature estuarine washover barrier occurs at the edge of a coastal lagoon (Fig 9). Moderate to well sorted sands are flanked by salt marsh clays and silts on the one side, and a mixture of mud and sand in the nearshore submerged area of the coastal lagoon on the other. The estuarine washover barriers landward of Rehoboth Bay are much smaller in geographic extent and are more discontinuous than those of the

ESTUARINE WASHOVER BARRIER

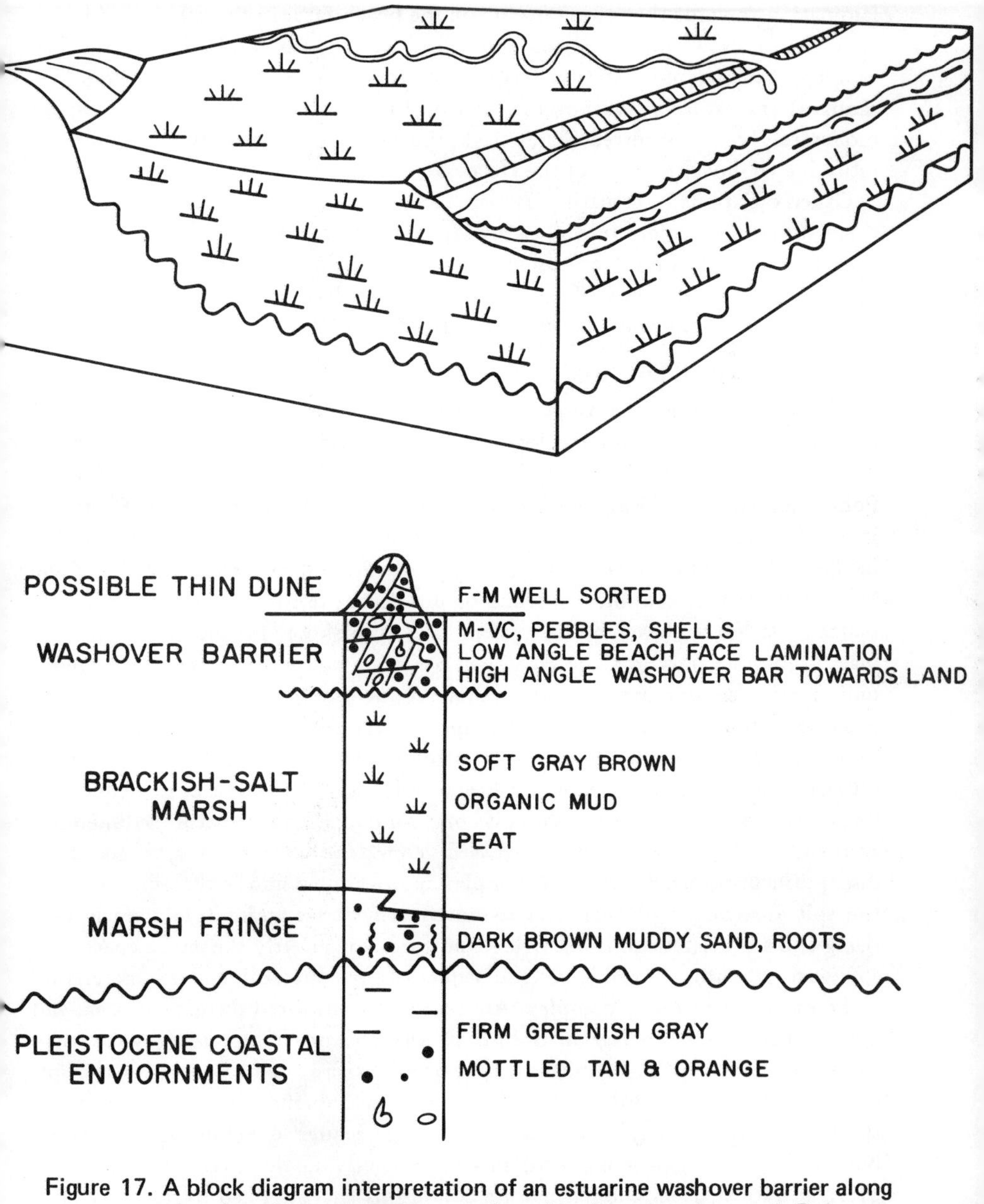

Figure 17. A block diagram interpretation of an estuarine washover barrier along the broad transgressive marshes fringing the drowned Delaware estuary. The thin, estuarine washover barriers migrate landward and upward in space and time across the marsh area which is also growing upward from tidal sedimentation of mud. A typical vertical sequence as encountered by the drill is shown.

Delaware Bay to the north. However, they exhibit similar features. High angle cross-lamination, dominantly in a landward direction with angles from 9-30°, occurs throughout the washover barriers. Low angle, less than 5° parallel lamination towards the lagoon occurs on the beach face. In the submerged beach area, the sands are poorly sorted and mixed with marsh and lagoonal mud. The eroding *Spartina* mud section frequently crops out at the low tide line. The sands of the estuarine washover barrier on the shores of the coastal lagoon are moderately to well sorted. They lack the high gravel content typical of the estuarine washover barriers of the Delaware estuary.

Other variations in estuarine barriers can occur because of the substrate that these estuarine barriers transgress across. The pre-transgression topography can vary from the Piedmont crystallines in the northern Delaware Bay (Fig. 9), to pre-Pleistocene sediments of eastern Chesapeake Bay, to the wide expanses of Holocene marsh in lower Delaware Bay (Fig. 9 and 10).

IV. PROGRADING SPIT TRACTS

The most complex transgressive barrier systems occur in the areas of prograding spits. The Cape Henlopen spit, marsh, dune and lagoon tract between Rehoboth Beach and Lewes, Delaware (Fig 19), as well as the spit at Fishing Point, Chincoteague, Virginia (Fig 12) are undergoing intensive study. Figure 20 is a schematic cross section of the Holocene sedimentary sequence that has infilled the deeply incised topography of Pleistocene sediments at Cape Henlopen. The geographic sequence of pre-Holocene topography, *Spartina* salt marsh, washover beach-berm system and nearshore marine sands of the marsh-shoreline tract at Cape Henlopen is not indicative of the vertical sequence underlying the washover barrier system at Whiskey Beach (Fig 9) north of Rehoboth Beach, Delaware. In the line of section of Figure 20, the vertical sedimentary sequence includes the pre-Holocene surface and unconformity, an estuarine-shallow marine section, a beach and spit sand and gravel sequence, and washovers or dunes. The amount of complexity in vertical sedimentary sequences has, accordingly, increased considerably. The major sequence disruption introduced by a spit complex in a transgression is that shorelines in the spit area might advance over shallow marine areas while in fact sea level is rising and a transgression is occurring landward along nearly the entire coast.

Figure 20 shows possible vertical sequences and the sediment characteristics to be expected in a spit complex. After a spit has migrated through an area, and a back barrier coastal lagoon is formed, the onward transgressing ocean may form a vertical sequence similar to that of a barrier system. The tip of the spit complex, itself, would more likely show a vertical sequence such as illustrated on the left of Figure 20. This vertical sedimentary sequence will include beach and buried shoreline against highland situations overlying the basal unconformity, overlain by estuarine or open marine conditions characterized by fine to coarse sand with moderate to poor sorting and an abundant molluscan fauna. The estuarine-open marine environment may be terminated by the deposition of a

Figure 18. An air photo oblique showing the Island Field-South Bowers estuarine washover barrier on the shoreline of mid-central Delaware Bay. Note the outcrop of marsh mud in the surf area at low tide on the bayward side of the washover barrier.

Figure 19. An air photo oblique showing the migrating coastal elements of the spit complex at Cape Henlopen. The vertical sedimentary sequences in the spit-tip area are "regressive" in the sense that land or shoreward environments are migrating over shallow marine area.

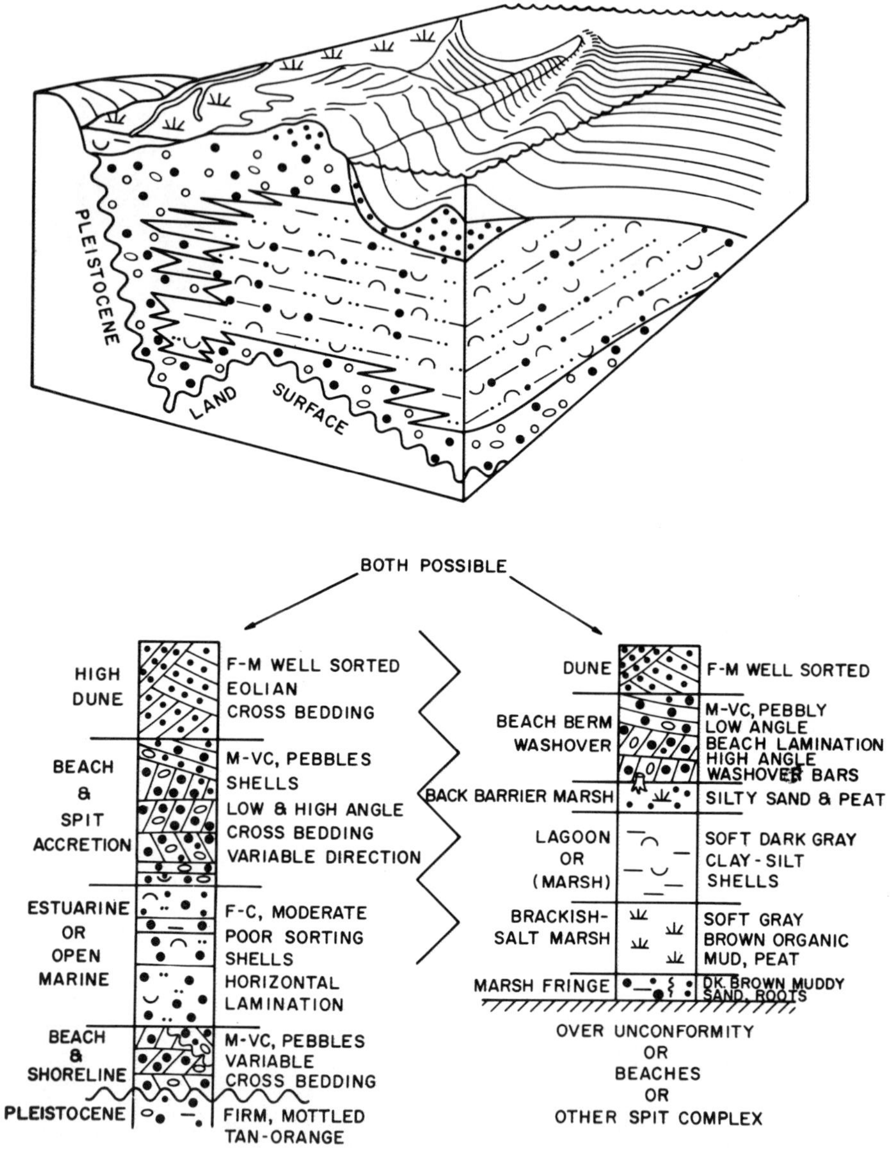

Figure 20. A block diagram showing the Holocene sedimentary lithosomes of a migrating spit complex at Cape Henlopen, Del. Vertical sedimentary sequences in spit complexes are highly variable. As shown, a "normal" transgressive sedimentary sequence of coastal beach-dune sediments over lagoon and marsh is possible on the one hand as at Whiskey Beach (right above), whereas, on the other hand, a "regressive" vertical sequence occurs when the spit migrates over the shallow marine area and coastal spit and beach sands and gravels overlie estuarine and shallow marine sediments (left above).

beach and spit accretion sequence of highly variable lithologies as the spit progrades and infills the shallow marine area. Sediment characterizing this section includes medium to very coarse sands, pebbly sands, and gravels with varying amounts of detrital shell material and of high and low angle cross-bedding with variable dip. The entire sequence may be overlain by a fine to medium well sorted eolian crossbedded sand.

The Cape Henlopen area is one of a rapidly eroding Atlantic shoreline with large beach-washover barrier (Whiskey Beach, Fig. 21) moving landward across an infilled lagoon, now the site of a coastal marsh. Longshore or littoral drift here is to the north. The sand and gravel in motion in the longshore drift reaches the tip of the spit and most is dumped into the relatively deep estuarine-shallow marine area of the mouth of Delaware Bay (Fig. 19). A smaller portion of the sediment is swept around to the lee of Cape Henlopen spit forming large scale flood oriented swash bars that migrate across the tidal flat thus adding sediment to the estuarine side of the spit (Halsey and Kraft, 1971). As the spit progrades and the ebb and flood tidal currents impinge upon the tip of the spit, a highly selective winnowing action occurs. The ebb tide current removes the fine sand fraction and selectively transports it into a narrow linear ebb tide shoal extending toward the southeast for about 10 mi. This shoal historically can be shown to have always been associated with the tip of the spit. Accordingly, it

Figure 21. An air photo oblique showing the large beach-berm washover barrier transgressing across the Lewes Creek marsh at the southern end of the Cape Henlopen spit complex at Whiskey Beach. In this part of a migrating spit complex, "transgressive" vertical sedimentary sequences occur as the shoreline berm and washover sands and gravels migrate landward across lagoonal and marsh sediments at the leading edge of the Holocene transgression.

follows that the formation of this shoal is controlled by the development of the spit, Cape Henlopen.

The large recurved spit at Fishing Point, Va. near Chincoteague (Fig. 11 and 12), also shows the complex superposition of environments similar to Cape Henlopen, Del. Here 45 ft of fine to coarse sand has migrated over lagoonal muds underlain by muddy shelly sands of a large open bay, possibly once the size of Pamlico Sound, N.C. Sediments interpreted as Pleistocene lagoon or marsh underlie this shallow marine unit.

Due to the progradation of the recurved spit of Fishing Point, a protected bay, Toms Cove, has formed causing various transgressive elements of the barrier system vertical sequence to develop on, around, and because of, the progradation of the regressive units of the spit. Lagoonal sediments are forming in Toms Cove while back barrier marshes are growing and modifying the shores of the cove. Storms and times of high wind activity cause extensive areas of washover sands, dune migration, and beach truncation. Therefore, in areas of prograding spit tracts, multiple vertical sequence of regressive spit sequences alternate with transgressive sequences of barrier systems making the interpretation of drill holes very complex.

CONCLUSIONS

In summary, there are four elements to the Delmarva Peninsula model of coastal transgressive barriers: "The barrier system" includes barrier islands, baymouth barriers and barrier island chains, all characterized by strikingly similar vertical transgressive sequences, which are identical to the horizontal sequence in the direction of the transgression. The second element, "the beach against highland" is the simplest vertical sequence model. Here the beach encroaches on a low lying pre-Holocene highland of Pleistocene or pre-Pleistocene sediments which erode and provide a source of sediment to the littoral drift system. This situation may be ephemeral, and not occur in the stratigraphic record although some local evidence suggests a beach system preserved against the underlying pre-Holocene highland erosion surface.

The third element, "the estuarine washover barrier" occurs on both the larger and smaller estuarine bodies in the area of study. Estuarine washover barriers along the western shoreline of Delaware Bay are narrow, extremely thin, but continuous for great lengths, whereas estuarine washover barriers on the southwest sides of the coastal lagoons and bays are in miniature and extremely short and discontinuous. It would be difficult in the subsurface to differentiate between estuarine washover barriers from a small body of water and a larger body of water. In fact, problems might occur in differentiating between the stratigraphic record formed by estuarine washover barriers and the normal washover barrier-beach systems of the open Atlantic Coast. However, ocean barrier systems are much larger in size and overall dimension than the estuarine washover barriers. Sedimentary sequences may sometimes be differentiated between estuarine washover barriers and ocean barrier systems in Holocene sections when one can clearly identify lagoonal microfaunal and marsh grass fragments in the underlying mud. Where extremes of lithification and leaching have occurred in the geologic record, it may be impossible to separate the two.

The fourth element, "spit complexes", are anomalous. One might expect normal washover barrier and barrier systems intermixed with vertical sedimentary sequences that include beach and spit sands and gravels migrating over open marine and estuarine sediments. So long as three points of vertical sequence control are available, it appears feasible to identify the nature of the barrier systems of a spit and whether or not a transgression or a regression has occurred. However, with a single point of control, it may be extremely difficult to make an objective interpretation of a vertical sedimentary sequence in the coastal environment.

The determination of the Holocene transgressive barrier sequence models has many potential applications to geology and geography. Detailed study in an area such as Delmarva affords other workers a model to use in study of their coastal areas. With the increased emphasis on environmental study and the population crush in coastal areas it is necessary for geologists, geographers, engineers and planners to understand the types of geomorphology that are peculiar to coastal areas. Such problems as coastal erosion; water resources, engineering structures such as high rise condominiums, bridges and causeways; dredging for landfill and beach nourishment; and maintenance of navigable waterways all hinge on intelligent understanding and use of the coastal zone. However, intelligent and proper use of the coastal zone depends heavily on the information and data geologists and geographers can dispense to the people responsible for building and development.

The data scientists gather about our present (Holocene) time has a great application to predicting the types of landform change and development into the short and long term future. Geologists base assumptions on Hutton's Law of **Uniformitarianism** –"the same processes acting today have been acting throughout geologic time and will continue to do so". Therefore, the data collected on coastal processes and morphology of the present ongoing transgression have use not only in interpreting ancient coastal areas but in predicting the future so long as the present marine transgression continues. Studies of rate of change of coastal morphology can prove useful in precise delineation of Late Holocene paleogeology and paleogeography. These paleogeomorphic studies are now proving critical in understanding the archaeology of coastal areas in the recent past, as well as through neolithic time. In addition, paleogeographic analysis forms the basis for accurate projection of coastal morphology into the short term "environmental problem future" as well as into the longer term end of the Holocene Epoch of geologic time (Fig. 22).

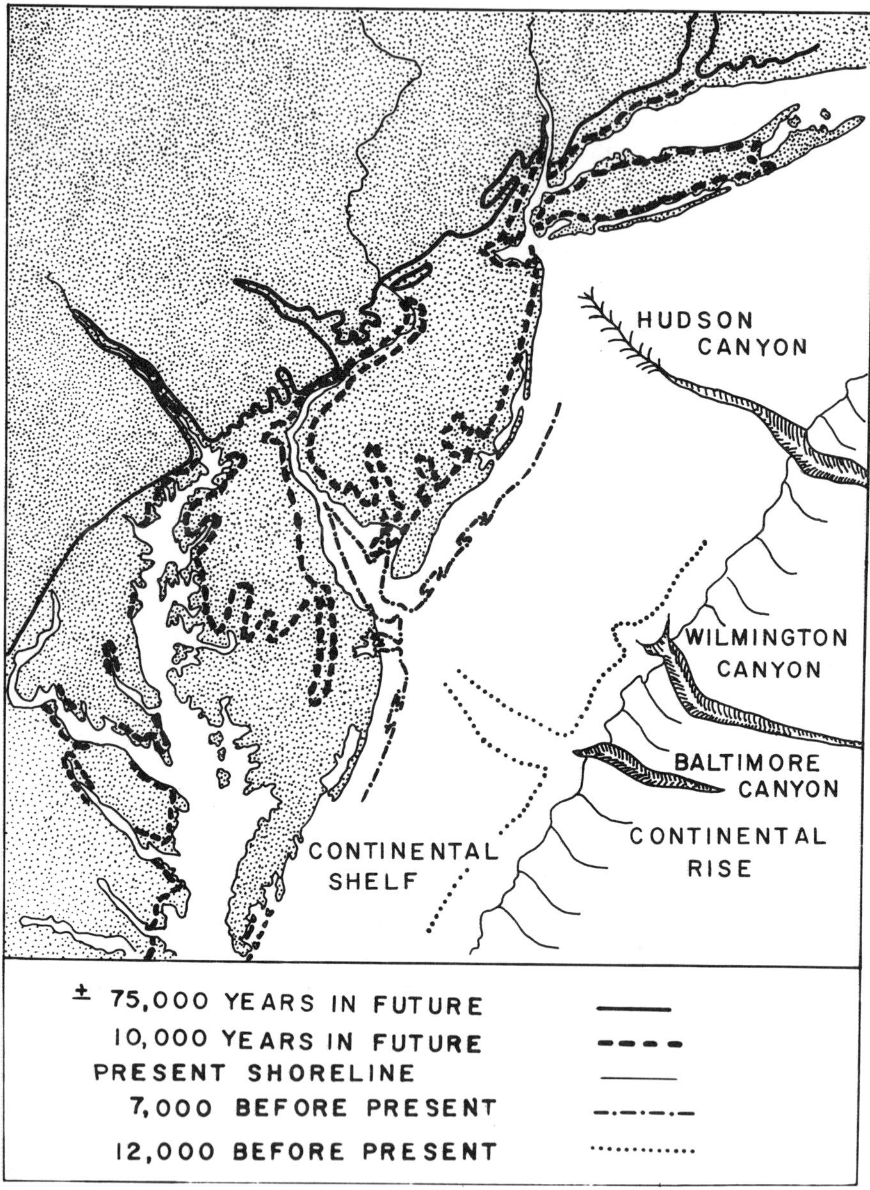

Figure 22. Paleogeographic and future projection of coastal morphology of the Delmarva Peninsula and New Jersey area. Projection is based on paleogeographic analysis (Kraft, 1971c) and the assumption that present processes of coastal change and sea level rise will continue into the short term geologic future of the Holocene Epoch.

REFERENCES

Biggs, R.B. 1970. The origin and geological history of Assateague Island, Md. and Va: p. 8-41 in Assateague Ecological Studies, Final Report, Part I, Univ. of Maryland, Natural Resources Inst. Cont. No. 446, 426 p.

Coastal Research Group. 1969. Coastal environments: Northeast Massachusetts and New Hampshire: Field trip guidebook, Eastern section, Society of Economic Paleontologists and Mineralogists, Coastal Research Group Contrib. No. 1 CRG Dept. of Geology, University of Massachusetts, 462 p.

Disney, L.P. 1955. Tide heights along the coasts of the United States: Proceedings, Amer. Soc. of Civ. Eng., v. 81, p. 6661-6664.

Fisher, John J. 1967. Origin of barrier island chain shore lines: Middle Atlantic States: p. 66-67, in Abstracts with Programs, Annual meeting of Geol. Soc. of Amer., New Orleans, Special Paper No. 115.

Gawne, C.E. 1966. Shore changes on Fenwick and Assateague Islands, Md. and Va.: B.A. Thesis, University of Illinois, 140p.

Halsey, S.D., and Kraft, J.C. 1971. Anatomy of a spit: Cape Henlopen, Del.: Abs. with Programs, Geol. Soc. Amer., Blacksburg, v. 3, p. 314-315.

Harrison, W., Malloy, R.J., Rusnak, G.A., and Terasmae, J. 1965. Possible Late Pleistocene uplift Chesapeake Bay entrance: Jour. Geol., v. 73, p. 201-229.

Hicks, S.D. 1972. Note on rising sea level: National Oceanic and Atmospheric Agency release—in Amer. Geol. Inst. Report, v. 5, n. 1, Jan. 3.

Kraft, J.C. 1971a. Sedimentary sequences and facies patterns in a middle-late Holocene spit complex: Program with abstracts, VIII International Sedimentological Congress, 1971, p 53.

Kraft, J.C. 1971b. Sedimentary facies pattern and geologic history of a Holocene marine transgression: Geol. Soc. Amer. Bull., v. 82, p. 2131-2158.

Kraft, J.C. 1971c. A guide to the geology of Delaware's coastal environments: Field trip Guidebook Geol. Soc. of Amer., College of Marine Studies Publications, Univ. of Delaware, 220 p.

Krumbein, W.C., and Sloss, L.L. 1963. *Sedimentology and stratigraphy:* Freeman and Co., San Francisco, 2nd. ed, 660 p.

Newman, W.S. and Munsart, C.A. 1968. Holocene geology of the Wachapreague lagoon, eastern shore peninsula, Va: Marine Geology, v. 6, p. 81-105.

Newman, W.S. and Rusnak, G.A. 1965. Holocene submergence of the eastern shore of Va.: Science, v. 148, p. 1464-1466.

Shepard, F.P. 1960. Gulf Coast barriers: p. 338-344 in Shepard, F.P. and others, ed., Recent Marine Sediments, Northwest Gulf of Mexico, AAPG, Tulsa, Okla.

Shepard, F.P. 1963. *Submarine Geology:* 2nd ed., Harper and Row, New York, 558 p.

Shepard, F.P. and Wanless, H.R. 1971. *Our changing coastlines* : McGraw Hill, New York, 579 p.

Stuiver, M., and Daddario, J.J. 1963. Submergence of the New Jersey Coast: Science, v. 142, p. 951.

Swift, D.J.P. 1969. Inner shelf sedimentation: processes and products: p. DS 4-1-46, in Stanley, D.J., *The new concepts of continental margin sedimentation: application to the geologic record:* Amer. Geol. Inst., Washington, D.C., 276 p.

Truitt, R.V. 1969. High winds—high tides: a chronicle of Maryland's coastal hurricanes: Educational Series No. 77, Natural Resources Inst. Univ. of Maryland, 35 p.

Chapter 16

COASTAL DUNES OF YOUNGER BERMUDA

Len Vacher

ABSTRACT

Younger Bermuda (Sayles, 1931) is the collection of large dune-shaped hills along Bermuda's external shoreline. These lithified calcarenitic dunes retain their original depositional morphology and are easily distinguished from altered, older dunes which lie inland.

The stratigraphy of these young calcarenites is presented in terms of multiple systems of stratigraphic classification. The Paget Formation is the rock-stratigraphic equivalent of the morphostratigraphic unit, Younger Bermuda. The Paget is divisible in its eolian facies into upper and lower members that reflect two distinct Late Pleistocene submergences of the shallow platform surrounding Bermuda. The marine (beach and shallow-water) facies includes long beach wedges transitional with eolianite. A soil-stratigraphic unit separates the two members and represents an interval of platform emergence.

The Bermudian dunes were large accretionary mounds that extended inland no more than about 0.5 km from their source beaches. As the mounds grew, they merged laterally to produce a transverse ridge parallel to the shoreline. The growth direction of the individual dunes is reflected by their long axes or, equivalently, by the symmetry axes of sets of horizontally convex-leeward foresets that represent the individual mounds. Apparently this growth direction was determined largely by strong winds and gales. The consistent pattern of horizontally convex-leeward foresets curving around the high points of the ridge is one of several ways the accretionary transverse dune bodies of Bermuda differ from migratory transverse dunes.

Two source-sink systems leading to the formation of calcareous dunes in Bermuda are inferred from stratigraphic relations and present-day sedimentary patterns. The first is a reef-to-dune, source-sink system for shorelines close to the platform margin where reefs are established during platform submergences. The resultant dunes (South Shore) consist of clasts of reef-dwelling organisms, and, apparently, the size of the dunes reflects reef productivity. The second system is a lagoon-to-dune, source-sink system for shorelines separated from

---

Department of Geology, State University of New York at Binghamton; and Consultant Geologist/Hydrologist, Public Works Department, Hamilton, Bermuda.

platform-margin reefs by a wide lagoon. Under "normal" conditions of platform submergence (like today), the reefs act as a submerged breakwater that filters out the waves capable of moving lagoonal calcarenite onto the shoreline. The lagoon becomes a source for dunes (North Shore) upon a rapid rise of sea level that nullifies the breakwater.

The rapid, 15 m rise in sea level about 125,000 yr ago that led to the formation of early Paget dunes along Bermuda's lagoon-facing shoreline may represent a large surge of the Antarctic Ice Sheet.

INTRODUCTION

Bermuda is a well-known group of limestone islands located about 1,000 km east of Cape Hatteras, the nearest point of land. The islands number over 150, with five of them accounting for nearly all of Bermuda's 20 sq. mi. The islands are near the southern margin of a 650 km^2 platform, presently submerged to depths of 20 m. The platform is the site of the northernmost coral reefs in the North Atlantic.

The exposed limestones are Pleistocene calcarenite and are representative of a widespread coastal Pleistocene lithofacies. Over 90 percent of the calcarenite is eolian (Fig. 1). The remainder formed in beach and shallow-water environments. The calcarenites are interstratified with paleosols, some of which are red (terra-rossa) and have downward protruding soil pipes. Other well-known examples of calcareous eolianites and associated rocks are in the Bahamas (Ball, 1967), Australia (Fairbridge and Teichert, 1953), and along Mediterranean shorelines (Butzer and Cuerda, 1962; Wright, 1962; Yaalon and Laronne, 1971).

The youngest calcarenites of Bermuda occur around the periphery of the islands. To a very large degree, these young rock bodies retain their depositional morphology – large, dune-shaped hills, the sum of which Sayles (1931) called Younger Bermuda. The Younger Bermuda surface is one of the most striking features of the geomorphology of these small oceanic islands.

This chapter concerns the depositional history of the dunes comprising Younger Bermuda. The key questions are:

—What were the conditions necessary to their formation?

—Given that dunes would form, what factors determined their morphology?

—What was the environment of Bermuda when the dunes formed?

—What facets of Bermudian geologic history – recorded by the dunes – reflect events of regional or global scale?

Obviously from the historical nature of these questions, answers must be found in the stratigraphy.

STRATIGRAPHY

General

The stratigraphic framework of this paper (Table 1) is patterned after the Illinois Geological Survey's system of multiple schemes of stratigraphic classification (most recently summarized by Willman and Frye, 1970). A morphostratigraphic unit (Willman and Frye, 1970, p. 43) "... (comprises) a body of rock that is identitied primarily from the surface form it displays ... (and) may or may not be distinctive lithologically ...." Rock-stratigraphic units are (p. 40) "... defined on the basis of observable lithology without necessary regard to biological, time, or other types of criteria. ... (and are) sufficiently distinctive to be recognizable by common field ... methods." A soil-stratigraphic unit (p. 42) "... is a weathered zone ... defined on the basis of the stratigraphic position of the top of the soil."

One new formal rock-stratigraphic unit, the Paget Formation, is used here. It embraces everything above the Shore Hills Soil of the stratigraphic column of Land, *et al* (1967) (see Table 2). The Paget is the rock-stratigraphic equivalent of the morphostratigraphic unit, Younger Bermuda, as that term was used by Sayles (1931). The Paget includes an eolian facies and a volumetrically minor marine (beach and shallow-water) facies. The formation is divisible in its eolian facies into upper and lower (informal, but objective) members. Each member consists of areally disjunct, informal map units, called eolianites or eolian map units, which are lithologically indistinct, but not necessarily precisely time-equivalent. Marine map units are essentially tongues enclosed by Paget eolian calcarenite. They are not included in any particular member, for to do so would commonly destroy the lithologic identity of the member.

The Paget Formation is underlain by and interlaced with paleosols. Two within the Paget and the one beneath the formation constitute useful informal soil-stratigraphic units.

Time-stratigraphic units are inferred from the objective rock-stratigraphic units and are informal. References to "Paget time," "early Paget," and "late Paget" mean the time spanned by, respectively, the Paget Formation, and the Lower and Upper Paget members.

Figure 1. Paget eolianite at Horseshoe Bay. Foresets and overlying windward cross beds.

The stratigraphic framework of Table 1 represents a revision of earlier stratigraphic work in Bermuda (Table 2). Revision, at least in names, is necessary because of the loss of stratigraphic identify following the redefinition of type localities. The Pembroke Formation (an eolian unit) of earlier reports is an example. The type Pembroke of Sayles (1931) in part grades into the type Belmont of Land, *et al* (1967). This beach unit is in turn overlain by their type Shore Hills (an extension of which overlies Sayles' type Pembroke), and their own type Pembroke.

The revision is more than a change in names, however, because it represents the abandonment of a theme which has prevailed since Sayles (1931): that the various movements of Pleistocene sea level *must be reflected* in the stratigraphic column. Thus the Bermudian column has become a set of time-stratigraphic units that are not mappable on the basis of objective criteria. This was clearly recognized by the most recent authors, Land, *et al* (1967), for they state (p. 1005), "... (the units) cannot be differentiated on the basis of lithologic or paleontologic criteria alone.... The geometric and temporal limits of Bermudian formations, i.e., their boundaries, are dictated by the conceptual scheme into which they are placed – the rise and fall of the Pleistocene sea." The role of major sea-level oscillations and of fresh-water diagenesis in the formation of the Bermudian aggregate of lithified beach and dune calcarenites and interspersed soils clearly would not have been appreciated without the Land, *et al* (1967) column and those from which it grew. However, the questions now being asked of Bermudian stratigraphy demand documentation of how the individual rock bodies (lithosomes such as eolianites and beach wedges) fit together in a three-dimensional mosaic. This documentation must be a geologic map, which in turn demands a stratigraphic column consisting of objectively mappable units.

The mappable column in Bermuda is a succession of major bundles, each *internally* a complex mosaic of lithosomal packages – eolianites, shallow-water calcarenites, and paleosols. It is the overall succession that is treated by the objective stratigraphy. Thus the objective units are broader-scale, or more coarse-grained, than the units of preceding columns; the longer objective units contain multiple limestone-soil cyclothems (as they are called by Gould, 1960), and, in the time sense, embrace multiple sea-level oscillations. The details of the internal mosaics are handled by the myriad informal map units. Use of the two together – objective stratigraphic units and a host of informal map units – deals with both the overall succession and the internal complexities simultaneously. This chapter concerns the uppermost bundle, the Paget Formation, in five of Bermuda's nine parishes (Fig. 2).

Morphostratigraphy: Younger Bermuda

Sayles (1931, p.445) first recognized the two striking surfaces in Bermuda, called them Older and Younger Bermuda, and used the terms in the

TABLE 1. STRATIGRAPHIC FRAMEWORK

A. Multiple systems of classification (eolian facies).

ROCK- Formal objective units	ROCK- Informal	ROCK- Informal map units	/ MORPHO-	/ SOIL-	/ TIME-
PAGET FORMATION	Upper Paget member	Saucos Hill, Southlands	YOUNGER BERMUDA		late Paget
				intra-Paget	
	Lower Paget member	Barker's Hill, Spice Hill, Watch Hill			early Paget
				sub-Paget	

B. Paget Formation: eolian and marine facies.

North Shore marine facies	eolian facies	South Shore marine facies
St. Catherine map unit	Upper Paget member	St. Catherine map unit
i n t r a - P a g e t s o i l		
Barker's Hill beach unit	Lower Paget member	Devonshire map unit
s u b - P a g e t s o i l		

TABLE 4. PAGET LITHOLOGIC IDENTITY

A. Typical Paget (eolianite altered in the intermediate-aerated zone).

Individual grains	Grains of aragonitic organisms recognizable (mineralogically non-inverted). Not chalky as in upper units of sub-Paget. Moldic porosity rare. Constituent-particle morphology like that in sand of modern beaches. <u>Homotrema rubrum</u>, in Upper Paget, pink as in modern South Shore beaches; somewhat bleached in Lower Paget; generally colorless in sub-Paget.
Cement	Loosely bound and consolidated. Point-contact cement in Upper Paget. Point-contact and rim cement in Lower Paget.

B. Qualifications

facies depositional	facies diagenetic	
beach	beach-rock cemented	Not recognized in Paget rocks. Sub-Paget examples are like typical Paget, except they are more tightly cemented (Land, <u>et al</u>. 1967); such examples can normally be traced landward into eolianite of typical sub-Paget lithology.
marine	not beach-rock cemented	Commonly less altered and cemented than overlying eolianite. Thus some sub-Paget examples are lithologically like typical Paget; however, they can be identified if the overlying eolianite is sub-Paget.
eolian	phreatic	Generally more altered and cemented than contemporaneous eolianites of vadose diagenesis. Relationship well documented in sub-Paget rocks by Land (1970). Phreatic Paget not identified in surface exposures, and should be limited to elevations below maximum of Paget sea level.
eolian	upper-aerated zone	Commonly more leached or tightly cemented than underlying eolianite of intermediate-aerated zone diagenesis.

morphostratigraphic sense. "Older Bermuda (forms) a partially submerged core and (is) flanked on both the north and south by younger dunes. He noted that the 'relations may be well observed from Gibb's Hill lighthouse, which is built in a region of pronounced dune topography (along the southern shore). A mile to the north . . . the subdued, rolling topography is in striking contrast to the dune topography of the coasts. Two miles to the north are the islands of the Sounds, and they likewise have a low, rolling surface. In the vicinity of the city of Hamilton (to the north) similar conditions prevail." Around Hamilton "are areas with subdued surfaces." North is the North Shore which "portrays a typical dune topography."

Figure 3 shows the typical topographic distinction between Younger and Older Bermuda: details of the dune morphology preserved in Younger Bermuda and smoothed out in Older Bermuda. Such relations, as seen in large-scale topographic maps, are apparent throughout Bermuda.

There is no question that the two surfaces are of different age. Younger Bermuda is equivalent to the Paget Formation. Older Bermuda is sub-Paget. Thus the lithologic distinction (reflecting largely a different degree of diagenesis) between Paget and sub-Paget calcarenites applies to the two morphostratigraphic units as well. The major unconformity marked by the sub-Paget soil that lies between Paget and sub-Paget calcarenites also separates Younger and Older Bermuda. There is also little doubt that the Older Bermuda surface evolved from an initial surface like that of Younger Bermuda, for sedimentary-structural relations in the two morphostratigraphic units are identical. The evolution is the result of geochemical processes attending the passage of water through the calcarenites (Plummer *et al,* in prep). The processes affect both the lithic properties (through mineralogic stabilization and cementation) and the surface morphology simultaneously – thus the parallel distinction between morpho- and rock-stratigraphic units.

Rock Stratigraphy: the Paget Formation.

The Paget Formation is here defined from a type section along the Paget/Warwick Parish boundary, where the formation is entirely of eolian facies. This definition and type section supercedes the manuscript definition of the Paget Limestone by Vacher (1971). The type section is briefly described in Table 3, which also includes two selected reference sections (Barker's Hill and Grape Bay sections) to cover the depositional facies variations pertinent to the paper.

Figure 4 shows the areally disjunct eolianite map units that compose the Upper and Lower Paget members, defined in the Paget/Warwick section. Four eolianites (Spice Hill, Watch Hill, the Rocky Bay lobes, and the eolianite component of the Barker's Hill beach-dune complex) comprise the Lower Paget member. Two eolian map units (Southlands, Saucos Hill) comprise the Upper Paget member. Together these six rock bodies constitute the eolian facies of the Paget Formation.

TABLE 2. BERMUDA STRATIGRAPHIC COLUMN ACCORDING TO VARIOUS AUTHORS

Sayles, 1931	Bretz, 1960	Land, et al, 1967		This paper*	
Southampton eol.					
McGall's soil					
Somerset eol.	later eolianites	Southampton Fm. (eol.)	PAGET FORMATION	Upper Paget member	(S.H.)
Signal Hill soil					
Warwick eol.					
St. George's soil	St. George's soil	St. George's Soil		intra-Paget soil	
	Spencer's Point marine ls.	Spencer's Point Fm. (mar.)			(D.)
Pembroke eol.	Pembroke eol.	Pembroke Fm. (eol.)		Lower Paget member	(R.B.)
Harrington soil	Harrington soil	Harrington Fm. (soil)			(G.B.)
Devonshire (marine) ls.	Devonshire marine ls and eolianite	Devonshire Fm. (mar.)			(D.)
Shore Hills soil	Shore Hills soil	Shore Hills Soil		sub-Paget soil	
Belmont (marine) ls.	Belmont limestone	Belmont Fm. (mar.)			
		Soil?			
Walsingham eol.	Walsingham eol.	Walsingham Fm. (eol.)			

* The last column equates the Land, et al, stratigraphic units, at their type sections, with map units of this paper. Abbreviations: S.H., Saucos Hill; D., Devonshire; R.B., Rocky Bay lobes; G.B., Grape Bay soil.

TABLE 3. PAGET FORMATION: TYPE AND REFERENCE SECTIONS

A. Type section. Paget/Warwick boundary. Horizontal section from South Shore to Cobb's Hill Rd./Ord Rd. intersection.

5. Upper Paget member; Southlands eolianite. Forms ridge south of South Shore Rd.
4. intra-Paget soil. White to pink regosol. Exposed on South Shore Rd., 150 ft. east of Harvey Rd.
3. Lower Paget member; Spice Hill eolianite. Forms ridge north of South Shore Rd; topographically distinct from Unit 5.
2. sub-Paget soil. Terra-rossa. Exposed on Cobb's Hill Rd., 450 ft. south of Ord Rd.
1. sub-Paget eolianite. Windward beds.

B. Reference section. Vertical section in cliffs facing Grape Bay.

5. Upper Paget, Spice Hill eolianite. Foresets.
4. Grape Bay soil. White regosol.
3. Devonshire map unit. Shallow-water calcarenite; conglomeratic and with large marine fossils in lower part. Unit 8-ft. thick and resting on bench cut in sub-Paget. Top of unit variable elevation, generally about 12 ft. above msl.
2. Unconformity, locally marked by truncated soil pipes of sub-Paget soil.
1. sub-Paget beach-rock-cemented beach calcarenite. Grades into sub-tidal facies. Top of unit a bench, at about 4 ft. above msl.

C. Reference section. Horizontal section along Barker's Hill Rd. from North Shore to Vesey St.

4. Barker's Hill beach wedge. Beach calcarenites grading into Unit 3. Confined to strip north of North Shore Rd.
3. Lower Paget, Barker's Hill eolianite. Windward beds just south of North Shore Rd., overlying and rolling into foresets at old RR overpass.
2. sub-Paget soil. Terra-rossa. Exposed on Vesey St., 250 ft. east of Barker's Hill Rd.
1. sub-Paget eolianite. Windward beds and foresets.

Table 4A summarizes the Paget lithologic identity, in terms of features that are observable in the field. Figure 5 shows the mineralogy of a number of map units. This figure is drawn from data generously provided by B. L. Ristvet who has begun an important study (see Ristvet, 1972) on the spatial and stratigraphic variability of various diagenetic features in Bermuda. As shown by Ristvet's analyses (and, in a general way, by Land, *et al*,1967), stratigraphic variation in mineralogy parallels that of other lithologic features seen in the field: Paget calcarenite is distinct from sub-Paget calcarenite; Upper Paget eolianite is less altered than Lower Paget eolianite, but there is considerable overlap in the two units.

Since mineralogy and the lithologic features of Table 4A are related in degree of diagenesis, there must be certain qualifications (Table 4B) on the "Paget lithologic identity." In Bermuda, lithologic variation within an iso-time reflects the overlap of two, non-congruent facies mosaics, depositional and diagenetic; i.e., the diagenetic mosaic is an overlay on the depositional mosaic (see Land, 1970). The Paget lithologic identity applies to one of the "volume-fields" of the overlapping mosaics. This field is calcarenites of (a) eolian depositional facies, and (b) altered in a single hydrologic-diagenetic environment, the intermediate-aerated zone of hydrologists. This hydrologic environment is above the water table and below the surficial layer penetrated by roots (upper aerated zone) and, in Bermuda, is an environment predominantly of vadose seepage (cf., Thrailkill, 1968). This intermediate-vadose diagenetic facies probably does not constitute a significant part of the total volume of Paget-age calcarenite in Bermuda. However, in that portion of the Paget bundle exposed in cliffs, roadcuts, and quarries, the largest part, by far, is the eolian depositional facies and the intermediate-vadose diagenetic facies, and it is this part that is important to surface geologic mapping.

The Paget marine deposits are summarized in Tables 5 and 6. There are two kinds of units (Vacher and Mackenzie, 1970; Vacher, 1971): depositional-coastline calcarenites that represent the source beaches for the dunes, and erosional coastline deposits that formed along cliffs and pocket embayments and do not grade into eolinite.

The Paget marine deposits group into three map units (Table 6). The Barker's Hill beach unit, along the North Shore, is a beach wedge that grades into the Barker's Hill eolianite and is demonstrably transgressive. The Devonshire map unit, a discontinuous line of pods along the opposite shoreline, is a regressive (Bretz, 1960; Land *et al*; 1967; Vacher, 1971), erosional-coastline deposit. The Devonshire onlaps the Watch Hill eolianite and is in turn overlapped by the Spice Hill eolianite. Thus both these marine map units are related in time to Lower Paget eolianites.

The St. Catherine map unit is a collection of poorly exposed, small, isolated pods at a few, widely separated localities along Bermuda's external shoreline. The St. Catherine appears to be an erosional coastline deposit time-related to Upper Paget eolianites. The St. Catherine is less altered than Lower Paget

TABLE 5. TYPES OF MARINE DEPOSITS

	Depositional-coastline	Erosional-coastline
Defining characteristic	Wedges of largely foreshore-beach cross strata that grade landward into eolianite.	Lithosomes that have a distinct landward termination; i.e., they do not grade into eolianite.
Typical features	Commonly continuous for several tens of meters along strike. Calcarenite. Gently-inclined, seaward-dipping cross beds in wedge-shaped sets. Keystone vugs, parting lineation, and other bedding-plane features indicative of intertidal (swash-backwash) deposition. Large marine fossils rare.	Linear collection of isolated pods and lenses. Conglomeratic. Large marine fossils common. Associated with prominent erosional surface. Two varieties: a. Bench-veneer. Characteristic vertical, intergradational succession of (i) basal cobble conglomerate with subtidal fossils, (ii) gently-inclined beach calcarenite, (iii) supratidal, poorly cemented calcarenitic soil. b. Cliff plaster. Fossiliferous conglomerate resting against cliff. Vertical contact.
Modern analog	South Shore (e.g., Horseshoe Bay) where beaches grade landward into dunes.	North Shore, a line of cliffs indented with small pocket embayments.

TABLE 6. PAGET MARINE MAP UNITS

Barker's Hill beach unit	Depositional-coastline wedge that grades into Barker's Hill eolianite. Transgressive: unit is 10 m thick, and intertidal throughout; locally (e.g., Blackwatch Pass) it onlaps the contemporaneous eolian facies.	Extends the length of lagoon-facing shoreline of Pembroke and Devonshire Parishes; confined to strip north of North Shore Rd.
Devonshire	Erosional-coastline deposit.	South Shore of Devonshire and Smith's Parishes. Similar deposits outside map area.
	Bench-veneer. Regressive, as shown by seaward offlap of supra-, inter-, and subtidal facies. Overlies sub-Paget; underlies Lower Paget eolianite (e.g., Spice Hill).	Southern portion of outcrop (Grape Bay, Hungry Bay, Rocky Bay).
	Cliff-plaster. Overlies Lower Paget eolianite (Watch Hill); underlies intra-Paget soil.	Northern portion of outcrop (Saucos Hill, Spencer's Point).
St. Catherine	Small isolated pods of coarse-grained, essentially unaltered calcarenite. Basal contact not exposed.	Widely scattered on external shoreline (Ft. St. Catherine, Church Bay, Elbow Beach).

TABLE 7. PALEOSOLS

A. Facies (based on locality descriptions of Ruhe, et al, 1961).

Regosol	Terra-rossa
White or brown uncemented calcarenite of weak weathering zones or profiles.	Reddish, clayey layers of strong profiles.
Sub-soil horizons generally unaltered.	Sub-soil horizons leached. Common soil base (1-2 cm thick). Deep, downward protruding pockets and pipes.
Little or no clay or structural aggregation. Horizonation by slight changes in color. Organic content, as sparse carbonaceous coatings in upper horizons. Iron oxide, as thin pale-brown coatings in lower horizons.	Structural aggregation moderate to strongly developed. Horizonation evident in vertical distribution of clay content. Iron oxide a major constituent.
Abundant unaltered land-snail shells.	Land snails, commonly a hash of fragments.

B. Soil-stratigraphic units.

unit	overlying unit	facies	importance
intra-Paget	Upper Paget eolianite.	white regosol in Paget, Warwick Parishes; locally approaching terra-rossa. Terra-rossa in Smith's Parish,	separates Upper and Lower Paget members.
Grape Bay	Spice Hill eolianite or Rocky Bay lobes, where either overlies Devonshire map unit.	white regosol; locally pink.	back-beach facies of Devonshire pocket-bay deposits.
sub-Paget	Devonshire or basal Paget eolianite (Watch Hill, Barker's Hill); Spice Hill where it overlaps Devonshire.	generally terra-rossa. Regosol common at buried topographic highs.	separates Paget Formation from lower units.

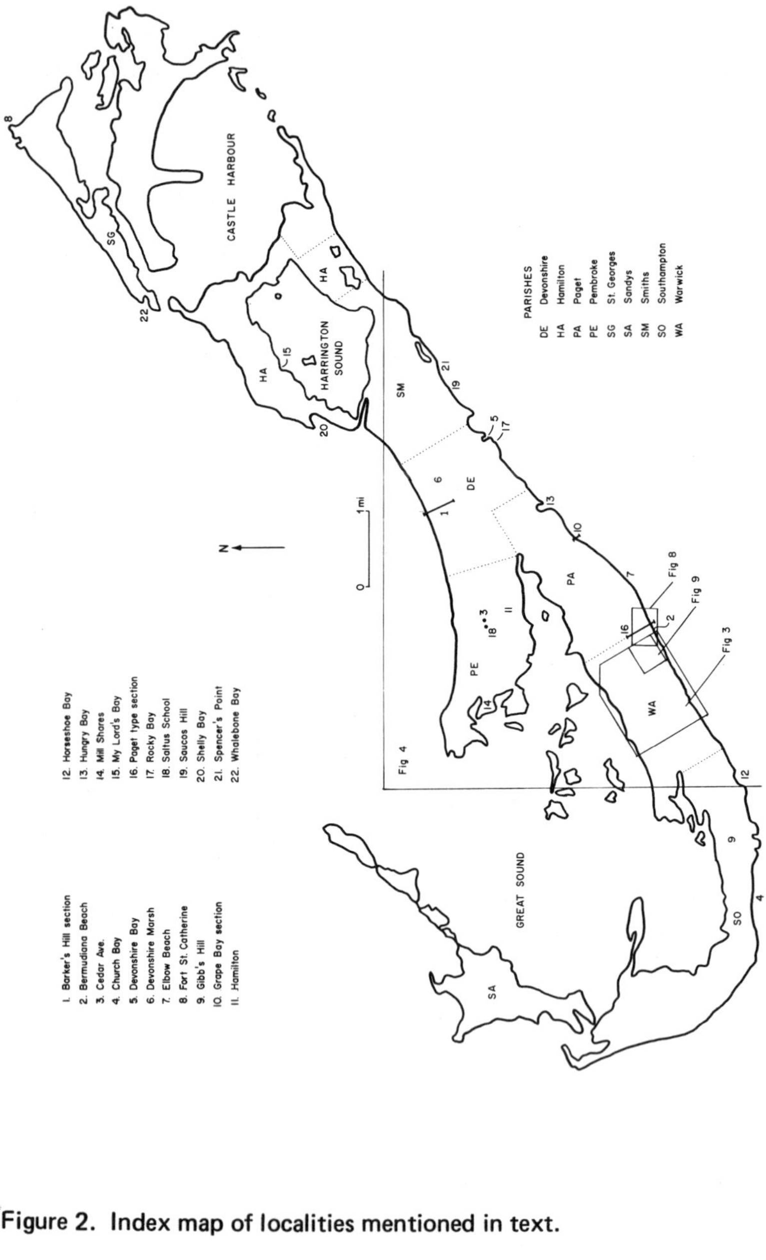

Figure 2. Index map of localities mentioned in text.

eolianites (Table 4A). Ultastructure (by scanning electron microscope) of *Siderastrea* from the St. Catherine (at Fort St. Catherine) is diagenetically altered from that of Recent *Siderastrea,* but considerably less so than that of *Siderastrea* from the Devonshire map unit (J. E. Sorauf, pers. comm., 1972).

The interpretation of the Devonshire in this paper differs from that of earlier papers. Land, *et al* (1967 also Vacher, 1971) considered that where the Devonshire was onlapped by a Lower Paget eolianite (e.g., the Spice Hill), the marine unit was distinctly older than where it onlapped a Lower Paget eolianite (e.g., the Watch Hill). Thus they thought in terms of a single eolianite (Pembroke) dividing two marine units (their Devonshire and Spencer's Point Formations). This chapter presents a case for the alternative: one South Shore marine unit between two temporally disjunct Lower Paget eolianites. The earlier deposited Lower Paget eolianite (Watch Hill) provided headlands for cliff plaster (seen as the Land, *et al* Spencer's Point at Saucos Hill) at the same time that pocket-bay calcarenites (Land, *et al* Devonshire) were deposited in the reentrants along lower portions of Older Bermuda to the south.

Soil Stratigraphy

Ruhe, *et al* (1961) gave useful detailed descriptions of Bermudian paleosols at the type localities of Sayles' (1931) soil-stratigraphic units. They recognized two soil varieties (see Table 7): red, clayey terra-rossa and white to brown regosols ("essentially raw eolianite;" Ruhe, *et al,* 1961, p. 1137).

Mapping of rock-stratigraphic units has revealed a spectrum of soil types and definite facies relation between the two end-members, terra-rossa and white regosols. Regosols commonly grade downdip from buried topographic highs into terra-rossas in buried topographic lows. The sub-Paget soil (see Table 7B) below the Barker's Hill eolianite north of Hamilton is an example. In a nearly continuous exposure, the layer changes from (pinkish) regosol at Cedar Avenue (elev. 30 ft) to terra-rossa in the Saltus School playing field (elev. 5 ft), a distance of 400 ft. Downdip facies changes from white regosol to terra-rossa can be seen within a single exposure in two places (Mill Shares and My Lord's Bay); in each case, the layer is stratigraphically below the sub-Paget of this paper.

The three soil-stratigraphic units of this report are summarized in Table 7B. The sub-Paget separates Paget limestones from older limestones, and the intra-Paget separates the two Paget members. The paleosols are hence valid soil-stratigraphic units in the sense employed by Willman and Frye (1970) – paleosols that conveniently mark surfaces of non-deposition.

The Grape Bay soil is in some ways a rock-stratigraphic unit and in others a soil-stratigraphic unit. It is a contemporaneous supratidal facies of the Devonshire map unit and analogous to the grass-covered, crab-burrowed, accretionary sands just beyond high-tide mark of pocket beaches (e.g., Whalebone Bay, Shelly Bay) of Bermuda's present-day erosional coastline. The Grape Bay built out over the Devonshire during a slow regression. Thus the Grape Bay is like any facies-restricted rock-stratigraphic unit that obeys

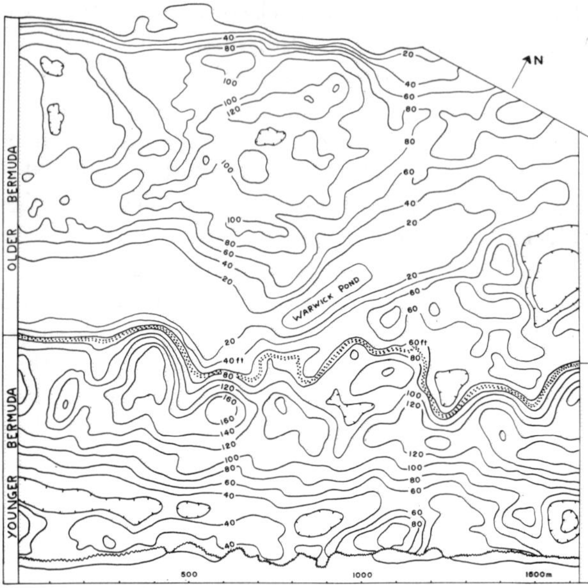

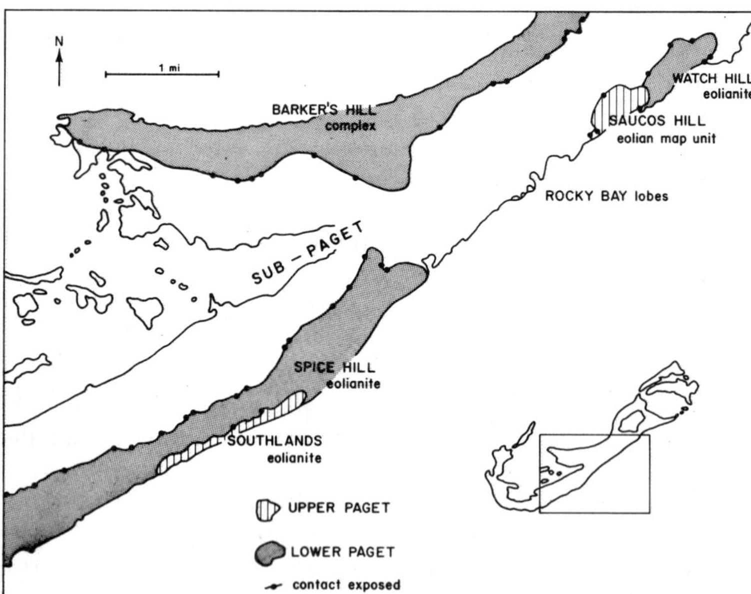

Figure 3. Topography of Younger and Older Bermuda. Warwick Parish

Figure 4. Paget eolian lithosomes in the central five parishes.

Walther's Law. However, the paleosol also marks a hiatus that terminated when the overlying Lower Paget eolianite overlapped the Devonshire by landward accretion of lee-side cross beds. In this sense, the Grape Bay is also a soil-stratigraphic unit.

Although the three soil-stratigraphic units of this chapter correspond (Table 2) to the three soils of the Land *et al* (1967) column, the three do not represent all the soil-stratigraphic units in Bermuda. Four distinct buried, inter-eolianite soil layers, each largely of terra-rossa facies, extend east-west across the city of Hamilton. These four are below the sub-Paget soil of this paper.

Time Stratigraphy

The simplest and most direct interpretation of the lithic relations described above is shown in Figure 6. This interpretation holds that:

1. The coastal dunes of Younger Bermuda were deposited during two broad time intervals, early and late Paget, separated by an interval of non-deposition represented by the intra-Paget soil.

2. The early Paget interval contained three temporally distinct dune-building events represented, in order of decreasing age, by the Watch Hill, Barker's Hill, and Spice Hill eolianites.

3. The early Paget interval was punctuated by a highstand of sea level above its present position. During the rise, dunes (Barker's Hill) formed along the North Shore. During the highstand, the Devonshire map unit, and no dunes, formed along the South Shore. Dunes formed along the South Shore while sea level was below that of present, both before (Watch Hill) and after (Spice Hill) the excursion of sea level above present datum.

4. The upper Paget interval was similar to the early Paget, because sea-level highstand was above present datum, with probably temporally disjunct dune-building events in different parts of Bermuda.

This time stratigraphy involving multiple dune-building events within a single (early Paget) platform submergence is consistent with:

1. Radiometric dates. The uranium-series dates on corals reported by Land, *et al* (1967) for the cliff-plaster (Table 5) Devonshire and the bench-veneer Devonshire are shown in Figure 6. The dates suggest a single highstand. The dates also decrease with decreasing elevation, a relation consistent with the overall regressive nature of the Devonshire.

2. Erosional-coastline paleogeomorphology. The modern erosional coastline of the North Shore is a line of cliffs indented with pocket bays and beaches. The coastline would be seen in the rock record as a map unit like the Devonshire – a discontinuous chain of cliff-plaster and pocket-bay deposits.

3. Land-snail occurrences. According to locality descriptions and snail identifications of Gould (1969), the Watch Hill eolianite contains *Poecilozonites cupula cupula* and other South Shore Paget eolianites contain *Poecilozonites bermudensis zonatus*. The interpretation of Fig. 6 suggests these subspecies have temporally disjunct ranges along the South Shore. Moreover, all of Gould's

Harrington-Pembroke (see Table 2) localities where he found other subspecies of *Poecilozonites cupula* are below the Paget of this paper.

4. Eolianite mineralogy. The one Watch Hill sample analyzed by Ristvet (1972) is mineralogically the same as the most altered of his 15 Spice Hill samples.

5. Paleosol facies. According to evidence accumulated by Bricker and Mackenzie (1970), the iron-oxide and clay of Bermudian paleosols is derived largely from atmospheric dust; thus ignoring the effects of post-settling redistribution, unconformities should be increasingly red the greater the lapsed time. The intra-Paget soil is predominantly of regosol facies in Paget and Warwick Parishes where it overlies Spice Hill eolianite, and entirely of terra-rossa facies in Smith's Parish where the time represented includes that of the Spice Hill eolianite and Grape Bay soil.

An alternative to the interpretation of Figure 6 is represented by the Land, *et al* (1967) stratigraphic column and holds that there was one early Paget dune-building episode (i.e., Pembroke Formation) separating two temporally disjunct early Paget sea-level highstands (i.e., "Devonshire" and Spencer's Point Formations). The following argues against this alternative.

1. No single locality has been found that convincingly shows the "Devonshire" to Spencer's Point succession in direct superposition.

2. The radiometric dates of Spencer's Point and "Devonshire" are in inverted order (Fig. 6). Taking account of the error, the Spencer's Point could be as young as 115,000 and the "Devonshire" as old as 128,000 yr. This would imply that within a time interval roughly 13,000 yr long, sea level dropped from +5 m to below present sea level, rose again to over +10 m, and while sea level was below its present position a dune ridge formed and was lithified to the extent it could resist erosion and be cliffed by ensuing sea-level rise. The implied complete sea-level oscillation would have involved very large volumes of ice, more than is currently stored in the Greenland cap.

3. Gould (written comm., 1972) has traced Harrington-Pembroke containing *P c cupula* from the "Devonshire" type section to within a couple hundred meters of *P b zonatus*-bearing Harrington-Pembroke of the Spencer's Point type section. He found no evidence of interspecific hybridization like that known in other parts of Bermuda and has concluded it is not likely the two Harrington-Pembroke localities are time-correlative.

Thus the balance of evidence supports the hypothesis of multiple, early Paget dune-building events, instead of multiple, early Paget sea-level highstands. However, the hypothesis of Figure 6 is not strictly demonstrated, nor can it be along the South Shore. The convincing proof would be the tracing of the Devonshire map unit continuously from the Spencer's Point – Saucos Hill localities where the marine unit rests against "Pembroke" eolianite, to the Devonshire Bay – Grape Bay localities where the unit underlies "Pembroke." This is unfortunately futile, because the Devonshire at and between the northern and southern areas is discontinuous (an inherent property of erosional-coastline

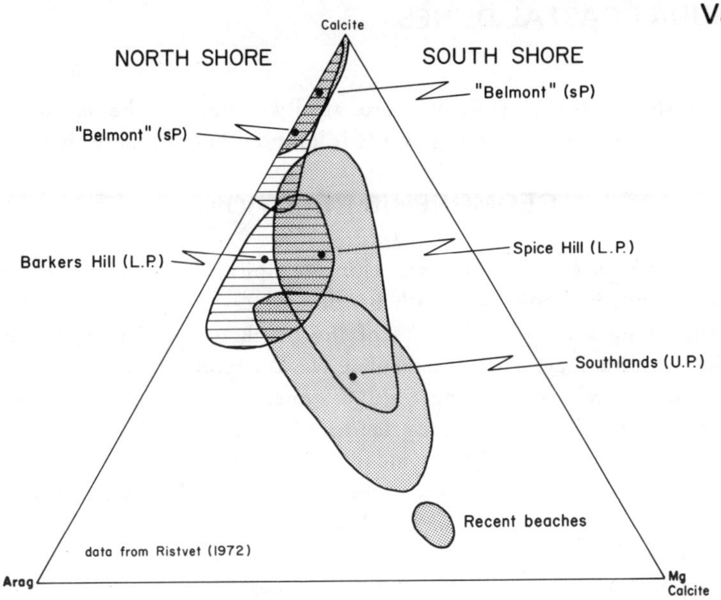

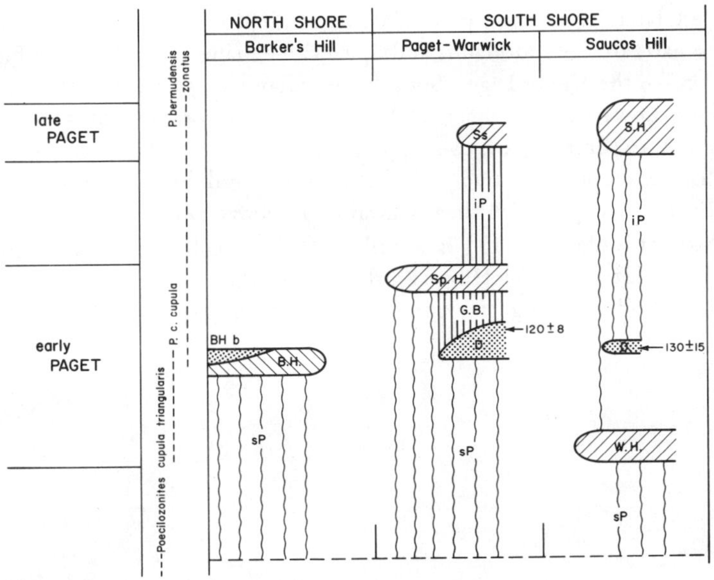

Figure 5. Vadose mineralogy of eolianites in Paget, Warwick, and Pembroke Parishes.

Figure 6. Paget time stratigraphy. Key: Time-space volume of (1) eolianites and (2) marine lithosomes; unconformities marked by (3) terra-rossa and (4) regosol. Abbreviations: Eolianite map units of Figure 4 (B.H., Sp.H., W.H., Ss, S.H.); Barker's Hill beach unit and Devonshire map unit (BHb, D); Grape Bay, intra-Paget, sub-Paget soils (GB, iP, sP).

units) and the pertinent eolianites are areally disjunct. The hypothesis of multiple early Paget dune-building events is adopted here because, it fits all the data at hand.

GEOMORPHIC RELATIONS

As shown in Figure 4, eolianites are elongate units, with long axes subparallel to the shoreline. For eolianites with beach sources (e.g., Barker's Hill of the Paget), their long axes parallel strike of the beach wedge. Younger eolianites onlap the seaward margin of older eolianites, reflecting a pattern of seaward accretion of progressively younger dune units. Large high-angle cross beds (foresets, as in Fig. 1) dip generally landward. Thus, in a general way, (a) the dune-forming winds were onshore winds; (b) leeward (at the time of deposition) is synonomous with landward (now), and windward, with seaward; and (c) the elongate units were formed as transverse bodies.

Topographically, an eolianite is a ridge with relatively gentle (5-15°) windward slopes and steep (30-40°) leeward slopes. The ridge-line is undulatory (Fig. 7). The ridge is a row of large hills, the lobate dunes (hereafter called lobes) that merged laterally as they grew (Mackenzie, 1964a; Vacher, 1971). Lobes comprising parts of the Southlands and Spice Hill eolianites are shown in Figures 8 and 9. Unlike the Upper Paget, Southlands eolianite, a single row of lobes, the much larger Lower Paget, Spice Hill eolianite is a complex of successively onlapping and, to some degree, overlapping lobes.

An eolianite consists of two sedimentary-structural units (Fig. 10): (1) a windward-topset wedge which lies seaward and above (2) a foreset wedge. The windward-topset wedge(Fig. 11) is a collection of trough-shaped, scour-and-fill structures. The foreset wedge (Fig. 12), the bulk of the eolianite, is a complex of laterally interfingering foreset lobes, each a set of convex-upward,

Figure 7. Ridge-line of the Barker's Hill eolianite. View from Devonshire Marsh East.

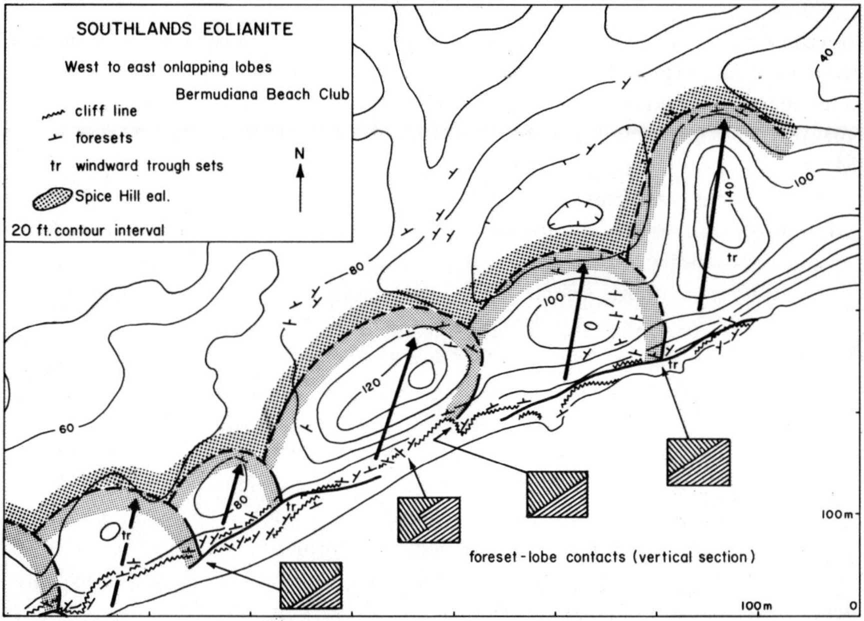

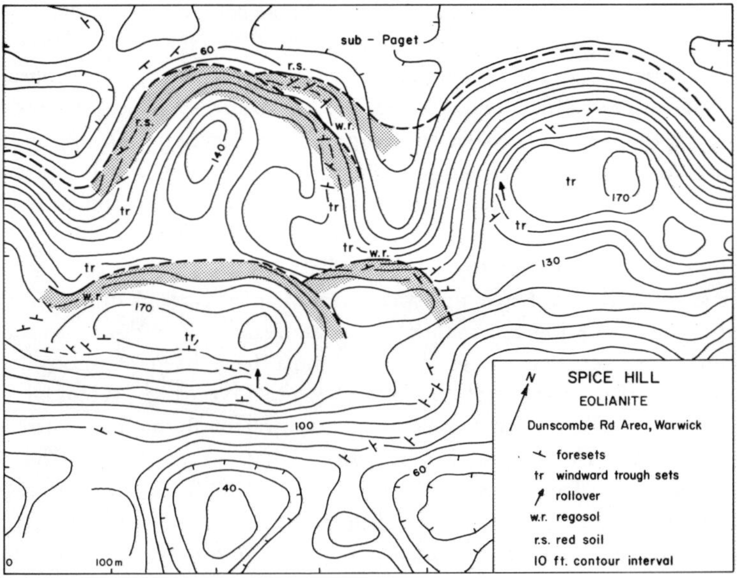

Figure 8. Lobes of Southlands eolianite. Arrows are foreset-lobe vectors, from axis of symmetry of curved foresets.

Figure 9. Lobes of Spice Hill eolianite.

convex-leeward, high-angle foresets. In windward parts of the eolianite, the two units are superposed, with the contact a pronounced seaward-dipping, sedimentary-structural unconformity (Fig.1). At the leeward extremity of the eolianite, the two units are intergradational in a "drape-over" or "rollover" (Mackenzie, 1964a) structure in which seaward-dipping beds of the upper wedge rotate and merge with foresets (Fig. 13).

The foreset lobes that interfinger to form the bulk of the eolianite rock body are the sedimentary-structural manifestation of the lobate hills that line up to define the topographic ridge. This correspondence of surface morphology and internal structure is fundamental (Figs. 8 and 9). Foresets strike with the contours around the landward part of the hills; foreset-lobe contacts are at the swales between adjacent hills. Since the cross-bedding surfaces record successive positions of the calcarenite-air interface, the simple sedimentary-structural anatomy shown in Figures 10 and 12 portrays the progressive growth of the dunes to their present volume. The dunes were *accretionary* mounds. They grew upward and landward with the additional sand delivered to the shoreline. The growth resulted from new layers being conformably draped over the brink (cf. Bigarella, *et al*, 1969, p.38). The ridges grew to great heights, generally more than 100 ft, and not uncommonly (in the Spice Hill) to 200 ft.

The Bermudian calcareous dunes did not migrate. They advanced inland but the advance was a landward progradation through the lee-side accretion that accompanied upward growth. They did not advance by the reworking of sediment in the entire dune body, as is the case of classic migratory dune ridges. In Bermuda, successive lithified eolian ridges are of stratigraphically different ages (e.g., Fig. 3); they are not frozen contemporaneous waves in a sea of sand. In the case of the Barker's Hill eolianite, where the contemporaneous source beaches occur above sea level, the landward extremity of the eolianite is generally within 0.5 km of the beach wedge. As emphasized by Bretz (1960), the Bermudian dunes were closely tied to their immediate source.

The importance of the evidence that the Bermudian calcareous dunes were not migratory but accretionary cannot be overstressed. For dunes to be deposited on the present island, the shoreline must have been at about its present position, and hence sea level must have been relatively high, submerging the platform. The corollary is that eolianite deposition in Bermuda occurred during Pleistocene interglaciations (Bretz, 1961; Land, *et al,* 1967). These statements are obviously demonstrable for the Barker's Hill. They must be true also for the other eolianites (e.g., Spice Hill, Southlands) in which foresets plunge into the sea (Fig. 1) and contemporaneous source beaches are not exposed; sea level was slightly below its present position.

Pleistocene calcareous eolianites can be interpreted to mean low sea level coincident with glaciation. The question of migratory *vs.* accretionary dunes needs evaluation in other areas. In terms of the large volume, the lobate morphology, and the internal sedimentary-structural anatomy of Figures 8-13, are the Bermudian eolianites representative of the widespread lithofacies?

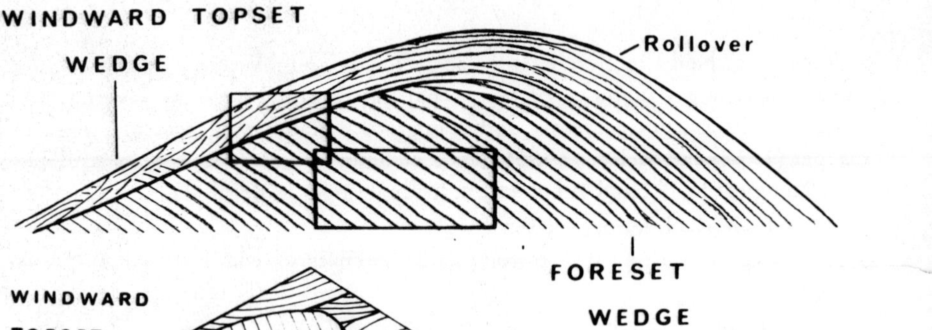

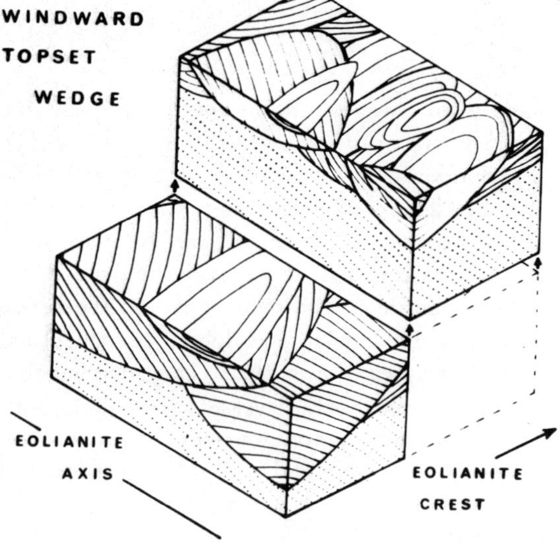

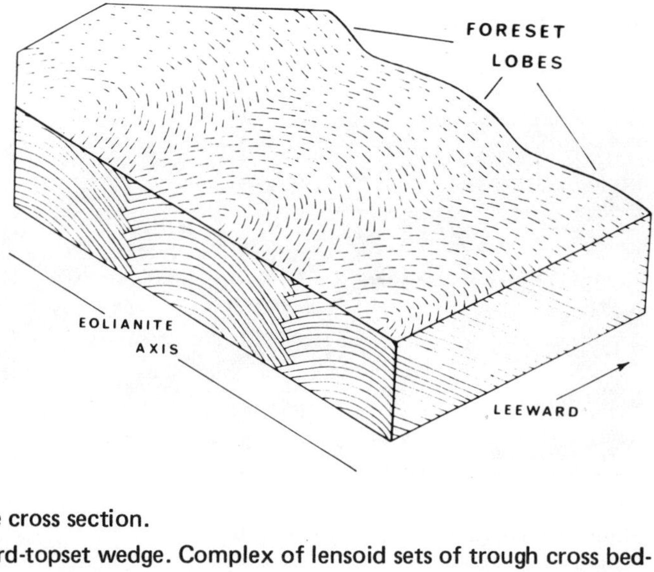

Figure 10. Eolianite cross section.

Figure 11. Windward-topset wedge. Complex of lensoid sets of trough cross bedding.

Figure 12. Foreset wedge.

Regarding the distinction between accretionary and migratory transverse dune ridges, there are a number of ways that the lithified accretionary ridges of Bermuda differ from presently active migratory transverse ridges described by Cooper (1958) and McKee (1966). These include:

1. The great height and volume of individual ridges in Bermuda.
2. Their stratigraphic separation.
3. The preservation of a considerable thickness of windward-dipping cross beds.
4. The common drape-over structure.
5. The over-steepened foresets (Mackenzie, 1964b).
6. The horizontally convex-leeward foresets that contour around the highpoints of the ridge. Apparently migratory transverse ridges tend to be barchanoid; that is, sliding is commonly "centripetal," with horizontally concave-leeward slipfaces opposite the high points of the ridge.

The Bermudian dunes, although considerably larger, are identical both morphologically and in cross-bedding relations to the Brazilian coastal dunes described by Bigarella, *et al* (1969). In particular, Bigarella, *et al* (1969) describe "linguoid protuberances" like the Bermudian foreset lobes, oversteepened foresets, and the prevalence of convex-upward cross beds. They emphasize a drape-over structure. Thus (p. 36) " . . .at least in some levels of a given dune, both the topset and foreset are made up by the same continuous set of strata, and not by different truncated sets as is usually the case;" and (p. 37) "the bent part of the strata represent the former dune's brink at the time of deposition." The dunes described by Bigarella *et al* are stationary (p. 38), the ridge having formed by (p. 7) "the coalescence of small linguoid shaped dunes against the vegetation line."

Figure 13. Drape-over structure. Sub-Paget eolianite; island in Great Sound.

The Bermudian accretionary ridges are reminiscent of the large "precipitation ridges" (Cooper, 1958) along the Oregon coast. According to Cooper (1958, p. 69), "If no effective vegetation is present the sand, if in good supply, moves inland in [migratory]transverse ridges. If the flat behind the beach is forested, the wind moving against it loses velocity, drops its load at the forest front, and produces a precipitation ridge parallel to the shore. . ." The precipitation ridge is (p. 60) "the ultimate stabilized form." The profile of the ridge (p.71) ". . .is similar to that of the ordinary [migratory] transverse ridge, with a gentle windward slope and a typical slipface . . . . Its mass in full development is, however, much greater than that of any [migratory] transverse ridge. Its advancing front may equal or even overtop the forest standing in its way. Height of the crest varies from place to place. Where activity is most pronounced, bulges project to leeward." In a locality description (p. 79), Cooper states that a particular precipitation ridge " . . . is strongly lobate, [suggesting the] merging of adjacent parabola structures."

The striking feature of the maps of foreset-lobe vectors (Figs. 8 and 14) is that vectors are oblique to the long dimension of the eolianite body. Thus in NE-trending South Shore eolianites (Southlands; eastern part of the Spice Hill), the foreset-lobe vectors consistently point N or slightly NE. This pattern apparently holds for all the Paget eolianites from Warwick Parish to the islands that close off Castle Harbour, and for the mapped sub-Paget eolianites of Warwick and Paget Parishes as well. The conclusion is that in this part of the island, southerly-derived dunes were supplied from NE-trending beaches (roughly parallel to the present South Shore) and prograded northward, not normal to the beach. The same conclusion is reached by the systematic SW-dipping foreset-lobe contacts in the Southlands (Fig. 8); a landward prograding lobe onlapped its neighbor to the northeast as they grew landward.

A similar *en echelon* pattern of foreset-lobe vectors holds for northerly-derived Paget eolianites. The Barker's Hill dunes of Pembroke and Devonshire Parishes and the Paget dunes of northern St. George's Island (Vacher, 1971) advanced SE from E-W trending beaches.

The maps of foreset-lobe vectors are consistent with the hypothesis that (a) it was gales and strong winds that influenced the growth direction of the dunes and (b) these winds in Paget times were, in terms of directional frequency, similar to those at present. The directional frequency of modern prevailing winds and of gales is shown in Figure 15, The obvious difference between the two is that gale directions are not uniformly distributed in the eight octants, whereas prevailing winds are. Dunes advancing under prevailing winds, therefore, should be generally normal to the strike of the source beaches; dunes advancing under gales should be strongly directional, generally oblique to the shoreline.

The Spice Hill illustrates the point. Figure 14 shows the frequency of modern gales that impinge on various segments of a "South Shore" parallel to the curving axis of the Spice Hill. In the eastern part of the unit, the preponderance of modern onshore gales impinges on parallel southern shores, comes from the south, and the foreset-lobe vectors point due north. Further west where the three temporally disjunct eolianites comprise a single, diagenetically uniform (vadose part) member, the Lower Paget. Because of the diagenetic uniformity, the entire member probably correlates with a single reef (i.e., the 127,000 yr B.P. Second High Cliff; Mesolella, *et al* 1969) of the Barbados section. The Barbados reefs recur, in this part of the section, at 20,000 yr intervals (Mesolella, *et al* 1969). Hence the Lower Paget eolianites must have each formed within only a very few thousand years. To account for the deposition of the Spice Hill volume in this amount of time, the annual flux of sediment onto the beach must have approached 1 m^3/m of shoreline; and, because of frequent gales, sediment was quickly swept into the dune areas. Under this relentless onslaught of sand, dunes were repeatedly buried and had to build upward.

The nature of winds that formed the large accretionary dunes of Bermuda is revealed by areal cross-bedding relations. Foreset-lobe axes (individually, the horizontal trace of the symmetry plane through the body of curved foresets) are about the same as the long dimension of the morphological lobes defined by contours (account must be taken of recent truncation by cliffs). These axes are made into foreset-lobe vectors by the addition of an arrowhead indicating the general direction of foreset dip. These foreset-lobe vectors are the important directional property of the eolianites, for the vectors should indicate the growth direction of the individual dunes that comprise the rock body. Foreset-lobe vectors are shown in Figures 8 and 14 for the Southlands and Spice Hill eolianites.

There are a number of reasons why the Bermudian dune ridges were accretionary and did not migrate. All of the following probably contributed.

1. The grains are calcium carbonate, thus reactive and subject to rapid incipient cementation (Mackenzie, 1964a) and case-hardening in the Bermudian environment of alternating rainy and sunny days.

2. The dunes advanced upon a forested landscape. The native Bermudian flora was dominated by forests of the Bermuda palmetto and Bermuda cedar (now, respectively, replaced by introduced flora and extinguished by a scale-insect blight). In the (few) uninhabited areas of Bermuda, the gentle sea-facing slopes of Paget eolianites are densely covered by the twisted remains of the cedar and testify to the kind of landmass that would have been presented (a few years ago) to a new rush of dunes. The evidence that the lithfied dunes formed during interglaciations means they formed in a similar climatic regime and hence encountered similar forests. Direct evidence is afforded by striking molds of palmetto leaves found in the foresets (Sayles, 1931; Bretz, 1960).

3. The dunes were built quickly under a tremendous influx of marine-derived sand, and in an environment, if like the present (Macky, 1956), experiencing a gale 36 days a year. According to the time-stratigraphic interpretation (Fig. 6), parallel shoreline changes strike so that it can pick up the frequent westerly gales, the foreset-lobe vectors have an eastward component. For comparison, the vector sum of the frequency of modern prevailing winds impinging on

NE-trending "South Shores" is shown in Figure 14. If the orientation of the lobes was due solely to the prevailing winds, the foreset-lobe vectors would have a westward component.

The case is stronger (see Fig. 15) for North Shore eolianites. Regardless of the precise orientation of the North Shore beaches, the growth direction of the dunes, if determined by gales, should have been strongly southeast, just so long as the source beaches faced the NW somewhat. From reconnaissance mapping, it appears that North Shore eolianites are composed of SE-directed lobes, regardless of whether the eolianite axes trend E or NE.

DUNES AS SINKS OF SOURCE-SINK SYSTEMS

The Paget eolianites were built from calcarenite produced offshore and transported to the shoreline during Paget times. The dunes, therefore, can be considered as the sediment sink in source-sink systems involving calcarenite production and transport on the platform surrounding Bermuda. Two source-sink systems leading to the formation of Pleistocene calcarenite dunes in Bermuda can be formulated, and a third rejected, from interpretation of Paget stratigraphic and sedimentologic data in the light of sedimentary patterns on the present-day platform.

Present-day platform.

As shown in Figure 16, the Bermuda platform consists of four physiographic-ecologic provinces. (The following synopsis is drawn from Upchurch, 1970, unless otherwise indicated.)

1. The reef-front terrace extends from the platform edge to the reef tract and lies at a depth of some 10 fathoms.

2. The main reef tract virtually encircles the Bermuda Islands. The reefs are of two basic types: (a) coral-algal reefs, and (b) algal "boilers" formed largely of an intergrowth of encrusting forms, *Lithothamnion, Millepora,* and the vermetid, *Dendropoma* (Ginsburg, *et al,* 1971). The boilers are just off the southern shore and occur intermittently along the outer extremity of the reef tract. The coalesced flats of the northern reef tract lie at depths of 1-2 fathoms (Garret, *et al,* 1972) and are broken by channels as deep as 9 fathoms.

3. The North Lagoon lies between the northern reef tract and the islands. The lagoon floor averages 6-8 fathoms, and has a maximum depth of 12 fathoms. Reefs decrease in frequency southward into the lagoon, where they are typically small, scattered pinnacles.

4. The islands themselves form a NE-trending chain near the southern edge of the platform.

Upchurch (1970) has shown (by correlation analysis of constituents of over 150 grab samples) that platform calcarenites are mixtures of two populations, which he called the reef suite and lagoon suite. The former consists predominantly of fragments of reef-dwelling organisms, principally *Homotrema,*

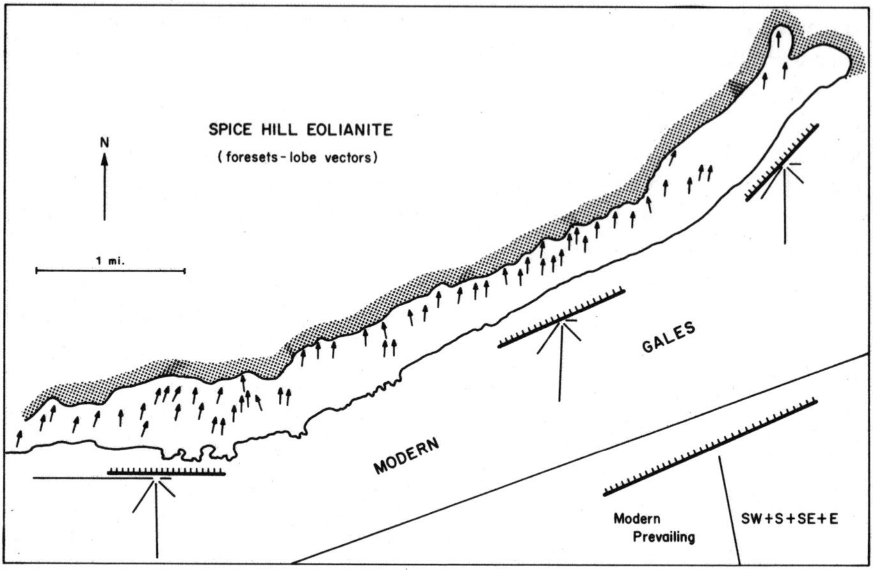

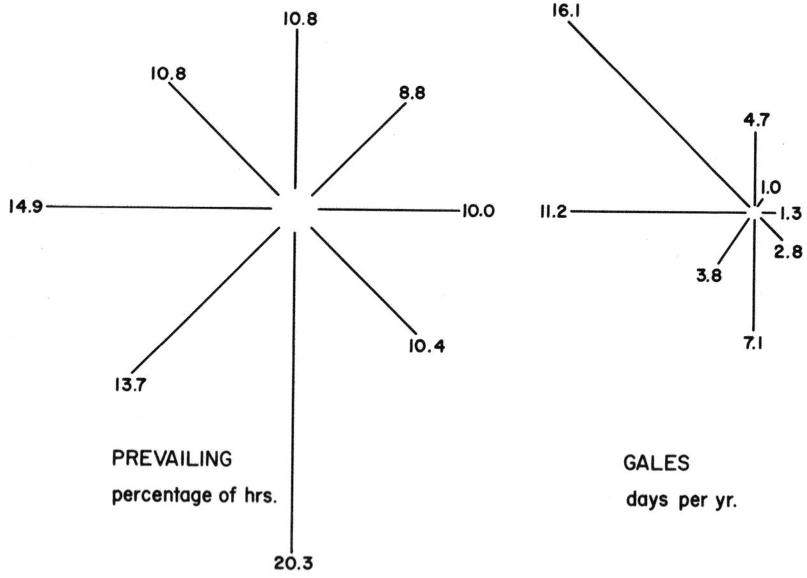

Figure 14. Foreset-lobe vectors of Spice Hill eolianite, may include Upper Paget eolianite in western part.

Figure 15. Modern winds. Date from Macky (1956).

gastropods, some varieties of coralline (red) algae, and coral. The lagoon suite consists of lagoonal organisms, principally the green alga *Halimeda* and small infaunal bivalves such as *Crassinella.*

The distribution and mixing of the two populations, as generalized from Upchurch's maps, is shown in Figure 17. Calcarenites of the lagoon are nearly entirely of the lagoonal suite of consituent particles; the exceptions are near the large patch-reef buildups of the inner shoals. The reef-tract calcarenites are a mixture of the two suites, with the reef suite predominating on the outer portion of the tract. Calcarenite of the inner portion of the reef tract is composed mostly of taxa living in the protected areas within the tract (Upchurch, 1970).

The distribution of the two suites, the distribution of individual constituents (Upchurch, 1970; Figs. 14, 15), and the rapid lagoonward decrease of *Homotrema* clasts (Mackenzie, *et al*, 1965), all show there is little landward transport of sediment in the North Lagoon. This must be due to the reef tract itself. Though the North Lagoon certainly should not be considered restricted (see Garret, *et al*, 1971), the reef tract does act as an energy barrier (or filter) to the ocean-fetched waves; that is, waves capable of moving calcarenite at 20 m are filtered (by breaking) at the shallow, platform-margin reefs (called "breakers" by many Bermudians).

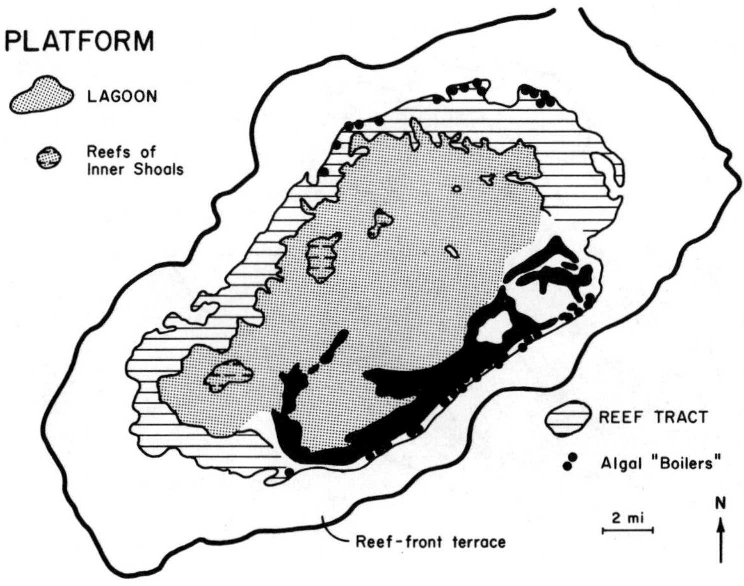

Figure 16. The Bermuda platform (adapted from Upchurch, 1970; Ginsburg, *et al* 1971; Mackenzie, *et al* 1965). The reef tract and islands enclose a central lagoon that lies at depths of about 20 m.

The effect on Bermuda's coastline is shown in Figure 18. The islands's lagoon-facing shoreline is erosional, a line of cliffs, with a few pocket embayments. Virtually no calcarenite occurs along the strand; that which does is in the small beaches at the head of pocket bays, and is derived largely from headland erosion (Mackenzie, *et al,* 1965).

The island's South Shore, on the other hand, is partly a depositional coastline. Here the famed pink Bermudian beaches occur, and many of them are backed by dunes. The beaches are pink because of large numbers of fresh *Homotrema* clasts (ca. 10 %; Mackenzie, *et al,* 1965), which indicate predominance of the reef suite, and hence a reef source. The other elements of the reef suite – particularly coralline algae – are abundant. The reefs are within 1 km of the shoreline.

The simple map relationship between the proximity of reefs to the shoreline and the distribution of erosional *vs.* depositional coastlines, the minimal landward transport of calcarenite in the North Lagoon, and the predominance of reef-suite consituent particles in the dune-supplying beaches of the South Shore, all establish a reef-to-dune, source-to-sink system in present-day Bermuda. The lagoon is obviously not a source for dunes today.

The present South Shore coastline shows also that the development of the reef-dune, source-sink system is not instantaneous upon platform submergence. The South Shore is in transition between an erosional coastline cut in Paget

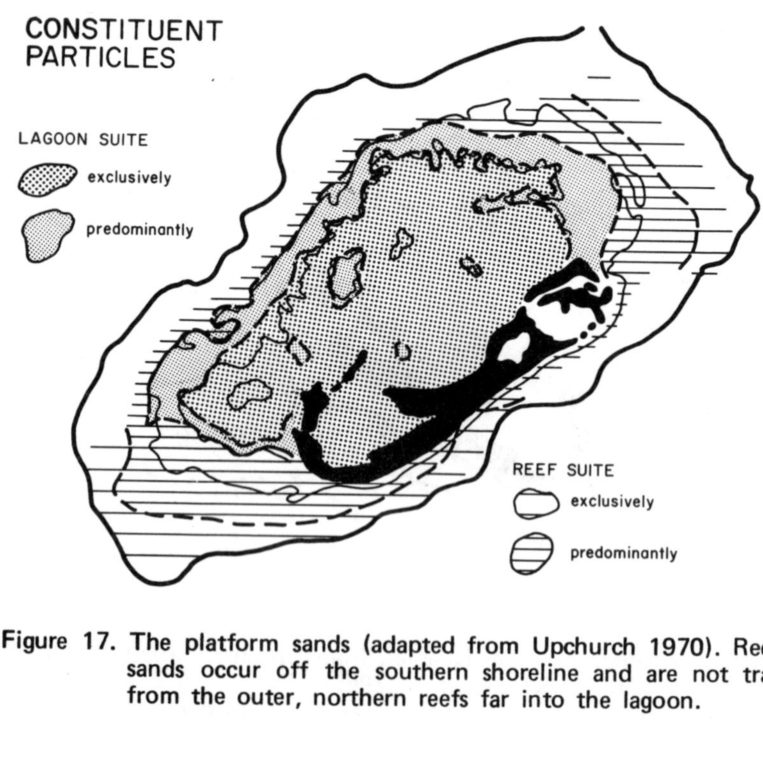

Figure 17. The platform sands (adapted from Upchurch 1970). Reef-derived sands occur off the southern shoreline and are not transported from the outer, northern reefs far into the lagoon.

rocks, and a depositional coastline like those that must have prevailed during the major dune-building events of the Paget (i.e., very long, continuous beaches supplying a long dune line). Much of the South Shore is simply cliffs with no significant offshore detritus (e.g., Saucos Hill). Along other segments (e.g., Bermudiana Beach Club), offshore-supplied beaches are impacted directly against cliffs. In other areas (e.g., Horseshoe Bay), the cliff cut in Paget is well inland of the strand and the intervening area is covered with dunes; these areas, obviously, are the most abundantly supplied.

The development in Bermuda of a reef-supplied depositional coastline during early stages of platform submergence can thus be inferred. Initially, cliffs are cut in the lithified eolianites, while the reefs are just getting established. As productivity in the reefs increases, they build up to sea level and become an increasingly important sediment source. The calcarenite carried to the shoreline accumulates first as a beach against the cliff. With continued supply, the beach progrades, allowing room for dunes which form from the ongoing supply. The development proceeds at varying rates along the shoreline, depending upon the local productivity of the reefs and the nearshore current and wave patterns. Eventually the source-sink system is well established; the isolated beaches coalesce; and the dunes grow large and override the cliffs. This last stage, of course, is not the case in Bermuda today.

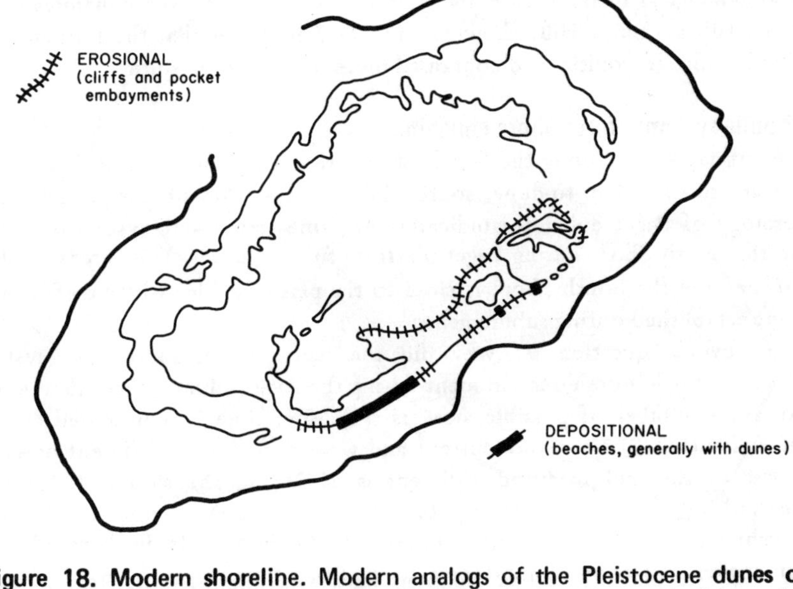

Figure 18. Modern shoreline. Modern analogs of the Pleistocene dunes occur along only short segments of the South Shore. Sediment is not brought to the North Shore today.

Source of Paget eolianites.

The source of the Paget eolianites can be identified from Ristvet's petrographic data. The Paget eolian calcarenite is like the modern calcarenite that lies directly offshore. Thus, according to Ristvet (written comm., 1972), constituent particles of the North Shore Barker's Hill eolianites are 21% molluscs and 19 % *Halimeda* (both indicators of Upchurch's lagoon suite) and only 8 % *Homotrema* (reef-suite indicator). In contrast, constituent particles of South Shore Paget eolianites (Spice Hill and Southlands) are 7 and 4 % mollusc and *Halimeda*, respectively, and nearly 30 % *Homotrema*. There is little doubt that (a) the South Shore eolianites were derived from reefs, just as the modern South Shore beaches and dunes are, and (b) the North Shore Paget eolianite was derived from the North Lagoon, which today does not supply sediment to the shoreline.

The same conclusion is reached from Ristvet's mineralogical analyses (Fig. 5) and appears to apply to sub-Paget eolianites as well. This is possible because of the mineralogical disparity between Upchurch's lagoon and reef suites. The lagoon suite is aragonitic, and the reef suite is a mixture of magnesian calcite and aragonite. The S-derived eolianites define a band on the mineralogical diagram (Fig. 5) from sub-Paget through Upper Paget, and back into the area of modern South Shore beach samples. Thus the original mineralogy of South Shore Paget eolianites was essentially identical to that of the present-day South Shore beaches; that is, these eolianites are largely of the reef-derived reef suite. The mineralogical differences between N-derived and S-derived eolianites (e.g., Barker's Hill *vs.* Spice Hill) shown in Figure 5 indicates that the former were originally more aragonitic, and contained more of the lagoon suite.

Reef-dune system: South Shore eolianites.

Present-day dunes along the South Shore are derived from reefs; that is, the dunes are sinks in reef-to-dune, source-sink systems. Constituent particles and mineralogy of Paget eolianites indicates the same source-sink system operated along the South Shore during Paget platform submergences. This is reasonable *a priori,* because the South Shore is close to the platform edge where reefs would become established during submergence.

The obvious question is: why did the reef-to-dune, source-sink system produce so much more dune sediment during the Paget submergences than now? There are a number of possible answers. (1) More time is represented by the Paget eolianites. (2) Nearshore current and wave patterns are different now, so that less of the reef-produced sediment is carried to the shoreline. (3) The source-sink system is just getting started (i.e., a necessary time lag for the establishment of the sediment source). (4) Modern reefs in Bermuda are producing less sediment than did reefs during Paget time.

The last is an interesting possibility, because it implies a stratigraphic parallel with Barbados reef volumes. In Bermuda, the early Paget eolianites represent by far the greatest amount of sediment brought to the South Shore during the last

130,000 yr; calcarenite brought to the shoreline during the late Paget was less, and that of Recent times very minor. Hence during the last 130,000 yr, reef productivity in Bermuda may have been greatest during early Paget (125,000 yr ago); less during late Paget (80,000?); and, in comparison, very small today.

Mesolella, *et al* (1969, p. 266) summarize the volumes of fringing reefs in Barbados, "Whereas the 125,000 B.P. reef tract and, to a lesser extent, the 105,000 and 82,000 B.P. reef tracts formed around much of Barbados, Recent reef growth of any significance is restricted to discontinuous fringing reefs along the west coast of the island. Furthermore, *Acropora palmata* was a major framework builder in the 125,000 B.P. reef tract and only slightly less important in the 105,000 and 82,000 reef tracts. However, the modern Barbados reefs are virtually devoid of *A. palmata.* Much of the present-day coastline is unprotected by reefs, and the margins of the coral cap are being heavily attacked by wave abrasion."

Lagoon-dune, transgressive system: Barker's Hill eolianite.

The North Lagoon is not a source of strandline sediment today because of the "breakwater" effect of the outer, platform-margin reef tract. During early Paget times, the source was tapped, and large quantities of lagoonal calcarenite were delivered to the shoreline to form the voluminous Barker's Hill eolianite which consists largely of clasts of lagoon-dwelling organisms. The facies relations of the Barker's Hill beach-dune complex, which represents the only Lower Paget eolianite with source beaches above present sea level, are demonstrably transgressive; the beach wedge is internally transgressive, and it onlaps the penecontemporaneous eolianite.

Thus the second source-sink system can be formulated. The source is a partially reef-fringed lagoon; the transport to the shoreline occurs during a rise in sea level. The system is shown in Figure 19. Initially, reefs are established on the platform margin, and act as a submerged breakwater; that part of the wave spectrum that could move sand from lagoon to shore expends its energy on the reefs; the lagoonal sediments quietly accumulate. With a rapid rise of sea level, the reefs are submerged, and more of the wave spectrum is transimitted across the breakwater. For some critical amount of reef submergence, transmitted waves are capable of moving sand on the lagoon floor, even though it lies at greater depths. It is these transmitted waves that sweep lagoonal sediment to the shoreline; beaches form, and from them, a calcarenite dune ridge.

The obviously important element of the lagoon-to-dune, transgressive system is the rapidity of the sea-level rise. Rate of sea-level rise must have exceeded the rate at which reefs in Bermuda could build upward.

Such a reef-submerging, sea-level rise would have an effect on the reef-to-dune, source-sink system that operates on shorelines close to the platform margin. The effects are seen in the early Paget stratigraphy of the South Shore (Figs. 6, 19). The same sea-level rise that produced a dune ridge along the lagoon-facing North Shore, terminated the operation of the reef-to-dune system of the South Shore. The reefs were submerged, and no longer sediment souces; erosional coastlines were cut along the South Shore.

Lagoon-dune, regressive system.

Platform-derived calcareous eolianites are commonly interpreted to have formed during drops in sea level. Calcareous eolianites are called "regressional eolianites" in a well-known textbook (Butzer, 1964), and the idea permeates the literature of the Mediterranean Pleistocene. The hypothesis is that freshly deposited, marine sand is progressively bared by the retreating shoreline, and, as a result, is blown into dunes. Thus the regressional eolianites are interpreted to correlate with the advance of the Pleistocene ice sheets.

The Barker's Hill unit refutes the regressional-eolianite hypothesis in Bermuda. The 10-mi wide North Lagoon, the predominance of NW gales, and the early Paget highstand combine to provide a seemingly perfectly designed natural experiment to test the hypothesis. As sea level dropped progressively lower from its early Paget maximum, the strandline moved N across the lagoon; increasingly more lagoonal calcarenite was exposed to the gales; eventually the entire 10-mi wide platform lay bare. According to the regressional eolianite hypothesis, lagoon-derived calcareous eolianites should offlap the deposits of the early Paget highstand. Instead, it is the Barker's Hill beach unit that occurs along the present shoreline; there is no overlying eolianite. The last depositional episode recorded by lagoon-influenced Paget calcarenite occurred when sea level was above its present position; the ensuing regression produced no lagoon-derived eolianite on Bermuda's lagoon-facing shoreline.

There are probably a couple of reasons why the sea-level drop from the early Paget highstand produced no dunes from the exposed sediments. First is the likelihood of rapid cementation of the lagoonal sediments upon subaerial exposure; the presence of muddy (carbonate) sediments would increase the likelihood. the second reason is that only a little sediment would be exposed per unit time; the shoreline would not experience the continual delivery of the volumes of sand needed to overwhelm the lagoonward march of vegetation. As stated by Wright (1962, p. 530; unfortunately, in leading up to the regressional-eolianite concept), "(Coastal) dune formation represents a contest between vegetation and sand accumulation." The vegetation appears to have won in Bermuda during the post-Barker's Hill regression.

The product of a regression across a lagoon thus may be little different than the product of a regression from large, sediment-producing pocket embayments like those represented by parts of the Devonshire map unit. No dunes – not even small ones – extend out from the Devonshire. The landward facies is simply the Grape Bay accretionary soil, analogous to the grass-covered, crab-burrowed, back-beach soils along parts of the present-day North Shore.

PAGET SEA-LEVEL HISTORY

As emphasized by Land, *et al* (1967), the history of Pleistocene sea-level changes is the controlling variable of stratigraphic and facies relations in Bermuda. Oscillation of sea level above and below the platform brink produced the general alternation of limestones and paleosols (Bretz, 1960; Land, *et al,* 1967). Reefs were established at the platform margin during platform submergences, and these reefs supplied nearby shorelines with calcarenite for dunes (reef-to-dune, source-sink system). A rapid, reef-submerging rise in sea level was the factor responsible for dune formation along other shorelines separated from reefs by a wide lagoon (lagoon-to-dune, source-sink system).

One of the leading objectives of stratigraphic and facies analysis in Bermuda has been (Land, *et al,* 1967; Vacher, 1971) to plot the controlling variable, relative sea level, through time. The goal has been to document and establish a standard sea-level curve, the chart from a mid-ocean "tide-gauge" (Land, *et al,* 1967). No evidence has been found to remotely suggest – and, *a priori* there is no reason to think – that Bermuda has not been stationary throughout the accumulation of the exposed limestones (Land, *et al,* 1967; Vacher, 1971). The curve of relative sea level in Bermuda *vs.* time must be an indicator of "absolute sea levels", if that term has any meaning at all.

Figure 20 shows the Bermudian sea-level history that emerges from the data and interpretations of this paper. Within the last 130,000 yr of the Pleistocene, there were two distinct submergences of the Bermuda platform. During each submergence, sea level rose above its present position. At least for the early Paget submergence, sea level did not simply rise monotonically to and fall monotonically from its maximum position. Within the interval of submergence, there was an abrupt rise of sea level to above its present position. The sea-level high stand itself appears to be a second-order feature of the curve. The first-order feature is the relatively long interval of sea-level above the edge of the platform but below present datum.

The most annoying paradox of Bermudian stratigraphy is removed by the conclusion that the early Paget highstand is a second-order feature of the interglacial sea-level curve, and that the significant buildup of reefs occurred while sea level was below its present position (Fig. 19). It has been known for some time that the Bermudian eolianites were derived largely from reefs, that the early Paget of this report included a sea-level highstand, and that this highstand was marked by dune formation from source beaches preserved above present sea level. The obvious problem has been that early Paget reefs have not been found. According to interpretations of this paper, the early Paget reefs probably did not reach above present datum and the eolianite of the North Shore was supplied by a different source.

The early Paget sea-level curve is of more than local interest. The "normal" position of early Paget sea level slightly below present datum may mean that Northern Hemispheric ice volumes were not significantly less during the

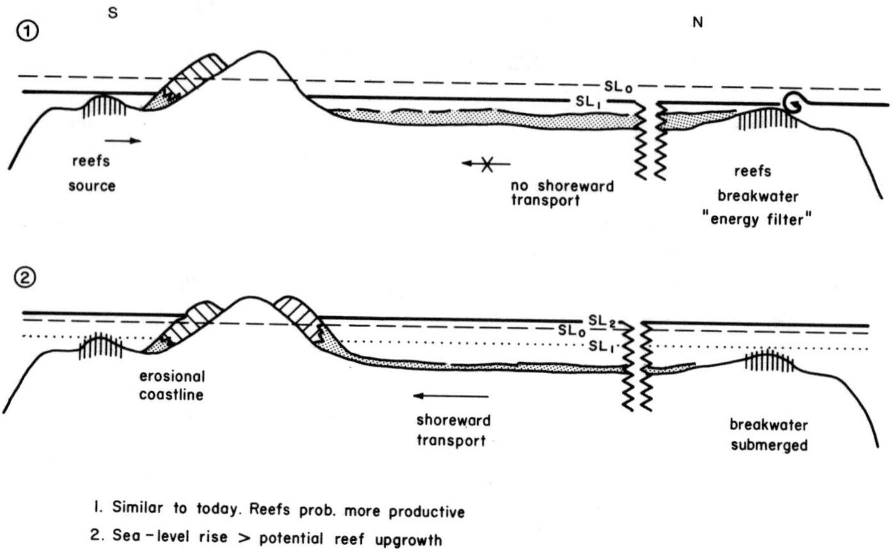

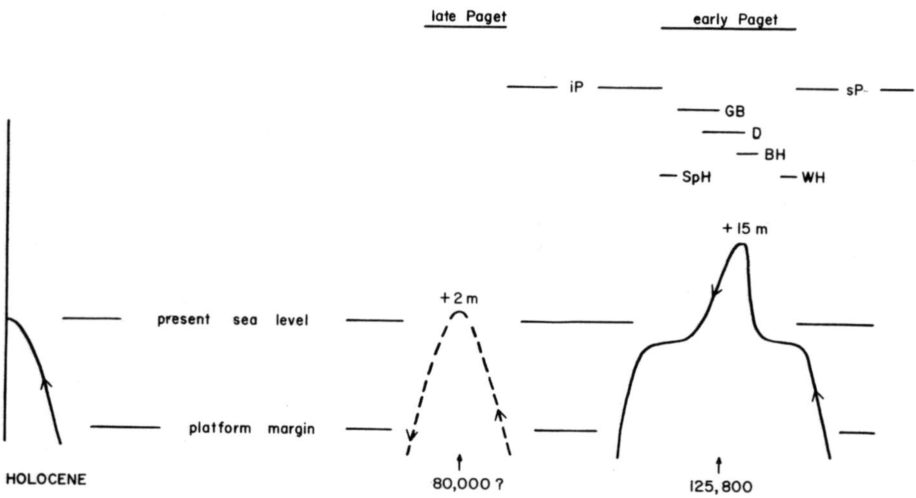

Figure 19. Source-sink systems during early Paget submergence of the platform. During initial phase (sea level below that of today) reefs become established on both platform margins. The reefs supply sediment for dunes along South Shore and prevent shoreward transport of sand from the North Lagoon. With a rapid rise in sea level, reefs are submerged; supply to South Shore is terminated, and the North Lagoon becomes a source for dunes along the North Shore.

Abbreviations: SLo, present sea level; SL1, position at initial submergence; SL2, position after rise.

Figure 20. Paget sea-level history.

indicated interglaciation (Sangamonian?) than they are now. The second-order excursion above present datum, an excursion with a rapid transgressive phase, may signify a large, catastrophic surge of the Antarctic Ice Sheet – the kind of surge argued by Hollin (1969).

CONCLUSIONS

The main points can be summarized in terms of the questions posed in the introduction.

1. *Conditions necessary for the buildup of dunes* on Bermuda are (a) relatively high sea levels so the strandline would be near the island and not below the edge of the platform, and (b) a source of large volumes of clacarenite. There are two cases. The first has the source as the reefs which are concentrated near the platform edge. Productivity in the reefs goes largely to production of calcarenite which is transported a short distnace to the southern shoreline and is then blown into dunes. In the second case, the source is calcarenite deposited initially in the North Lagoon. This sediment is transported to the northern shoreline upon a rapid rise in sea level which nullifies the breakwater effect of the reefs surrounding the lagoon. A third source-sink mechanism, involving subaerieal exposure of earlier deposited calcarenite of the lagoon, is rejected as a cause of large platform-derived dunes.

2. *The factors determining dune morphology* are those that prevented inland migration of the dunes. The dunes took on the accretionary form of large lobate hills that joined laterally to produce a ridge immediately behind the source beaches. The trend of the source beaches is reflected by the axis of the ridge. The direction of strong winds and gales that impinged on the shoreline controlled the direction of growth of the individual lobes.

3. *The Bermudian environment during dune formation* was an interglacial environment like that of today. The platform was submerged. Carbonate-secreting organisms, including hermatypic corals, were productive. Directional frequency of strong winds was similar to that of strong winds today.

4. *The elements of Paget geologic history that are of general significance* involve the history of sea level during Pleistocene interglaciations. Two separate intervals of relatively high sea level are recorded by the Paget rocks. The earlier of the two was about 125,000 yr ago. For the most part, sea level remained below its present position during this early Paget interval. Sea level did rise above present datum during this interval; the rise was abrupt and rapid, and may signify a catastrophic surge in Antarctica.

ACKNOWLEDGMENTS

This paper is an outgrowth of a dissertation problem at Northwestern University. I am indebted to my advisors F. T. Mackenzie and L. L. Sloss, and to the Department of Geology for supporting funds. Continuation of fieldwork was possible through a Research Associateship at the Bermuda Biological Station

(BBS). Support for field study was provided by NSF Grants GA-901 and GA-13540 (to Mackenzie), the Bermuda Government, BBS, and the GSA (Penrose).

The paper was written while a Research Associate at the State University of New York at Binghamton. I thank J. R. Beerbower, Chairman, Department of Geology, and the geology faculty for providing the congenial atmosphere which led to the production of the paper.

I am especially grateful to Byron Ristvet, of Northwestern University, for the use of his mineralogic and petrographic data for the purposes of this paper; Fred Mackenzie and Paul Enos for helpful reviews of the initial manuscript and Mackenzie, also, for his continuing and encouraging interest through all phases of my work in Bermuda.

This paper is BBS contribution number 555.

REFERENCES

Ball, M.M. 1967. Carbonate sand bodies of Florida and the Bahamas: J. Sed. Pet., v. 37, p. 556-591.

Bigarella, J.J., Becker, J.B., and Duarte, G.M. 1969. Coastal dune structure from Parana (Brazil): Marine Geol., v. 7, p. 5-55.

Bretz, J.H. 1960. Bermuda: a partially drowned late mature, Pleistocene karst: Geol. Soc. Amer. Bull., v. 71, p. 1729-1754.

Bricker, O.P. and Mackenzie F.T. 1970. Limestones and red soils of Bermuda: discussion: Geol. Soc. Amer. Bull, v.81, p. 2523–2524.

Butzer, K.W. 1964 *Environments and Archeology*: Aldine Publishing Company, Chicago, 524 p.

Butzer, K.W. and Cuerda, J. 1962. Coastal stratigraphy of southern Mallorca and its implications for the Pleistocene chronology of the Mediterranean Sea: J. Geol., v.70, p.398-416.

Cooper, W.S. 1958. Coastal sand dunes of Oregon and Washington: Geol. Soc. Amer. Memoir 72, 170 p.

Fairbridge, R.W. and Teichert, C. 1953. Soil horizons and marine bands in the coastal limestones of Western Australia: J. Proc. Roy. Soc. N.S. Wales, v. 86, p. 68-87.

Garrett, P., Smith, D.L., Wilson, A.O., and Patriquin, D. 1971. Physiography, ecology, and sediments of two Bermuda patch reefs: J. Geol., v. 79, p. 647-668.

Ginsburg, R.N., Schroeder, J.H., and Shinn, E.A. 1971. Recent synsedimentarycementation in subtidal Bermuda reefs. p. 54-58 in O.P. Bricker, ed., *Carbonate Cements:*Johns Hopkins Press, 376 p.

Gould, S.J. 1969. An evolutionary microcosm: Pleistocene and recent history of the land snail P. (*Poecilozonites*) in Bermuda: Bull. Museum Comparative Zoology, v. 138, p. 407-532.

Hollin, J.T. 1969. Ice-sheet surges and the geological record: Canadian J. of

Earth Sci., v. 6, p. 903-910.

Land, L. S. 1970. Phreatic vs. vadose meteoric diagenesis of limestones: evidence from a fossil watertable: Sedimentology, v. 14, p. 175-185.

Land, L.S., Mackenzie, F.T., and Gould, S.J. 1967. The Pleistocene history of Bermuda: Geol. Soc. Amer. Bull., v. 78, p. 993-1006.

Mackenzie, F.T. 1964a. Geometry of Bermuda calcareous dune cross-bedding: Science, v. 144, p. 1449-1450.

Mackenzie, F.T. 1964b. Bermuda Pleistocene eolianites and paleowinds: Sedimentology, v. 3, p. 52-64.

Mackenzie, F.T., Kulm, L.D., Cooley, R.L., and Barnhart, J.T. 1965. *Homotrema rubrum* (Lamarck), a sediment transport indicator: J. Sed. Pet., v. 35, p.265-272

Macky, W. A. 1956. The surface wind in Bermuda: Bermuda Meteorol Office, Tech. Note. 7.

McKee, E.D. 1966. Structures of dunes at White Sands National Monument, New Mexico (and a comparison with structures of dunes from other selected areas): Sedimentology, v. 7, p. 1-69.

Mesolella, K.J., Matthews, R.K., Broecker, W.S., and Thurber, D.L. 1969. The astronomical theory of climatic change: Barbados data: J. Geol., v. 77, p. 250-274.

Plummer, L.N., Mackenzie, F.T., Vacher, H.L., Land, L.S., and Bricker, O.P., in prep. Geology and geochemistry of Bermuda ground water.

Ristvet, B.L. 1972. The progressive diagenetic history of Bermuda. Reports of Research, 1972, Seminar on Organism-Sediment Interrelationships: Bermuda Biol. Special Station Publication (in prep.).

Ruhe, R.V., Cady, J. G., and Gomez, R.S. 1961. Paleosols of Bermuda: Geol. Soc. Amer. Bull., v. 72, p. 1121-1142.

Sayles, R.W. 1931. Bermuda during the Ice Age: Proc. Am. Acad. Arts and Sci., v. 66, p. 381-467.

Thrailkill, J. 1968. Chemical and hydrologic factors in the excavation of limestone caves: Geol. Soc. Amer. Bull., v. 79, p. 19-46.

Upchurch, S.B. 1970. Sedimentation on the Bermuda Platform: Detroit, U.S. Lake Survey, Res. Rept. 2-12, 172 p.

Vacher, H. L. 1971. Late Pleistocene sea-level history: Bermuda evidence, Ph. D. Thesis, Northwestern Univ., 153p.

Vacher, H.L. and Mackenzie, F.T. 1970 (Abs.). Pleistocene calcarenite lithosomes of Bermuda: Am. Assoc. Petroleum Geologists, v. 54, p. 873.

Willman, H.B. and Frye, J.C. 1970. Pleistocene stratigraphy of Illinois; Illinois State Geol. Survey Bull. 94, 204 p.

Wright, H.E. Jr. 1962. Late Pleistocene geology of coastal Lebanon: Quaternaria, v. 6, p. 525-539.

Yaalon, D.H. and Laronne, J. 1971. Internal structures in eolianites and paleowinds, Mediterranean Coast, Israel: J. Sed. Pet., v. 41, p. 1059-1064.

AUTHOR INDEX

SUBJECT INDEX

HOMERTON COLLEGE LIBRARY

HOMERTON COLLEGE LIBRARY